NOMIC INFERENCE

NOMIC INFERENCE

AN INTRODUCTION TO THE LOGIC OF SCIENTIFIC INQUIRY

by

SALVATOR CANNAVO

MARTINUS NIJHOFF / THE HAGUE 1974

For my parents
Joseph and Francesca

ISBN 978-94-015-0266-5 ISBN 978-94-015-0788-2 (eBook)
DOI 10.1007/978-94-015-0788-2

TABLE OF CONTENTS

FOREWORD

Those who speak of the philosophy of science do not all have the same sort of study in mind. For some it is speculation about the overall nature of the world. Others take it to be basic theory of knowledge and perception. And for still others, it is a branch of philosophical analysis focused specifically on science. The present book is meant to be a study falling under this last category. Generally, such a study has two aspects: one, *methodological*, dealing with the logical structure of science, the other, *substantive*, dealing with scientific concepts. Our concern here is primarily methodological; and, where discussion veers at times towards substantive matters, this will be largely for the purpose of illustrating underlying methodological points. It should also be added that our considerations will be of a general sort, intended to apply to all of science with no special concern for any particular divisions. Except in an incidental manner, therefore, we shall give no primary attention to special problems in the methodology of the social sciences or in the philosophy of physics or of biology. And if we draw the larger portion of our examples from the physical rather than from the behavioral sciences, this is done merely for simplicity, succinctness, and similar conveniences of exposition rather than out of specialized concern for any particular area.

It is generally convenient in methodological discussion to distinguish two major phases of all scientific inquiry, namely: the *growth* of science and the *systemization* of results. The growth of science provides content and includes both the discovery and acceptance of scientific hypotheses. Systemization invests this content with logical order. It results, so to speak, in the finished scientific report. The major portion of our discussion concerns systemization (Chaps. IV-VII), although some consideration is given to scientific growth in both the early and final portions of the text (Chaps. III and VIII).

As regards general methodological outlook, we admit, straight away, to some heresy in at least two matters: First, we depart from the main line of tradition by siding with the minority opinion that there *can* be a measure of "logic" or method in discovery, and that therefore – besides falling under history, psychology, or sociology – the study of scientific discovery can also make the agenda of philosophy of science.

We also depart from another tradition, this time concerning scientific *explanation*. Except for a few recent dissenters, philosophers since Aristotle have traditionally regarded explanation as central to the structure of systemized science. Our quarrel with this tradition is that the centrality view is wrong. We shall contend that explanation, construed as an answer to the question, "why"?, fails to represent the structure of large and important portions of modern science. We propose, therefore, to speak more generally of *subsumptive* or *nomic inference*, and to regard explanation as merely a species of such inferences. When this modification is granted, explanation and prediction come out as overlapping rather than as mutually exclusive forms of nomic inference. Moreover, on this analysis, several mixed subsumptive patterns come to light which we take the liberty to baptize with new hybrid labels.

Nomic inferences are the structural units of *subsumptive science*. They are the logical means for the basic task of showing how given matters of fact may be regarded as special cases of others that are more general. The structure of nomic inference is our principal concern in Chapters IV and V where two modes of such inference are distinguished : *deductive* and *probabilistic*. The aim of discussion is to trace out various patterns of nomic inference. But the distinctions are drawn in such a manner as to apply equally to both deductive and probabilistic nomic inferences. This manner of exposition seems to be an advance over more usual approaches to the subject, which tend to put great stress on the primacy and fundamentality of deductive subsumption and to neglect notable aspects of isomorphism with the probabilistic form. More particularized discussion of nomic inferences relating specifically to the derivation of laws and to probabilistic nomic inference is left to Chapter V where we also consider two special topics pertaining to nomic inference – namely, the question of whether there are non-nomic explanations and the status of functional or teleological accounts. At this point, having examined the structure of nomic inference, it seems natural to go on to a consideration of the lawlike statements which serve as premises of nomic inferences. Thus, in Chapters VI and VII we take up the structure of laws, theories, and related nomic forms.

Our opening chapter offers a discussion of philosophic method for the purpose of orienting the reader to philosophy of science as an *analytic* discipline. But the reader who prefers to forgo it, may easily do so without loss of coherence by beginning instead with Chapter II which serves as an introduction to the main body of discussion.

Chapter VIII concludes our considerations with a brief *perspectiva* on relatively recent trends in methodology, not developed in the main text. The aim is to round off discussion with a short presentation of these developments, along with some attempt to say how they relate to the more traditional body of methodological opinion. As it happens, the currents of thought which we touch on have bearing on the *growth* ("ampliative phase") of science. And, though there are some implications also for *subsumptive systemization*, these closing reflections represent a shift of attention from our more central concern – a shift which, it is hoped, can serve to enhance perspectival balance for the overall study.

It is a hard and frustrating fact of our culture that positive science – a great *product* of culture – cannot be reached by most *students* of culture. Nor does science readily unveil its broader delineaments to the casual visitor even when he can read its cant. Popularized science helps little in this respect; for, at its best, it tends to highlight factual content rather than logical structure. Some burden of duty, therefore, may well fall on the philosophy of science which, like any account *about* a precisely formulated system of discourse, can (and in fact usually does) proceed in language that is far less exact and artificial than that of the discourse under study.

With considerations of this sort in mind I have presupposed, in the main body of discussion, no special knowledge of logic or mathematics, the discussion proceeding, therefore, with relatively little use of formal symbolic devices. But where technical details seem desirable, they are placed in the footnotes, so they may be easily bypassed without loss of continuity or coherence. I hope to have done this in such a manner as to invite the participation of the more generally oriented reader without disinvolving the professional student of philosophy.

In the way of acknowledgments, I should express, first, my indebtedness to the teachers, colleagues, students and friends who immeasurably influenced the growth and direction of my interests. In particular, I must mention Professors Sidney Hook, A. J. Ayer and Albert Hofstadter to whom I owe my first real lessons in philosophical analysis. I also recall with much pleasure my earlier teachers at Fine Hall and at the Palmer Physical Laboratory. Among these, I am especially grateful to Professors V. Bargmann and J. M. Jauch for the breadth and depth of their lectures in mathematical physics and its foundations.

The late R. Carnap, N. R. Hanson and H. Reichenbach, together with Professors Carl G. Hempel, Ernest Nagel and my friends, Adolf Grünbaum and John Hospers, have not been formally my teachers, but my debt to them will no doubt be obvious.

My thanks to Brooklyn College for supporting preliminary research with a six-month, paid leave in 1968 and to the University of Southern California for an enriching semester of specialized teaching and research in the methodologies of the physical sciences, 1970. Finally, I wish to thank my dear wife, Gaetana, and daughter, Francesca, without whose help and understanding this book could not have been.

New York, 1973 S. C.

I

ANALYTICAL PHILOSOPHY OF SCIENCE

A. WHAT IS THE PHILOSOPHY OF SCIENCE?

Many writers refrain from starting an introductory work with a definition of the subject and do so for good reason. An accurate and concise enough definition at the start of a work would probably have to be too general, abstract or technical to communicate anything of much value to the beginning reader; while one which avoids these features would tend to become so elaborate and burdened with illustrations as to seem redundant, if the main body of the work is to follow anyway. There is indeed a real difficulty here and its source lies deep. For talk about any subject is, in general, not part of that subject. A definition of science would not be science; just as a definition of music would not be music. Even a practicing scientist attempting to write *about* his subject would soon find himself using language not ordinarily called for when writing *in* his subject. Thus, in talking about physical theory he might, for example, take up such matters as the general structure of abstract deductive systems, the relation of such systems to physical subject-matter, the nature of scientific models, and much more – topics which have come to be thought of as more properly belonging to semantics, applied logic or, more generally, to the philosophy of science rather than to natural science itself. What is more, actual practice seems to tell us that a working scientist, in elaborating any particular theory or in presenting experimental evidence for it, need not (and often does not) become extensively involved in any considerations *about* what he is doing. Indeed, he seems quite capable of going about his business without giving sustained and systematic attention to a wide variety of foundational questions about his doings – questions which if carefully pursued tend to take him into specialized areas outside of science.

In the case of philosophy, however, one must somewhat modify the claim that talk *about* any discipline tends to fall outside that discipline.

Historically, it has been a major concern of philosophy to study the structures and foundations of other disciplines. But nothing in this tradition excludes *philosophy itself* as an object of study. In actual practice, philosophy can hardly avoid self-reference. For in talking about another discipline it immediately begins to refer to itself. This is so simply because in order to talk about another discipline it must carefully set itself outside of that discipline. But here, self-reference eventually becomes self-study. For study about disciplines is readily generalized to include the student discipline itself. This is only a natural extension for it requires no radical changes of methods and conceptual apparatus. An inquiry into the structure of philosophic discourse would still be philosophy, namely *philosophy* of philosophy. A similar inquiry about the philosophy of science, though of course not itself philosophy of science, would be nonetheless philosophy, viz., *philosophy* of the philosophy of science. Thus in its concern with other disciplines, be this a concern for their methods or for their conceptual content or for the way their results relate to the world, etc., philosophy can hardly remain insensitive to its own character. A total lapse in this respect would be about as unthinkable as a consciousness that never had the slightest inkling of its own activity. Philosophers, like other inquirers, can, and almost always do, plunge into their special tasks without first offering some systematic comments on the sort of activity they are about to engage in. But in the long run, even as a mere token to universality, this can be only a postponement rather than a permanent omission.

In our own considerations, we shall prefer to begin with some brief attention to the scope and character of what it is we are about. This is largely motivated by the fact that the headings, 'philosophy of science' and 'philosophical analysis', under which we pursue the present study, have been used for a variety of activities. It seems advisible therefore to say something, at the very outset, about the sort of meaning we have in mind in the use of these headings.

The term, 'philosophy of science', has been applied to least three quite different sorts of study. We will attempt to pursue only one in all that follows. But let us begin by taking a brief look at each of them :

1. We list first what may be described as the *sociological* conception of the philosophy of science. It includes such topics as the history of science, the relation of science to culture and society at large, and the behavior of working scientists – their motivations, influences upon one another, the way they come upon their discoveries, etc. The philosophic flavor which these subjects seem to have is partly due to the broad manner in which

they are often approached and to the somewhat unsettled nature of many of their conclusions. Where the history of science is specifically concerned, philosophic overtones are also, no doubt, produced by the logic which seems to be inherent in the subject. To note this, the term 'structural history' has sometimes been employed, the point being that though scientific ideas have a time-place history, unlike earthquakes and wars, they are usually tied to each other in a characteristically logical (rather than merely causal) manner.[1] Accordingly, this gives the study of such events a strongly analytical or, as we might say, *philosophical* (rather than empirical) character. And no doubt there is much truth in such a claim. For scientists are especially motivated by considerations which link their creations logically to the existing body of ideas, with the result that the psychogenesis of science is bound to be more logic-like than, say, political or economic history. But one who, for such reasons, calls these subjects philosophy, ought not to forget that the behavior of scientists and the genesis of their ideas are ultimately matters of contingent fact rather than of necessary logical relations, no matter how philosophical the overtones may be. To overlook this is to invite serious misunderstanding regarding the intended methods of approach to such subject-matter. Indeed, there are established modes of usage that seriously resist applying the term 'philosophy' to any inquiry into matters of natural fact, the term 'science', or more specifically, 'natural science', being preferred instead.

2. A second conception of the philosophy of science, which, for want of a better label, could be called the *continental* conception because it is strongly dominant in continental Europe and Spanish America, places the philosophy of science within epistemology and metaphysics. Here one finds the traditional concerns for the alleged foundations and presuppositions of scientific knowledge, often with special emphasis on the social sciences, sometimes referred to as the "humane sciences".[2] One also finds under this heading varying attempts to provide a speculative synthesis or a contemplative, world view (man included) based on the accumulated results of science. Representative topics are the "total" universe, its all-inclusive evolution from problematic beginnings to probable ending,

[1] G. Bergmann, *Philosophy of Science*, The University of Wisconsin Press : Madison, 1957, p. 8.

[2] For further references to this school of philosophy, see : G. Radnitsky, *Anglo-saxon Schools of Metascience*, ed. H. Törnebohm, Akademiförlaget : Göteborg, 1968, vol. I, preface, where the term, 'hermeneutic dialectic', is used for the continental conception of the philosophy of science.

along with the place this leaves for purpose and intelligence.[3] Also under this conception of the philosophy of science one often finds studies in many of the subjects of traditional metaphysics along with such topics as perception, memory and other parts of what is sometimes called "philosophical psychology". But despite the generality and fundamentality claimed for these studies, at least a part of what they seek to fathom seems to be of a fundamentally matter-of-fact character, as is the concern of science. Serious questions arise, therefore, regarding whether some of these subjects require special "philosophic" methods of investigation that are fundamentally different from those of empirical science. And, quite apart from their admittedly philosophic flavor, the question of whether (and in what fundamental sense) they are to be considered philosophy has been the occasion of some of the most divisive controversies in present-day thought, controversies which invariably reflect profound philosophic differences regarding the nature and grounds of human knowledge.

3. The third conception of the philosophy of science, which may be described as the analytical conception prevails largely in England, Anglo-America, and Australia and is the one which we shall adopt here. It views the philosophy of science neither as an empirical nor as a speculative study, but merely as logical analysis of the formal structure and conceptual content of science. Unlike the first two kinds of study, this version of the philosophy of science does not deal with time-space things, i.e., temporal events, substantial objects, their properties and changing relations. Nor does it concern itself primarily with what it is to perceive or to know such things. It focuses instead on the manner in which scientific results are expressed. More particularly, it concerns itself with the *methods* which give scientific knowledge its coherence and certification together with the conceptual content which gives such knowledge its scope and applicability. The term "method" here is not to be construed to refer to matters of experimental practice and technique (except, possibly, where these have a direct bearing on questions of validity). It refers, rather to a wide and complex range of logical features that give to scientific discourse its characteristic structure.

Scientific knowledge like all other knowledge is expressed and communicated linguistically. More particularly, it is expressed as sets of ordered statements which may take the form of literal sentences and other forms of expressions such as diagrams, graphic models, and mathematical expressions. The systematic aspect of the knowledge – and this means also

[3] An example of contemporary philosophy in this vein would be some of the writings of Teilhard du Chardin (1881-1955). Especially notable in this respect is *The Phenomenon of Man*, Harper & Row : New York, 1959.

its basis in experimental fact – is expressed by the order, as well as the content and the form of statements. Accordingly it is sometimes highly convenient to approach problems in the philosophy of science as analytical studies of scientific discourse or, simply, of scientific language. The technical advantages of this method result largely from the ease with which formal logic is sometimes applied to certain aspects of language. In such an approach talk about concepts becomes talk about the meanings of linguistic expressions (e.g., terms, statements, groups of statements, etc.); while the structure of science is nothing more than the logical form of such expressions and the various relations that occur between them.[4]

Under the analytical conception the scope of the philosophy of science is generally, though not entirely without exception, limited specifically to *scientific* knowledge, thus by-passing the broader epistemological questions relating to knowledge in general. Accordingly, in such a study, one does not usually find any consideration of topics like perception, common sense experience and the relation of these to our knowledge of things.[5] By virtue of such limitation the analytical philosophy of science has become somewhat less controversial and has thus accumulated a more stable content than one usually finds on the more general philosophic levels. Our own considerations will remain largely within the limits of this third conception of the philosophy of science.

B. METHODS OF ANALYTICAL PHILOSOPHY

1. But despite its relative stability, the philosophy of science, as an analytical discipline does have some basically problematic aspects which it inherits from philosophical analysis in general. Quite apart from their diverse philosophic techniques – dialectical, polemical, formal, logistic, etc. – analytical philosophers seem to be divided on fundamentally two approaches to their subject, which have come to be known as *ordinary language analysis* and *explication*. The favoring of one or the other of these approaches, though it usually goes with a preference for certain

[4] In our own discussion we shall, for idiomatic reasons and general ease of exposition, often speak of concepts, notions and of the meanings of terms interchangeably. In doing so, however, we do not wish to convey by the terms 'concept' or 'notion' anything like a mental state, or mental activity, or other psychological disposition.

[5] There are important exceptions to this. The problem of induction, for example, is basic to all empirical knowledge yet frequently becomes the specialized concern of philosophers of science.

techniques, generally reflects much more than such a preference. It often stems from far deeper differences regarding the nature of language, how it relates to things, and what it is that philosophical analysis is supposed to do. It seems appropriate therefore to pause, somewhat later in our discussion, for a brief consideration of each of these two approaches.

2. Philosophical analysis is no doubt as old as discursive activity itself and is amply illustrated throughout philosophy even as far back as pre-Socratic thinkers like Parmenides and Zeno. It owes much to its great modern progenitors, Descartes and Immanuel Kant, and finally, achieves its contemporary refinement in the writings of logical empiricists and of the Cambridge analysts.

Essentially, it is a method for clarifying concepts or, what is the same thing, for clarifying both the structure and content of linguistic expressions such as terms, statements, and even systems of statements. Broadly speaking, this is done by rendering that which is to be clarified in language which, compared to the original expression, is better understood or, as one might say, more perspicuous because of being either more familiar or more systematic.[6] The account aims at providing an explicitly articulated significance for the expression under consideration and usually requires attending not merely to the meaning (i.e., semantical) content of the expression – which is to say – its manner of application to possible kinds of circumstances but also investigating its formal (i.e., syntactical) grammar. In actual practice, the analysis may proceed in one of fundamentally two ways : (1) It may take the form of a *reductive* analysis which transforms the original expression into a second expression. (2) Or, it may take the form of a *discursive* analysis which instead of translating or otherwise transforming an expression proceeds to characterize it by talking *about* it. The account is given more or less abstractly in terms of some chosen framework of discourse – logical, grammatical, philosophical, etc. Examples of the first, i.e., the reductive mode, are one of C. L. Stevenson's models for ethical meanings, in which he suggests : 'He ought to do this' means "I disapprove of his leaving this undone; do so as well" [7] or, David Hume's analysis of the expression 'An object causing another' which he gives to be "an object precedent and contiguous to another, and so united with it in the imagination that the idea of the one determines the mind to form the idea of the other".[8] An example of the second, i.e., the discursive,

[6] For naturalness of expression, throughout all that follows, more traditional terms like 'concept' and 'notion' will often be used instead of the term 'meaning'.

[7] C. L. Stevenson, *Ethics and Language*, Yale University Press : New Haven, 1944, p. 21.

[8] D. Hume, *A Treatise of Human Nature*, Vol. 1, E. P. Dutton : New York, 1911, p. 169.

mode is Immanuel Kant's account of the term 'being' and its closely related variants 'is' and 'exists'.

"*Being* is evidently not a real predicate, that is, a conception of something which is added to the conception of some other thing. It is merely the positing of a thing, or of certain determinations in it. Logically, it is merely the copula of a judgment ... Now, if I take the subject (God) with all its predicates ... and say *God is* or, *There is a God*, I add no new predicate to the conception of God, I merely posit or affirm the existence of the subject with all its predicates — ...
Whatever be the content of our conception of an object, it is necessary to go beyond it, if we wish to predicate existence of an object. In the case of sensuous objects, this is attained by their connection, according to empirical laws, with some one of my perceptions" [9].

There is no disagreement over these two ways of doing philosophical analysis. In practice reductive and discursive analysis virtually never occur unmixed, and preference for one over the other seems to be little more than a matter of style and expository technique. In analyzing a passage we paraphrase, translate, reconstruct and otherwise transform it; but then we also examine the transformations and talk about them by characterizing, categorizing, explaining or generalizing the results. Serious contention, however, does arise as to whether explication or the ordinary language approach is the basic method of philosophical analysis; and to a consideration of these we must return in what follows – but not without first giving some passing consideration to a rather prevalent notion regarding the character of logical analysis in general.

It is a rather widespread, time-honored view that the central feature of any logical analysis is the "decomposing" or "unpacking" of a complex meaning into presumably simpler, or, as it is sometimes put, more "elemental" and, therefore more clearly understood, parts. Thus Descartes, early in his *Discourse on Method*, resolves "... to divide each of the difficulties under examination into as many parts as possible, and as might be necessary for its adequate solution". In this way he hopes to conduct his thought "... by commencing with objects the simplest and easiest to know" or, as he says later of straight lines, "... than which I could find no objects more simple, or capable of being more distinctly represented to my imagination and senses ..." [10]

[9] I. Kant, *Critique of Pure Reason*, J. M. D. Meiklejohn translation, Willey Book : New York, 1943, pp. 335-336.

[10] Descartes here is proposing the analytic method as a general problem-solving procedure, hence his reference to dividing a *difficulty*, rather than a meaning, into as many parts as possible. Obviously his considerations are easily construed specifically as a method of meaning rather than of problem analysis. R. Descartes, *Discourse On*

But such a characterization of analysis, however suggestive and widely held it may be, is much too vague for actual implementation except in the most highly systematic contexts where the notion of an expression (or concept or object) being or not being more elemental can be given some precise sense. One would, for example, be hard pressed to apply such a criterion to C. L. Stevenson's account, in our example above, where he seeks to analyze the meaning of the term 'ought' by reducing it to an approval (expressed by : 'I approve') and an imperative requiring others to approve also (expressed by 'do so as well'). How does an approval or an imperative count as more elemental than the *ought* which is being analyzed? Indeed some writers have maintained views which would seem to say that the term 'ought' has, itself, so elemental a meaning as to be incapable of being further unpacked. Thus, the elementality of whatever is designated by the expressions 'ought', 'I approve', and 'do so as well' would not, in this case, seem to be a matter of settled opinion.

In some cases, where that which is to be analyzed involves a geometric structure of some kind, it may seem natural to regard the components or parts of the structure as elements in terms of which to carry out the analysis. (Here Descartes' notion of "simplest objects" or "parts" seems more appropriate.) Thus, one may analyze the notion of a geometric plane by characterizing it as : *a surface such that a straight line joining any two of its points lies entirely on the surface*. In such a case it may seem appropriate to say that straight lines and points are more elemental than planes because lines and points are in a genuine sense parts of planes. Obviously, however, one cannot impose this structural sense of elementality on all analyses. For not all concepts have geometric structure as, for example, freedom, love, and obligation. But, even where such structure is involved, elementality in this last sense seems to remain inapplicable to a wide class of cases. Thus, in our example, the notion of *surface* seems to resist being compared in this manner to that of *plane*. Both notions are geometrical, the first more general than the second. But how does one decide whether one is structurally more elemental than the other? Corresponding difficulties arise with *joins*, *lies*, and other notions occurring in our example. Still another difficulty arises, stemming from the fact that we sometimes analyze a concept in terms of notions that, in the sense here involved, one might want to say are structurally *less* elemental, for example, in characterizing a *straight line* as the intersection of two *planes* or a *point* as the intersection of two *straight lines*.

Method, tr. J. Veitch, Open Court : New York, 1945, p. 19.

These considerations tend to caution against regarding it essential that philosophical analysis incorparate the feature of elementality in the vague, pre-systematic sense in which this feature has often been associated with analytical exposition. Modern logic provides us with a clear notion of what it is for a term to be *elemental* or *primitive*. But this requires such a term to be part of some highly formal system such as a logical calculus or artificial language. Without stopping to consider the logical structure of such systems we note merely that formal systems are, among other things, stipulated, i.e., "artificial", devices for analyzing terms and that whether or not a term in the system is elemental or not depends on the way the formal system is constructed. Now, what sorts of things can or cannot be expressed in such formal or artificial systems is a matter of some contention. What is not disputable, however, is that if anything at all can be so expressed or *formalized*, the system for doing this can be constructed in a variety of ways each taking a different set of terms as "elemental" or "primitive". Thus the elementality of any term, in this sense, is ultimately a matter of stipulation, that is to say, a matter of how we choose to construct our formal system. It is not a matter of the term's simplicity in some pre-systematic sense. But it would be futile to attempt to characterize all philosophical analysis as the "reduction" of concepts to the elements of some formal system. For many areas of discourse, no such systems are available; indeed, philosophical analysis goes on very extensively without them.

On the basis of these considerations it would hardly do to list elementality or simplicity, either in the formal sense just mentioned or in the pre-systematic sense of being highly familiar or conceivable, as essentially involved in any philosophical analysis. It is of course undeniable that in practice the analysis of concepts aims at increased comprehension and communicability of meaning. Hence, one may reasonably expect that an analysis will in general make use of terms whose meanings are more familiar – perhaps, more easily imagined, conjured, or, as it is sometimes put, "intuited" – than whatever is to *be* analyzed. And no doubt familiarity and conceivability of this sort figure heavily in what loosely passes for "simple". But despite the way such features may ease communication it would be baseless to make them canonical requirements of any analysis. Familiarity and intuitability are in large measure dependent on psychological dispositions which can vary from time to time and from person to person. What is more, the analysis of concepts in the history of thought has not always been a reduction to more familiar ones. Some basic scientific notions such as energy, particle, field, or electric charge in terms of which other notions have been rendered were far from familiar when they were introduced, though the ensuing accounts eventually added much to scientific understanding.

3. Traditionally, philosophical analysis, unlike mathematical analysis, has been a rather loose and individually stylized affair, attended as much by rhetoric as by logic. Basically, its aim has been to heighten general intelligibility, though its stratagems have often been polemical – exposing paradoxes, ambiguities, inconsistencies, logical gaps, circularity, and question begging in otherwise uncontested claims. But there are those who see in this rather unregularized procedure the alleged "danger" of not offering much methodological assurance against fundamentally two possibilities, i.e., distorting or losing the meaning which was to be analyzed and failing to provide sufficient definiteness in the analysis. These are two somewhat divergent concerns. And they have generated respectively two over-arching and somewhat opposed approaches to philosophical analysis, to which we now return, namely, (1) *ordinary language analysis* and (2) *explication*. Whether the mode of exposition is reductive or discursive, these two approaches represent divergent alternative conceptions of philosophical analysis. Both seek out meaning, the first by pursuing and capturing every one of its *actual* nuances, the second by systematically revising it.

(1) Ordinary language analysis in its contemporary form owes much to the early twentieth century, Cambridge philosopher, Ludwig Wittgenstein. Broadly speaking, it holds that philosophical investigation can be relevant only if it focuses on that which actually articulates and communicates experience, namely, ordinary usage.[11] Accordingly, philosophical analysis is essentially a reportative process whose underlying concern is the customary meanings of terms, no matter how subtly these may shift and change with context. All philosophic issues must first be squared with actual usage which in turn may be clarified by careful high-fidelity analyses based on representative instances or paradigms and put to test by possible counter-examples.

This approach to philosophical analysis has had a sweeping contemporary influence. Its proponents have applied it to a variety of traditional philoso-

[11] Wittgenstein, in some of his works, claimed much more. In effect, he maintained that only if the structure and composition of a language corresponds to a subject-matter is it applicable to that subject-matter. But ordinary language or usage is unquestionably applicable to everyday experience and hence to the world, which is the subject of that experience. Such usage, therefore, is a map of what there is and the analysis of this usage a key to reality – hence the importance of analyzing language ordinary. The claim

phic issues in order to show these to be pseudo-problems arising primarily when philosophers unwittingly distort and misuse ordinary language. This attempt at the "therapeutic dissolution" of old problems by analytic procedure has been one of the most dramatic and impressive thrusts of contemporary philosophy and marks ordinary language analysis as a striking and far-reaching development in modern intellectual history.

A possible objection to the *exclusive* use of this mode of analysis, however, is that it is much too modest in its aims. Despite its efficacy in the pursuit of customary meanings and its salutary effects on philosophic discourse, it does not in its most typical instances, bring into play powerful logical techniques of reconstruction now available for eliminating vagueness. Ordinary language analysis, as the objection might go, may indeed unveil meaning, but, in doing so, what it may also unveil is the vagaries and unclarities of everyday language. Nowhere does it stipulate linguistic changes for rectifying loose usage or for correcting a variety of other semantical defects in natural language.

To this an ordinary language philosopher might wish to retort that on both systematic and historical grounds philosophy ought to be concerned with unravelling the strands of ordinary usage rather than with *regenerating* language. Natural language provides the broad linguistic framework for basic conceptualization and communication, and these have been central philosophic concerns from the beginning. The artificial reconstruction of language on the other hand is a task properly left to the scientist who, as a specialist, is presumably best acquainted with the conceptual needs of his subject.

But, then, might one not answer that such a recommendation too severely limits philosophy which has traditionally been critical and reconstructive with respect to *both* ordinary and specialized discourse. After all, the reconstruction of language is, in part, the reconstruction of meanings or, more traditionally, the formulation or "definition" of concepts, a practice that, historically, seems to run throughout all of philosophy.

Another objection to ordinary language analysis calls attention to what may be regarded as one of its stylistic and therefore less intrinsic characteristics. Ordinary language philosophers, in their desire to adhere strictly to ordinary usage, do all their philosophizing not only *about* ordinary language but also *in* the idiom and style of ordinary discourse.

---

that linguistic structure reflects world structure or at least our own experience of it is, of course, not new with Wittgenstein, views of this sort having been expressed by both Aristotle and Kant.

Their writing therefore, though lively and often perspicacious is sometimes disadvantaged by the avoidance of powerful symbolic and formal devices that have been very successful in many areas where analytic thinking occurs, e.g., mathematics, logic and theoretical science, and that could well prove effective in those areas from which ordinary language philosophers seem to have banned them.

(2) The alternative to the ordinary language approach – perhaps better, an extension of it – is *explication.* It has been developed in contemporary times mainly by those empiricists who have had a predilection for logic and the more formal modes of exposition. Its beginnings, however, go back to the very earliest attempts to formulate clear and systematically tenable meanings for familiar but elusive terms like "knowledge", "justice", "beauty" or "virtue". As a method of philosophical analysis, it is usually employed where, for certain purposes, a key concept is insufficiently clear or, for certain purposes, systematically ill-suited. In effect, it seeks to replace the customary or pre-systematic meaning of an expression by precisely stipulated versions, formulated in more or less systematic context. Following the nomenclature of R. Carnap, the terms *explicandum* and *explicatum* have come into use to label the original and the revised meanings, respectively.

As a basic method of philosophical analysis, explication has played a strong role in the philosophy of science having been applied both to concepts arising primarily *within* science – for example, *cause, probability, space-time* and *particle* – and to concepts occurring largely in discourse *about* science – for example, *explanation, theory, law* and *model.* Our principal concern, however, is not ordinary language analysis and explication as approaches to philosophical analysis in general, but their more specific roles in the philosophy of science.

C. METHODS OF ANALYTICAL PHILOSOPHY OF SCIENCE

1. Ordinary Language Type Analysis of Scientific Discourse

How is the analytical philosophy of science to be done?

Broadly speaking two approaches present themselves paralleling ordinary language analysis and explication. One of these would be committed to return at every step to prevailing usage, in discourse both *within* and *about* science. The other would be more or less disposed to reconstruct such usage in order to achieve greater precision and systemization.

Which of these would be most appropriate?

There is no doubt some scientific terminology which may be characterized as natural or ordinary in the sense that it has not been explicitly defined or introduced by formal postulation. And for the elucidation of essential scientific content, this may indeed require non-reconstructive analysis paralleling the ordinary language type. Thus in the formally less developed sciences where theory is still too amorphous to shape and stabilize meanings, the prevailing dynamics of natural usage may require something like an ordinary language approach to reveal what is going on from context to context. Terms like 'subconscious', 'ego', 'person', 'social consciousness', 'gestalt', 'mental state', 'pain', 'hunger', 'will of the majority' and a host of others occurring in the behavioral sciences often call for elucidation on the basis of careful contextual study of the ordinary language type, the subject-matter in such a case being of course ordinary *scientific* language rather than ordinary everyday or common language. Even in the more matured and structured disciplines there can be deep, and sometimes shifting, layers of pre-systematic meaning barely receiving explicit recognition, yet so basic as to limit the scope of an entire science or blur portions of it. Among the more striking historical examples of this are terms like 'space', 'time', and 'simultaneity at a distance' whose ordinary "scientific" sense had to be made fully explicit before the science of mechanics could go on from where Newton had left it. The notorious clock paradox in connection with the special theory of relativity arises only when instead of adhering to new meaning conventions, one "unwittingly" or "naturally" slips back into customary usage. And, similarly, the air of paradox characterizing parts of modern physical theory is in some measure due to certain customary scientific senses which we persistently attach to terms like 'object', 'particle' and 'probability' in contexts where such senses are no longer appropriate. In all cases of this sort philosophical analysis of the ordinary language type can do much to bring underling usage to the surface in order to get at sources of error and perplexity.

2. The Role of Explication

According to the Wittgensteinian critique metaphysical puzzles are generated when philosophers unwittingly abandon customary meaning in favor of non-sense producing distortions. Bring customary meaning back into focus by ordinary language analysis and the puzzles may be dispelled. The historical development of scientific discourse, however, strongly suggests that perplexities having to do with science may be resolved not only by analytically unveiling natural or established scientific usage but also by refining or, as we say, by *explicating* it.

Philosophical discourse about science is either methodological or substantive. The first deals with structural aspects of science such as the nature of theory, models, scientific inference and explanation; the second with scientific content itself as for example with notions like space, time, particle and field. In the case of methodology, everyday language even if unravelled by the ordinary language approach, is of little interest. Pre-systematic methodological concepts are much too casually and loosely conceived for ready use in any study of the structure of science. Accordingly, it has been formal explication rather than any ordinary language approach that has figured heavily as a means of precisioning methodological concepts.

But, not only does methodology call for explication; to a large extent, so does substantive analysis. The history of science is in large measure a history of semantical innovations, or, as we might say, meaning reforms which replace pre-systematic "natural" usages with revised or "artificial" ones. The latter, far from being nonsensical distortions, often reflect considerable advances in both clarity and content. The continuing growth of both scientific fact and theory calls for evolving concepts, both scientific and methodological. Thus much room is generated for the analytical reconstruction of concepts or explication. And, despite the obvious dispositions of scientists towards remarkable precision and clarity, the results of science, as *ordinarily* articulated, often exhibit sufficient inexplicitness to warrant some reconstructive analysis. Thus, in statistical studies such terms as 'random' and 'relevant' at times seem too indefinite for a firm understanding. In quantum theory, the relation of measurement to the "measured" state, the appropriate interpretation of probability, the physical significance of such notions as *single event* or *individual particle* have been matters of semantical contention for scientists and philosophers alike, even after the sciences involved have enjoyed long and successful careers. For that matter, such seemingly well-planted terms as 'electron', 'atom', and 'physical object' – or 'space', 'time' and 'mass' are a challenge to anyone seeking to give intelligible accounts of their scientific significance. Well-defined and performable operations may indeed be coordinated with these terms, as for example, clock measurements with the term 'time' and ruler measurements with the term 'space'. It would, however, be naive to suppose that any list of such operations, even if it could be given exhaustively, would provide a full account of the desired meanings.

In general, scientific language may suffer from two fundamental forms of unclarity, sometimes labelled *semantical* and *syntactical*, respectively. The first of these refers to the occurrence of terms whose application, i.e., *referential* meaning is vague. The second refers to the case where the form of a term as well as its formal relations to other terms in the system are nowhere explicitly indicated. Thus a semantcal unclarity is vagueness regarding what a term designates. One may say, for example, that yesterday's rain was *improbable* without being sure of just what sort of attribute is being predicated of yesterday's rain by calling it "improbable". Syntactical unclarities have to do with uncertainties regarding the purely formal grammar of the term. The uncertainties arise because the formal characteristics in point have not been explicitly specified anywhere and are not entirely evident under the informalities of pre-systematic usage. Thus, for example, it may not be clear whether quantity is relational or not. The notions of up or down, of motion and location are prime historical examples. A failure to distinguish the referential or semantical content of a term from its formal or syntactical aspects can seriously hamper any attempt at clarifying its meaning. Some of the long polemics on the meaning of the term 'probability' or of terms like 'number', 'line' and 'point' have been complicated by a failure to distinguish these two aspects of language.

Explication as we have seen seeks to eliminate unclarities by refining and to some extent, therefore, revising the area of usage in question. To accommodate these revisions it is sometimes necessary to introduce revisions elsewhere in the general framework of discourse. Thus, an explication may entail basic and drastic enough linguistic innovations to effect an entire science, system, area of discourse, language, etc. Indeed, it is possible for such a reconstruction to be so thoroughgoing as to transform radically the entire realm of discourse. If the reconstruction is carried through in a logical mode of exposition, then the resulting system is said to be *formalized.* The *formal* mode of exposition, by its strict attention to logical form and order, attempts to bring full explicitness to the structure of discourse. Contemporary philosophical analysis has shown this mode of exposition to be best carried out by what is known as the *axiomatic* method. It would not be convenient to discuss this manner of structuring discourse in any detail, so early in our account. We shall therefore return to it in a later section. Having mentioned it at this point, however, in connection with explication, it seems appropriate to pause and state briefly the broadest aims of the method. Accordingly, we may say that in axiomatizing a system of discourse we seek to accomplish with maximum economy and explicitness, the following three things :

(1) Separation of what we choose to talk about and choose to assume about it from the logical consequences of these two choices. (This is more usually expressed in terms of the distinction between the assumed elements

or "vocabulary" of our system together with postulates on the one hand, and the resulting theorems on the other.)

(2) Formulation of *all* the conventions or rules which *alone* will govern the elaboration of whatever follows from the above choices.

(3) Separation, to whatever extent possible, of the purely structural (i.e., syntactical) from the meaning (i.e., semantical) aspects of the system of discourse in point.

These objectives make the formal mode of exposition the most sweeping manner of systematizing scientific discourse. Yet despite a long and distinguished mathematical tradition, empirical science has not, except in most recent developments, made much use of what we would now consider to be rigorously formalized modes of exposition.[12] Formalizations of even our most sophisticated and successful sciences have come only within the past few decades. There are to be sure an increasing number of notable instances, especially in the newer biological and behavioral areas, where axiomatic approaches, or close approximations to these, are being used as primary techniques, at the very outset of inquiry rather than at later stages as reconstructions. But these are still quite rare. Thus, formalization continues to be essentially a *post facto* analytic process which, for the present at least, is as much entitled to be called *philosophy* – more particularly, *explication*, as it is to be called *science*.

3. Possible Objections to Explication

From the viewpoint of a strictly analytical conception of philosophy explication seems to invite a certain sort of objection. This would be that the stipulative and prescriptive character of the method features a *synthesis* of new meanings rather than the *analysis* of established ones.

The nub of this objection is that such synthesis is presumably a task for the specialized disciplines as part of theory construction, but hardly the

[12] Euclid's formalization of geometry, often held up as a paradigm of the axiomatic method cannot count as a genuine counter-example to this claim. Though the Greek geometer and his contemporaries no doubt thought his system to be an idealized science of physical space, historically, it seems to belong more properly to mathematics than to empirical science. For that matter Euclid's monumental work is something of an anachronism even as pure mathematics, the rest of whose formal logic does not really begin until two thousand years later. It might also be added here that Euclid's conception of the axiomatic mode of exposition differs in rather fundamental respects from present-day notions about it. In particular, Euclid's reconstruction of geometry makes no explicit distinction between matters of form (i.e., syntax) and matters of referential meaning (i.e., semantics). This, as we shall note in later discussion, is an important feature in the more abstract, present-day conceptions of formalization.

concern of philosophical analysis whose business does not customarily go beyond the clarification of meaning already in use. The objection might go on to indicate that even the proponents of explication disagree on what they may or may not do, particularly on how far to go in revising the original meaning. At what point, it may be asked, does a reconstruction differ so widely from the original meaning that it becomes gratuitous and irrelevant as a philosophical analysis of that meaning?[13]

Admittedly an explication may so greatly transform a meaning as to amount to a meaning creation *ex nihilo*, as it were. But such an extreme case would not invalidate distinguishing between the clarification and origination of meaning. Explicating a given concept or meaning is in fact a quite different sort of affair from originating it in the first place. In explication the primary commitment is to the pre-established usage which motivated the analysis in the first place; in concept formation there is little or no commitment of this sort. The first process has an uneliminable reportative aspect; the second, by contrast, is a free invention subject only to the logical constraints of the context. A newly contrived scientific concept as, for example, the chemical notion of acid may be useful or otiose, rich or meager, central or peripheral. But it cannot be distortive. An explication, on the other hand, can, in a very definite sense be mistaken with regard to the explicandum in which it is supposed rooted. It is mistaken or wrong when, though it may be ingeniously and even usefully contrived, it is not in fact an explication of the meaning it is supposed to analyze.

The basically stipulative character of scientific concept creation is sometimes lost sight of because of strong similarity that such concepts show to others of ordinary discourse, very often the same word being used for both. Thus, one may note some proximity of meaning between the scientific senses of terms like 'work', 'impulse', 'acid', 'fish', or 'particle' and their

[13] Some writers, as for example, H. Reichenbach, favor the independence of the reconstructed meaning from the pre-systematic one. Others, as for example, A. Tarski, favor similarity. For a somewhat more detailed survey of conflicting views in this respect see : J. H. Hama, "An Explication of Explication", *Philosophy of Science*, LXV (1968) 28-29.

In this general connection H. Reichenbach writes : "Logic is interested not in what man means but in what he *should* mean, that is, in the meaning that, if assumed for his words, would make his words compatible with his actions". H. Reichenbach, *The Theory of Probability*, University of California Press : Berkeley, 1948, p. 372. By 'logic', here, Reichenbach means philosophical analysis, and by 'what a man means' he seems to have in mind systematic usage which, he cautions, may be based on psychological states such as images, fears, hopes, wishful thinking and other associations made with words.

everyday senses. But this is often in the nature of a loose analogy having little more than heuristic value. For all scientific purposes, it is about as important that the physicist respect the non-scientific meaning of the term 'work' as it is for the mathematician, to concern himself with non-mathematical senses of words like 'eccentricity' and 'irrational'. Whatever vestiges of ordinary content may remain in the newly formulated meanings is merely a matter of lexical history. In contrast to this, explication as a method of *philosophical* analysis never breaks free of the meanings it seeks to *reformulate* for greater precision and clarity. Radical as the revision may be, it is, at least in principle, still a revision and not a free invention. As a matter of fact, unlike concept creation, it often includes a preliminary stage consisting of non-reconstructive analysis, possibly resembling the ordinary language type, which provides the explicanda. In this way, the process can and does remain sensitive, even if not *literally* faithful, to established usage.

It should also be noted that any modification of meaning, that results from explication is more or less linked with linguistic effects in the wider system of discourse. Broadly speaking, the full significance of an expression depends upon the language in which it is couched. Explication, therefore, is not simply the free revision of isolated meanings. It is a matter of context-dependent process which in some case mandates linguistic changes elsewhere in the system of discourse and in all cases must be so governed that the explicated concept (i.e., the explicatum) is logically well-fitted to the rest of the systematic context. In this respect explication admittedly bears a strong resemblance to scientific meaning construction which of course is also tied logically to the theoretic context. But, once again, there are differences. The scientific process is in effect a part of theory construction – in the broadest sense, a part of the process of discovery. Explication is fundamentally reconstructive and, though it may keep step with scientific discovery and formulation, it remains basically *post factum* throughout, coming essentially after, rather than before, the initial scientific steps. The anachronistic tease that twentieth century philosophy is nineteenth century science, though methodologically crass, nevertheless has its point. *Qua* philosopher, the investigator of scientific discourse does not create such discourse, he examines and sometimes refines it. At its core, his explication of scientific meaning remains essentially reportative. Ultimately it must – even if only in a broad sense – square with science as it *is* and not with any desired conception of what it *might* be. Scientific usage is, to be sure, a largely deliberate affair, controlled enough to escape most of the semantical vagaries of common language. But it does have its "ordinary" or "natural" aspects. In a very important sense, there is, indeed something which parallels ordinary everyday usage and which may be called ordinary scientific usage. It consists of a more or less systematic blend of formal expression and natural idiom which expresses the established results of science and thus constitutes the main body of actual scientific discourse. Every explication, if it is to be an explication at all, remains subject to comparison with this body of discourse by means of representative or paradigm cases and possible counter-examples.

It seems appropriate to mention here that where pre-systematic usage is ambiguous it may yield several competing explicanda for a given term. That is, several distinct pre-systematic versions of the concept may be in actual use. A possible objection therefore might be that a philosopher may seek to grind his own philosophic axe by adopting a particular one of them as a basis for explication, thus abandoning the neutrality one ideally attributes to a purely analytical exposition. In this connection some might want to say for example that Hume's empiricism leads him to reject all rationalistic senses of the term 'cause'. Similarly Hans Reichenbach urges a single-concept interpretation of the term, 'probable', excluding other concepts as vestiges of *a priorism* fundamentally incompatible with empiricism.[14]

This sort of objection, however, need not keep us long. In choosing an explicandum as a basis for explication a philosopher need not discredit other possible choices for analytical consideration. Such choices do not logically presuppose any particular philosophic doctrine. When a philosopher chooses to explicate one prevalent shade of meaning to the neglect of others his work may be said to be, at worst, limited or incomplete. Any speculative claim he might want to make for the correctness of one choice over another would have to be argued quite separately from the explication itself. If Hume rejected all but an empiricist account of causation he did so on the basis of epistemological arguments and not of the ensuing explication itself.

D. THE ANALYTICAL ACCOUNT OF SCIENCE

The philosophy of science as an analytical enterprise may undertake any of several things :

1. It may attempt, in the reductive manner, to restate a *part* of science so as to render it intelligible in some other set of terms. These latter may be the technical language of philosophy and logic as when we explicate

[14] H. Reichenbach, *op. cit.*, pp. 371-372.

the term, 'probable', or axiomatize quantum mechanics, or re-state for greater explicitness the content of Galileo's "law" of inertia perhaps to determine whether it expresses a definition or a matter of fact. On the other hand the terms for the reduction may be those of common language so as to give a popularization of science.

When reductive analysis takes the form of an explication or a formalization we have what one may describe as a *reconstruction* of science. In reconstructive analysis, the philosopher's interests overlap considerably with those of the theoretical scientist who may also be engaged in the linguistic refinement of theory, during those times when he is neither inventing nor confirming it. For the purposes on hand, we have stressed the distinction between such reconstruction and the originative construction of scientific concepts. But in practice originative construction and *ex post facto* analysis are but two stages in the continuous development of scientific discourse. Cases falling on the borderline between these two are many with much done by scientists that could be called philosophy and *vice versa.*

2. The philosophy of science may undertake to study the distinguishing characteristics of some *part* of science e.g., the particular features (if any) which cam be claimed for the theories, experimental laws or explanations of some special science along with the particular nature of its investigational procedures, the distinctiveness of its vocabulary and so on. Thus it may be asked – whether in biology or in history there occur modes of explanation that are fundamentally different from those found in the physical sciences; or whether in speaking of purposes as causes the behavioral sciences are using causal language differently from the way it is used in the physical sciences; or, again, whether the notion of *function* in the social sciences marks these disciplines as "autonomous" and social phenomena as ultimately unexplainable in terms of non-human factors.

3. Finally, the analytical philosophy of science may concern itself with science in general. This in fact is its most typical concern and will constitute a major portion of all that follows.

As a major enterprise of such a study one might at once envision a vast reconstruction, perhaps by axiomatization, that would bring the entire content of science under one theoretic system. But, for the present, this is at best only a prospectus since no theoretical base for reducing all the sciences to one has, as yet, been found – nor may ever be found. In this respect a broad analytical study can, at best, bring under clearer focus those junctions at which some sciences are logically linked to others, especially those cases where some sciences have been subsumed under or, as it is often put, *reduced* to others.

On the other hand, the general philosophy of science goes beyond matters of theoretical unity to consider characteristic aspects of science. In doing this it may seek to explicate basic terms like 'probable' or 'random' which seem to belong to all science rather than primarily to certain ones. Or it may talk about certain wide classes of terms as, for example, those designating so-called "unobservable entities" which either because of their minuteness or other characteristics attributed to them cannot, in any ordinary sense, be perceived. Terms like 'atom', 'elementary particle' and 'subconscious' would be of this type. A philosophical consideration of such theoretical terms, as they are sometimes called, raises questions regarding their empirical significance or the existence of what they purport to designate.[15]

But even more typically the general philosophy of science addresses itself to the common methodological or, as we prefer to say, *structural* features of science. Thus, it might investigate such pervasive notions as those of space-time and the causal relation; inquire into the nature of laws and how these differ from accidental regularities; seek to ascertain the roles of deductive and so-called "inductive-type" inferences; reflect on the patterns of such modes of inference as explanation, prediction, retrodiction, and others; study the systematic features of measurement; and so on. In such an enterprise it is sometimes thought necessary to give special attention to the social sciences, and possibly even the biological sciences. But if so, this is done in the context of more general discussion serving only to draw comparisons of similarity or difference between these broadest of divisions in scientific inquiry. More specialized parts or branches of science may also be singled our for attention but, again, this time mainly in the context of the general considerations.

The general analysis of science has gradually yielded a vocabulary that includes terms like : *data*, *measurement* and *error*; *singular fact*, *law*, *theory*, *interpretation* and *model*; *chance*, *probability* and *relevancy*; *confirms*, *explains*, *predictive base*, *inductive*, *deductive*, *valid*, *sound* and many more. Terms such as these are much used by scientists. Yet, except for a few like 'probability' and 'error' they do not, very strictly speaking, belong to the vocabulary of science but rather to philosophical discourse *about* science. With such vocabulary one could, for example, say of a given science that some of its laws are more or less general or better confirmed than others; that certain of its statements confirm, explain or form a predictive

[15] The question of the existence of such theoretical entities is intended here as a purely logical one whose answer depends only on the evidence already offered us by the empirical sciences and on how the term, 'existence', is to be understood.

base for others; that some of its statistical practices are sound or unsound. Or one could claim that some sciences are more concerned, than others, with the implementation rather than the creation of explanatory theory; that one science is reducible to another; that the laws of some sciences are largely statistical rather than deterministic. Or, in speaking of all science, one could say that its central tasks are : discovery, explanation, and other forms of subsumptive inference, extrapolative and interpolative calculation, prediction and confirmation. Or, again, one could even venture to say that the scope of science with respect to explanation and factual knowledge is or is not limited in principle, and so forth.

But now, this "metascientific" terminology, that is, terminology in discourse *about* science, and employed by philosophers and, to some extent, also by scientists, has what we may call its proto-systematic stage of development, and even its "ordinary" or pre-systematic phase. At such levels of usage many key terms – 'generality', 'explanation', 'confirmation', etc., used in philosophical talk about science, may, themselves be in need, of clarification, often of the explicative type. In such a case the philosophy of science becomes reflexively engaged in analyzing its own language, potentially to yield a more or less precise framework for systematic discourse *about* science.[16]

E. THE PHILOSOPHICAL ANALYSIS OF SCIENCE AND A THEORY OF SCIENCE

1. The philosophical analysis of science leads to the sort of account that some have characterized as a 'theory of science',[17] but it is far from clear as to just what has been added by putting it this way. We can, no doubt, suppose that, in some one of its several nuances, the term 'theory' can be applied here. But what we should not wish to convey by such a term is the unqualified notion that the analysis of science is some sort of

[16] Some writers have introduced the term, 'metascience', to describe their studies *about* science. For example see : M. Bunge, *Metascientific Queries*, C. C. Thomas : Springfield, Illinois, 1959, *passim*; also G. Radnitsky, *op. cit., passim.*

Following the lead of present-day usage for terms like 'metamathematics', 'metaethics', and 'metalogic' these writers mean to suggest by the term 'metascience' the notion of a study *about* science rather than *in* science. As it happens, however, some of those who have adopted this term seem to include within its application an assortment of studies which diverge considerably from what would here be called analytical philosophy of science. Thus, we have refrained from any major use of this term in subsequent discussion.

[17] As for example, I. Scheffler in his *The Anatomy of Inquiry*, Alfred A. Knopf : New York, 1967, p. 6.

empirical enterprise having the "phenomenon", science, as subject-matter. Such a view could be highly misleading. The philosophy of science must, indeed, confront a definite subject-matter — namely, the "fact" of science. But there the term, 'fact', can be misleading unless it is construed in a rather special way. In some respects the occurrence of science is factual in much the same sense that the occurrence of lightning or of inflation are. But in other respects the "fact" of science is of a very different kind, and calls for a study that could not be called factual or empirical in any sense in which we would apply these terms to physics or economics. The study which we refer to here is not one that views science as an historical or cultural phenomenon happening in space-time. It is a study which focusses on science as a body of statements – statements which are to be regarded as essentially linguistic items. And this means items whose fundamental character is their *use* to express systemized knowledge. Thus, the subject-matter of the philosophy of science, though determinate in character, is not a set of "world occurrences" but rather a highly systemized body of discourse which purports to say something true about the world. Indeed, the array of statements constituting this body of discourse has its own time-place genesis, subject to causal factors much like other human products, or circumstances. Again, however, it is not the genesis but, rather, the logical structure, and conceptual content of the body of scientific statements – and this includes also its credentials – that primarily concern the philosopher of science. If it be asked whether logical structure and meaning are not themselves human creations and therefore natural objects or events that can be brought under empirical study, the answer may indeed be yes. But the philosophy of science, as we construe it here, has no part in such an empirical study. It does not attempt to predict how scientific knowledge (structure or content) will develop with time, what systematic preferences scientists will express, and what formulations will be made in the next few years. Nor does it try to give causal reasons for any such development. Thus it would not be concerned with why, in terms of causes, motives, etc., scientists prefer to speak of efficient rather than final causes; or tend to adopt theories that are logically similar to others already in use. It may well be the case that some alternatives are more effective than others for ordering our experience, and this the philosopher of science would certainly want to show explicitly. But by doing so he does not explain why the more effective ones are in fact adopted or even preferred. Much in the form or content of science may eventually change. But no such future change in what now passes for science can falsify any present account in the analytical philosophy of science; it can only call for another analysis.

The schemas and formal patterns in terms of which scientific discourse is explicitly analyzed include no time-functional laws describing the development of scientific knowledge. The analytical account of science, though essentially reportative and descriptive of what lies at the fundaments of scientific discourse, renders all its subject-matter in terms of time-independent relations which are, broadly speaking, logical rather than causal. Nor does the philosophy of science, as we understand it here, join with science in the empirical study of any other subject matter. Unlike other enterprises that sometimes go by the name of philosophy, it attempts to add nothing of factual significance to the account which science has already written about the world; it only attempts to turn more light on what may be called the "text" of that account.

2. Philosophers unlike scientists often feel compelled to furnish some pragmatic justification at the very start of any work. In particular the philosophy of science is supposed to yield a variety of practical benefits such as applicability in the teaching and learning of science, effectiveness in expository writing, especially of popular science, methodological utility to practicing scientists and more. No doubt something can be said in support of any of these items. But we feel no need here to indulge in this sort of business except, possibly, to touch on one matter that seems to have gotten insufficient attention frcm justification of this kind.

We refer here to the matter of sheer communication not only between scientists and the rest of society, but also between scientist and scientist even in the same fields of specialization. That science is rapidly ramifying under staggering technicalities of language is a familiar enough fact. But what is not sufficiently acknowledged is that : (1) within the same science, highly specialized scientific techniques (linguistic, computational, experimental, etc.) can seriously hamper the communication of results even for colleagues of the same discipline; (2) while the scientist has at his disposal an elaborate and precisioned language in which to articulate his accounts of phenomena, this language does not in general afford the facilities for talking *about* these accounts. Thus it would be gratuitous to expect that in talking *about* what he does a scientist can achieve the same levels of systematic strength which he exhibits in the doings themselves. Newton's disclaimer : *hypotheses non fingo* requires the imagination of an apologist to reconcile it with what he in fact did; while some polemics between particle and wave theorists in the history of physical optics leave notable room for clarity and explicitness. The problem of communication is much more acute on the interdisciplinary level. For here the task of communication often comes very close to the task of popularizing specialized results. In principle, any scientific statement, however abstract, could be reduced to every day language if enough words were used to do it. In general, however, sheer length or wordiness would rule out such an account as totally impractical. With increasing polarization between specialist and layman, this sort of difficulty seems to be approaching irresolvability except perhaps for the possibility of developing an appropriate system of discourse in which to establish such key methodological or logical distinctions as might help get behind some of the technical obstacles to communication [18] Philosophy of science can be the source of such a system, i.e., of a language not in which to do science but in which to communicate more precisely about it.

[18] A suggestion along similar lines is briefly made by S. Toulmin in his *The Philosophy of Science*, Hutchinson's University Library : London, 1953, p. 15.

II

WHAT SCIENCE IS : AN INTRODUCTORY CONSIDERATION

A. SCIENCE AND NON-SCIENCE

1. Some Areas Falling Under Science

The term 'science' is occasionally used loosely to apply to almost any controlled and well-organized activity as for example in saying that Mrs. Jones has her housekeeping down to a science. Or it may refer to any branch of systematic inquiry as when we say that mathematics is the science of quantity. or that aesthetics is the science of beauty, or that logic is the science of discourse. Here the term, 'science', will be used in a much narrower sense. This is the sense in which physics, biology and certain versions of psychology and the social studies are sciences, but logic, pure mathematics, philosophy and theology are not. This restricted sense of our subject is a well established and philosophically interesting one. It marks off a family of disciplines of incomparable distinction. The influence of these disciplines on human existence has been incalculable. Collectively these disciplines recognize no limitations on their applicability to the space-time world. And their methods claim unique validity, indeed, parity with sanity itself, in all such matters of inquiry. As a matter of custom these disciplines are often referred to as *empirical science*. The term is a good one and we shall on occasion find ourselves using it. The simple term, 'science', however, will be understood here in precisely the same sense and will, for the most part, be used instead. Moreover, our use of the term will generally apply, not to the physical procedures and techniques that are so essential to the pursuit and growth of the disciplines in point – but rather to the intellectual product or, as we might say, to the "knowledge" which results from these activities and is expressed in ordered sets of statements which we will sometimes refer to as *scientific discourse*.

But this is only an indication, largely by example, of what science is. It offers little in the way of giving the characteristics which set it apart from other areas of inquiry. Unfortunately, however, the framing of any synoptic definition that is reasonably clear seems to be a rather sticky matter tending to unleash questions that easily lead one into labyrinthine discussion usually falling under the general theory of knowledge and ontology and, for the most part, lying beyond the scope of the present work. Nevertheless, to heighten orientation and to illustrate some of these underlying questions we venture a brief excursion of this kind in the general considerations that follow.

2. The Beginnings of a Definition

In the way of a broad characterization one might want to say, as is sometimes done, that *science is a systematic body of matter-of-fact knowledge*. Reflection immediately raises several questions concerning such a definition. It seems to exclude fragmentary, factual "bits" of knowledge, all right; but how does it exclude that vast body of practical knowledge ranging from crude day to day know-how to rather high level technology, all of it obviously factual and, in some ways, undeniably systematic, yet none of it quite science? And how are we to exclude mathematics, philosophy and several other disciplines which seem not to be science in our intended sense. Mathematics, for example, a paradigm of systematic structure, is sometimes said to be about matters of fact, i.e., "facts" having to do with numbers, sets, functions, etc. Again, some branches of philosophy search out general "facts" about space, time, cause, substance, personal identity, fate, the universe as a whole, the ultimate, being in general, and much more. And logicians have claimed to be dealing with "facts" about sentences, propositions and meaning. Yet none of these disciplines counts as a science, although all of them claim to generate truths, i.e., "facts" about some sort of subject-matter, and to do so systematically. Perhaps we could settle the matter by differentiating science from everyday factual knowledge by the extent and kind of systemization which the former institutes. Mathematics, logic and other highly systematic non-sciences could then be excluded by specifying that the subject-matter of science, unlike that of the non-sciences, is existential, or – as it is sometimes put – that science tries to make true statements about the world? [1] To answer "yes" to this question is to invite still another question. What is it to be existential or real? Substances (apart from properties), causal connections

[1] The characterization of science as an attempt to make statements about the world is very common. For one of the simplest and most recent statements to this effect, see : A. C. Danto, *What Philosophy-Is*, Harper & Row : New York, 1968, p. 15.

(apart from events), color (apart from colored objects), etc. – sometimes referred to as "metaphysical entities" have at some time or other been said to "exist", Are they, in some determinate sense, a part of the world? Mathematicians and logicians sometimes attribute existence to some of their subject-matter – as for example – when they speak of the existence of sets, numbers, functions, a variety of relations and more.

Indeed, in an important sense, there is virtually no limit on the sort of things we may intelligibly say "exist". For, as has been cogently argued, it is all a matter of how we choose to talk, a matter of the system of discourse, within which we decide what sorts of *instances* the neutral term 'thing' can have.[2]

Thus, to say that scientific subject-matter is "existential" without offering further qualification seems to be insufficient for characterizing science differentially. Obviously science is not concerned with all that can, in some sense or other, be said to "exist". The appropriate suggestion here seems to be that scientific subject-matter must, somehow, show up in space-time. Moreover science is, to some extent, a search for causes. And causal relations seem, at the very least, to be constant space-time conjunctions of events. Indeed, some philosophers such as David Hume have argued that causal relations are no more than this. Admittedly, to suit the vagaries of discourse we may, with logical impunity, bloat our universe with all sort of "entities", but unless such items are *conceived* as capable of *space-time occurrence* and therefore of causal efficacy, they seem to go not to science but to other disciplines – some to logic, some to mathematics, some to metaphysics, and some, perhaps, to still other areas of concern. Thus it is that we exclude from scientific concern such metaphysical favorites as: causal linkages, considered apart from the space-time phenomena which they allegedly connect; "bare substances" stripped entirely of their attributes (including space-time ones); and even the universe as a whole which, though consisting of all space-time objects, cannot itself be in space-time because it would lack a frame of reference – itself a space-time object.

But once granted that all scientific subject-matter must be in space-time, we are confronted with new questions.

[2] In a well quoted passage W. V. O. Quine expresses approximately this same view by saying "To be assumed as an entity is, purely and simply, to be reckoned as the value of a variable". *From a Logical Point of View*, Harper & Row : New York, 1963, p. 13. Also : "... the objects we are to be understood to admit are precisely the objects which we reckon to the universe of values over which the bound variables of quantification are to be considered to range". *Word and Object*, The M.I.T. Press : Cambridge, 1960, p. 242.

3. Does Our Subject-matter Requirement Rule Out too Much? Theoretic and Mentalistic States

To begin with, it seems appropriate to mention an enigmatic but important class of scientific "objects" known as theoretical or hypothetical entities, e.g., atoms, electrons, photons, and many more, which are not observable in the direct manner that tables and chairs are, their detection requiring inferences generally mediated by great spans of theory. In what sense are such entities space-time objects? If this is not a sense similar enough to that in which tables and chairs are, then what sort of subject-matter do they and the sub-sensory events in which they ostensibly occur comprise? How can characterizing scientific subject-matter as spatio-temporal include such entities? Answers to these polemical questions are not easily attempted without making careful reference to the theories in which hypothetical entities are imbedded. They are important *substantive* matters in the philosophy of the theoretical sciences, but lie beyond our concern here.

A more immediate difficulty arises in connection with some versions of psychology whose alleged subject-matter is mental states, i.e., experiences, feelings, thoughts, memories, etc., not to mention more hypothetical aspects of mind itself such as certain *supposed* unconscious or sub-conscious "states". The unavoidable question is how, in view of all this, can one regard space-time as a distinguishing feature of scientific subject-matter and yet hold these versions of psychology to be genuine sciences. Mental states undoubtedly occur at definite times and have duration; that is, they have time location. But, serious question arises regarding how spatial location can be assigned to such states. As usually conceived, mental states seem to have neither geometric size, nor shape nor position in the sense in which these attributes are ordinarily assigned to physical entities or physical events. And though in day-to-day discourse we speak very naturally of such states, the question of how they relate to physical processes has generated some of the most troublesome and unsettled philosophical issues of all times. And here, the question of causal efficacy comes up once again. If the spacial character of mental states is dubious, then, so is their causal character. Particularly is this so on the Humean notion of causation – a notion which has found wide acceptance and which construes the causal relation as involving, among other things, spatial continguity.[3]

By 'to admit' Quine means 'to be included among what there is'.

[3] "The idea ... of causation must be derived from some relation among objects;

Thus it seems that, doubts about the spatial status of mental states raises serious doubt as to whether they qualify as *bona fide* scientific subject-matter. In various attempts to avoid difficulties of this sort some schools of psychology have attempted to analyze all talk about mental subject-matter in terms of definite space-time "operations" or "behavior". Others, in the philosophy of mind, have offered other approaches to a solution, one of the most interesting of which is the *identity* theory of mind which submits that mental states may in fact be regarded as identical with brain (i.e., physical) processes. Talk about mental states has a different grammar and therefore different meaning from talk about brain processes. But both modes of discourse refer to one and the same thing, just as 'evening star' and 'morning star' do, though the two expressions have different meanings.[4] But the identity theory of mind, as this view is sometimes labeled, raises objections. What are we claiming when we say that a certain mental state, M, and a certain physical or brain state P, are the very same thing, that is, are numerically identical? As must seem perfectly clear to anyone who understands the language involved, we are claiming that, in fact, *such an M and P have precisely the same attributes*. We may have two very different ways of talking about this one thing, i.e., M-talk and P-talk, but whether we use M-statements or P-statements, we cannot by means of these statements attribute a property or relation to the M and deny it of the P. Grammatical differences between M and P languages may indeed mask this identity of attribution but they cannot, on pain of contradiction, violate it. But ordinarily we do say that P's have spatial attributes, that is, have space location and take up volume, and deny that M's do. The identity theorist, however, would say that this last denial does not express a natural "fact" or law about M's but merely reflects a linguistic (i.e., logical) impossibility. It merely expresses the senselessness of speaking of spatial M's, since M-language does not provide the grammatical machinery for space-talk. Strictly speaking it

... I find in the first place that whatever objects are considered as causes or effects, are contiguous; and that nothing can operate in a time or a place, which is ever so little removed from those of its existence. Though distant objects may sometimes seem productive of each other, they are commonly found upon examination to be linked by a chain of causes which are contiguous among themselves ... We may therefore consider the relation of contiguity as essential to that of causation". D. Hume, *op. cit.*, Vol. 1, pp. 78-79.

[4] For the identity thesis see : J. J. C. Smart, "Sensations and Brain Processes", *The Philosophy of Mind*, ed. V. C. Chappell, Prentice Hall : New York, pp. 160-172, also U. T. Place, "Is Consciousness a Brain Process?" *op. cit.*, pp. 101-109, and J. Shaffer, "Could Mental States Be Brain Processes?" *Journal of Philosophy*, LVIII (1961) 813-822.

is just as senseless to deny spatial attributes of M's as it is to affirm them. The denial in point therefore is not a genuine factual denial, not, at any rate, the same sort of denial as denying that a certain M occurred at a certain *time*. Similarly we say that P's are public things but *ordinarily* deny this of M's. Once again the air of paradox in this vanishes when we note that the affirmation and denial here are not factual, They are mere logical truisms which follow from the way P-terms and M-terms are used, respectively, in P and M languages. Such affirmations and denials, therefore, though apparently incompatible, are not contradictory because they stem from different grammatical rules.

There are some, however, who would say that our inability to attribute spatial volume or public accessibility to a thought, as we do to a brain state, is more than a matter of linguistic rules and limitations. They would maintain that it also says something about the nature of thought and that one cannot, therefore, establish the spatiality and public accessibility of M's by merely providing linguistic adjustments. Be this as it may, the question of what mind is remains an elusive and controversial topic. Not only the identity thesis but also other approaches to the question continue to be matters of highly unsettled opinion. Accordingly, if we are to say that the subject-matter of science is characteristically spatio-temporal then the status of at least some versions of psychology as a scientific endeavor falls open to question.

4. Still More Questions : Science and the a Priori

Other questions come to mind in connection with the space-time characterization of scientific subject-matter. What do we say to those who claim that there are approaches to space-time reality through systematic disciplines other than the sciences? If such a claim were true then having space-time subject-matter would not suffice for distinguishing science from other systematic disciplines. Indeed, any account that seeks to distinguish science by its subject-matter is made additionally uneasy by the patent fact that, in actual practice, it is simply not true that the non-sciences deal exclusively with what is not spatio-temporal. As a matter of historical fact science owes much to mathematics and speculative philosophy not only for the deductive and expository techniques which it inherits from these disciplines but also for many factual principles, unmistakably intended as descriptive of the space-time world.[5] Even simple factual laws have

[5] Scientific principles admittedly espoused *a priori* because of their alleged self-evidence or their clarity and distinctness to reason have played incalculable roles in the development of science. In some cases such principles have held key places in the theoretical structure

sometimes been tied to speculatively based, pre-scientific doctrine.[6] And philosophers are not the only ones that have ventured into matters of fact. Religious doctrine on its own sort of grounds has, in the past, had much to say about what was, is, and will be. Great mystics have offered not only alleged truths about God but also prophecies about the course of world events, and revelations about a "deeper" reality which despite its depth is, according to those who claim to know it, no less capable of space-time manifestations than tables and chairs. One could, perhaps, regard the factual assertions of the non-sciences as merely unessential vestiges from a long tradition in which distinctions between science, mathematics, logic, philosophy and, in some cases, prophecy were faint and, often,

---

for long periods before they were put to critical experimental test. Perhaps the most striking example is Euclidean geometry all of whose axioms were almost unexceptionally regarded self-evident truths about the world and remained virtually unquestioned for more than two millenniums. Similarly regarded were several great principles of symmetry, economy of nature, conservation and continuity, none of which were challenged until recent times. One may expand the list further to include absolute space and time; the mutual independence of space, time, and mass; and more.

[6] The law of optical reflection, for example, which says that the angle, r, at which light is reflected from a surface is equal to the angle, i, at which it impinges has, by some, been regarded as being, in part, a consequence of an old teleological principle that says : Nature acts only to achieve a purpose. This is understood to imply that nature achieves its purposes in certain minimal ways. For, to do more than the minimum is to do what is not required for the purpose. But, assuming no other purpose to be operative, doing more than the minimum would then be doing something for no purpose hence violating the original principle. As it happens, for a fixed reflecting surface the path from point A, before reflection, to point B, after reflection, is the least path for getting from A to B by means of reflection from the given surface, M, if and only if angles, i and r, are equal.

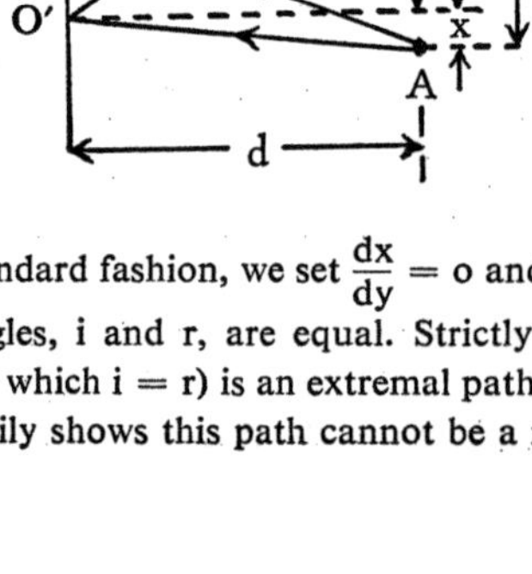

Thus (see figure) AOB is the least path from A to B via reflection from surface M. A non-demonstrative way of suggesting this is with a few scaled constructions to show that representative alternative paths are all greater than AOB. An analytical account might go as follows : We wish to find those conditions which will guarantee that AO'B (= AO' + O'B) is a minimum. It is, however, simpler to do this by determining when $\overline{OA'}^2 + \overline{O'B}^2$ is a minimum, since the conditions for both minima are the same. Now $\overline{AO'}^2 + \overline{O'B}^2 = x^2 + d^2 + (s\text{-}x)^2 + d^2 = y$. Proceeding in standard fashion, we set $\frac{dx}{dy} = o$ and get $2x - 2(s\text{-}x) = o$, which yields that $x = \frac{s}{2}$ or that angles, i and r, are equal. Strictly speaking, this guarantees only that AOB (the path for which i = r) is an extremal path, i.e., a minimal or maximal one. Reflection, however, easily shows this path cannot be a maximal one; hence it must be a minimum.

little more than a matter of special emphasis. More recent cases, of this would then be exceptional and spurious incursions by the non-sciences into areas where they have no business at all. But this of course would be based on the unexplained supposition that science alone *legitimately* deals with matters of fact. It therefore essentially begs the question. In the face of counter-examples, the hard and fast demarcation of subject-matter areas for the purpose of saying what is or is not science tends to appear arbitrary and open to question unless a firmer basis for the choice of area boundaries is offered.

5. Science and Sense Experience

One way of responding to this is to express serious doubt that any non-scientific manner of "seeing" into matters of fact, even if it led to some true assertions, could ever serve as acceptable grounds for such assertions; that is to say – such "clairvoyance", as those objecting might wish to call it, could never serve as the grounds of factual knowledge. Religious experiences and revelations about a "deeper" reality could be understood as psychedelic distortions of ordinary experience induced either by unusual physical circumstances, by drugs or by more subtle psychological influences. Prophecies, in turn, could be viewed as either blind guesses or else as common-sense appraisals unconsciously formulated on the basis of vaguely entertained principles and factual evidence, essentially unconnected to any unusual or non-sensory mode of perception.

The first part of this answer, though at first blush plausible, spawns its own difficulties. How, in assessing the experience of one who claims to have had a revelation do we decide that he has had a distortion rather than a "truly novel" experience, capable of serving as grounds for knowledge of some new kind? The distinction between "distorted" and "novel" seems itself to be troublesome; and qualitative novelty can be claimed for experiences differing from others in virtually any manner whatever.

The second part of the answer could be something of a philosophic evasion if it stemmed from a play on the term, 'knowledge'. What, it may be asked, is to be understood by the term, 'knowledge', when one denies that prophecies are knowledge? Is the term being used in one of the usual senses or is language being "stacked" in favor of the conclusion? A procedure of the latter sort would have no significance beyond instituting or assuring a certain usage for the term knowledge. Accordingly, prophecies even if based on some alleged, extraordinary perception, and even if they turned out to be true would not count as "knowledge", (and therefore would not be science) simply because they have been excluded by definition – in this case, by a stipulation regulating the use of the term 'knowledge'. Surely anyone denying the status of knowledge to prophecies must be engaged in something more than linguistic reform. If so, what could such a denial be saying? Let us this time approach the question by asking, what is a prophet claiming when he says he *knows* that there will be a disaster before the years end. Reflection indicates that he must be saying something like the following : (1) that there will in fact be such a disaster; (2) that he is convinced there will be; and (3) that he has an adequate basis for his conviction. Deciding on whether a prophecy meets conditions (1) and (2) would present no special problems. But, having satisfied these, the charge that prophecy is not knowledge amounts to saying that it fails with regard to (3). This means either that prophecies are totally baseless, meaning that the "grounds" they offer are of the wrong kind; or it means that though, perhaps, appropriate in kind, they are not sufficient for knowledge. The question of sufficiency when the grounds are of the right kind is a difficult and potentially technical one. It will not be profitable to pursue it here. But saying that science rather than prophecy is genuine knowledge suggests *scientific* grounds as being of the right kind. Our concern with differentiating scientific from prophetic or other, non-scientific types of discourse then becomes a concern for the nature of scientific grounds. And what seems most apparent in this respect is that the procedures of the scientific disciplines contrast rather sharply with those of the non-scientific approaches to reality, in at least one major respect. The sciences conduct laboratory experiments or engage in systematic field investigations; the non-sciences do not. Accordingly, one might want to suggest *experience* as the sort of basis required in order that true factual convictions be knowledge. But here we must pause for a precaution. The non-sciences which, as was noted, have sometimes claimed to be sources of factual knowledge, have also offered "experiences" of diverse sorts as grounds for such knowledge. The history of such claims is replete with references to intuitions, clarity and distinctness, self-evidence, revelatory visions, insights, extrasensory perceptions, new qualities and perceptual "dimensions", "altered consciousness", and so on, all of which are sometimes associated with or spoken of as "experiences "of one kind or another. If the non-scientific disciplines (even when they claim to generate matter-of-fact knowledge) are to be excluded, then the term, 'experience', must be further qualified.

The observational methods of science, experimental or other, obviously suggest *ordinary sense experience, rather than alleged "experiences" of any kind whatever, as the basis required for the having of knowledge about space-time subject-matter.* Let us, for convenience, then, put the term, 'fact', and its derivatives such as 'matter of fact' to a specialized use and apply them only in connection with *truths about space-time subject-matter.* Where other types of subject-matter are concerned, reference will be made simply to *truths* as, for example, in *truths of logic* or *truths of arithmetic.* We may now characterize science as *a systematic body of matter-of-fact knowledge* and thus exclude those disciplines which claim to offer "knowledge" about space-time subject-matter on the basis of alleged non-sensory "perceptions". This exclusion, however, has been achieved by denying the status of knowledge to any belief regarding matters of fact if it is not based on sense-experiential grounds or – as it is often put – if it is not empirically grounded. In more traditional terms, this denial is expressed as the impossibility of knowing any *synthetic* (i.e., matter-of-fact) statement to be true *a priori* (i.e., on non-sensory "grounds"). On such a view, for example, any claim to *know* the axioms of Euclidean geometry as matter-of-fact *truths of reason* would be considered simply false. Despite any truth these axioms may have when they are interpreted as statements about the physical world, they cannot be said to be known when the only basis for asserting them is their alleged "self evidence" or "clarity and distinctness" or some other intuitive "perception" of their truth.[7]

But here some very pressing questions arise – pressing because their answers seem to be required for a proper understanding of this *sensory* theory of knowledge. What sort of claim does one make when he says that the sole basis of factual knowledge, and therefore of science, is ordinary sense experience? Is it merely a behavioral report stating what sort of material scientists and men engaged in practical affairs actually take as a basis for their factual judgments. If so, the claim is merely a reportative statement arousing little philosophic concern. As a matter of fact, construed as such the statement would seem to be false. Scientists have not, in practice, always held to sense experience as the sole basis of what they have been willing to count as knowledge in matters-of-fact.[8] Modern science right

[7] This denial of the synthetic *a priori* is not to be confused with the sort of argument often leveled against the claim that certain statements as, for example, 'all colored things are extended', are synthetic, i.e., factual, but known *a priori.* For cases of this latter kind, the usual strategy is to argue that such statements are not synthetic at all, and that if "true" they are only logically true, i.e., analytic, and therefore factually empty. For a lucid introductory account of the major lines of controversy in this matter see : J. Hospers, *Philosophical Analysis*, Prentice Hall : Englewood Cliffs, N.J., 2nd edit., 1967, pp. 179-186.

[8] Rationalism is the philosophy of knowledge which claims that some matter-of-fact

up to the present century has held to some principles for which it is extremely doubtful that any experimental evidence was seriously sought.

Perhaps, then, we should understand our sensory principle to be not a report on precisely how scientists and others actually behave but rather a claim that sensory perception is the only *sound* basis for matter-of-fact knowledge, regardless of the fact that scientists have, as it seems, sometimes deviated from the principle. *A priorism* in science could then be regarded as a hangover from the past two thousand years of rationalism, a hangover which science has been trying to "kick" since the seventeenth century. Thus, it could plausibly be maintained that some *a priorism* in science is only apparent rather than actual. For, many scientific principles which seem to have been adopted without experimental basis are, in effect, implemented not as *a priori* conclusions, but as tentative, informal, perhaps even subconscious generalizations from day-to-day experience, representing

---

statements (i.e., statements about the time-space world, in some cases very important and theoretically fundamental ones) can be known *a priori*. Examples are the axioms of plane Euclidean geometry, understood not as uninterpreted assumptions of pure mathematics, but as principles of the physical world.

That rationalism was the dominant theme of Western thought prior to seventeenth-century science no one would seriously dispute. Its intellectual grip on subsequent science, holding on through the last quarter of the nineteenth century, has also been recognized by many scholars. Particular interest has centered on Galileo who, though generally hailed as the father of modern experimentalism, seems to have been deeply influenced by scholasticism and Platonism, both of which are variants of rationalism. (For divergent views on which of these two rationalistic traditions form the stronger background of Galileo's thought see : Pierre Duhem, *Les Précurseurs Parisiens de Galilé*, Paris, 1913, and Ernst Cassirer, "Galileo's Platonism", *Studies and Essays*, ed. M. F. Ashley-Montagu : New York, 1944, pp. 279-296, also, E. A. Burtt ,*The Metaphysical Foundations of Modern Science*. Doubleday : Garden City, New York, 1954, pp. 74-78.

The influence of rationalism on Galileo's thought is most obvious in his views on the crucial role of mathematics where scientific knowledge is concerned. This has been referred to as his mathematicism or Platonism. Some passages in his *Dialogo* suggest that he regards mathematics not merely as a logical tool – an analytical "key" to the universe – but as a body of fundamental and *necessary* truths holding, though only with approximation, for the actual world. Salvatio, who speaks for Galileo says, "... I say that human wisdom understands some propositions ... perfectly and is as absolutely certain thereof as Nature herself; and such are the pure mathematical sciences, to wit, Geometry and Arithmetic. In these, Divine Wisdom knows infinitely more propositions, because it knows them all; but I believe that the knowledge of those few comprehended by human understanding equals the divine, as to objective certainty, for it arrives to comprehend the necessity of it, than which there can be no greater certainty". See : Galilei, Galileo, *Dialogue on the Great World Systems*, Salsbury Translation, University of Chicago Press, 1953, p. 114.

at most postponement rather than a forgoing of rigorous experimental test.[9]

But even so, we are again compelled to ask : What sort of thesis is it that says that sensory experience is the only sound basis of factual knowledge? Is it a "truth" resulting merely from the way language is used? If a part of what we mean by 'factual knowledge' is *that which can be grounded only in expereince*, then the sensory thesis is only an empty truism – a tautology having no content beyond reflecting how certain words are used. If, on the other hand, it means to express more than a logical relation between meanings then it must be a statement about some "external" subject-matter, in particular, about the nature of factual knowledge and what constitutes appropriate grounds for such knowledge. Viewed in this manner, the thesis in question is known as *empiricism*, a

[9] Germane to the point is a restricted form of Galileo's law of free falling bodies which says that such bodies will fall with the same acceleration regardless of their weight. Associated with this law is the famous experiment from the Leaning Tower in which Galileo was supposed to have dropped two bodies of different mass and similar density to show their simultaneous arrival at the ground. There is good reason to doubt that the great experimentalist ever bothered to drop the two missiles. For had he done so it would most likely have been to convince others rather than himself. The principle was obvious to him as he himself indicated. See : *Le opere di Galileo Galilei*, ed. Alberi, Firenze, 1854, p. 49. To see why, consider a falling object to be subdivided into contiguous (i.e., closely touching) but smaller and therefore lighter parts each then falling as independent objects while they remained in fixed contact throughout the fall so as to constitute *one* large object, falling with the same acceleration as any of its parts. This consideration might lead one to think the law to be self-evident and therefore *a priori*. Implicit in such a judgment, however, is the assumption of another law, namely that whether or not two bodies of the same density are joined in close contact (thereby constituting a heavier body) has no effect whatever on their rate of fall, for negligible air resistance. Though this latter law seems *obviously* true, it is by no means *a priori*. For day-to-day sense experience seems to assure us that whether two bodies falling freely with the same speed touch or not has no effect on the speed with which they will continue falling. This latter law is not a truth of reason but a virtually unconscious generalization from ordinary sense experience.

When Galileo's law is extended to objects of varying densities its truth is, of course, not as obvious, and the foregoing considerations do not apply.

Galileo has been quoted as having written that "... in order to demonstrate to his opponents the truth of his conclusions, he had been forced to prove them by a variety of experiments, though to satisfy his own mind alone he had never felt it necessary to make any". (This quotation of Galileo is attributed to J. J. Fahie by A. E. Burtt, *op. cit.*, p. 76) There seems to be much in Galileo's thought to indicate that he might have regarded *a priori* mathematics as a key to the universe, meaning that parts of pure mathematics had factual content in the sense of the synthetic *a priori*. Nevertheless, examples like the one just discussed would not strictly imply this of Galileo's thought. His strong expectations regarding the outcomes of experiments like the legendary one from the Tower might have been grounded in ordinary experience with the phenomena in question.

doctrine which, as historically presented, seems to have been intended as something decidedly more than a tautological truism.[10]

If empiricism is not a thesis made necessary merely by the meanings of the terms involved, then its falsity is logically possible. How, then, does one decide whether it is true or false? What sort of subject-matter are *knowledge* and *grounds* and how do we decide general truths about them? Or, as we might say : What are the *grounds* for believing that only sensory experiences are adequate grounds for factual knowledge? How do we rule out non-sensory perceptions in this respect? One might wish to reply here that determinations made on the basis of sight, sound, touch, etc. are objectively confirmable, that is to say, are for the most part reproducible, with public agreement and under controllable conditions. Accordingly, whenever individuals disagree on any judgment based on sense experience it is in general possible to resolve the disagreement by means of more sense experience. This, as the argument might go, is not the case where judgments based on other purported modes of perception are concerned. Intuitions, revelations and other alleged non-sensory bases for judgment often lead to incompatible conclusions on diverse matters of fact. Furthermore, where such disagreement arises, there seems to be no specifiable procedure for calling up the appropriate non-sensory "perception" and resolving the disagreement.

But though this reply has much force, it does not quite seem to settle the matter. It is not at all clear how any public consensus in some mode of perception is decisive for regarding that mode as the exclusive basis of factual knowledge. If a class of individuals – let us say individuals highly trained in mathematical and scientific areas – demonstrated the ability to make unexceptionally correct judgments on certain matters of fact without either experimenting or making sound inferences from established knowledge, they might claim to be able to "see" or "intuit" the truths in question

[10] The great British empiricist Hume writes, "There is no single phenomenon, even the most simple which can be accounted for from the qualities of the objects, as they appear to us; or which we could forsee without the help of our memory and experience". D. Hume, *op. cit.*, Vol. 1, p. 73. Moreover, our acquaintances with ("ideas" of) things are also derived from experience (impressions). In Hume's words "... all our simple ideas in their first appearance are derived from simple impressions". *Ibid.*, p. 13. Hume seems to offer such statements not as truths "by definition" that is, by mere word usage, but instead as factual generalizations based, perhaps, on a close examination of his own experiences and modes of awareness. Similar passages may be found in the other two great empiricists, Locke and Berkeley. In all cases what seems to be intended for communication are "facts" about the character of human knowledge – not mere terminological relations.

after appropriate reflection. We might imagine such intuitions to require a high degree of advanced training in the area concerned, a qualitative familiarity with the particular subject-matter and a good amount of careful "reflection" upon it, but no observations relevant to the phenomenon involved. In this manner true assertions might, for example, be made about previously unfathomed configurations of atoms, subatoms or even cosmic objects – configurations which always turned out to be experimentally correct. Or, consider a sizable group of mystics remarkably manifesting unanimous agreement on the content of individual and independent revelations which, let us suppose, they could reproduce under pre-specifiable circumstances (e.g., fasting in isolation, or taking psycho-active drugs, or long exposure to psychedelic colors, sounds, etc.). Would assertions based on such "intuitions" or revelations then qualify as science? In this respect, some might even argue that while scientific and other sensory determinations may be highly reproduceable they are not the only ones having this virtue. Certain non-sensory judgments such as mathematical and logical ones seem to be assured with a certainty and regularity perhaps exceeding that of judgments based on ordinary sense experience.

We see then that any attempt at a synoptic differentiation of science from other disciplines which claim to be knowledge-gathering invites a plethora of considerations usually falling under basic theory of knowledge and much beyond the scope of the present work. Our brief excursion into some of these questions was taken merely to serve the more general reader with some introductory perspectives on the sort of issues that may underlie more specific considerations regarding the nature of science and possibly the philosophy of science as a whole.

6. The Scope of Science

It seems appropriate before ending our peripheral excursion to consider one other question about scientific subject-matter. We ask : What is the basis for limiting such subject-matter to space-time things? Is it merely a convenient way of keeping science down to manageable proportions, or is there a deeper reason? Science is man's attempt not only to know, understand and anticipate his circumstances but also to control and to some extent even add significant novelty to them. But this requires a knowledge of causes and when it comes to causes there seems to be no alternative to what Hume pointed out, namely that we have no clear idea (i.e., concept) of causation apart from (the constant conjunction of events in) space and time. Accordingly, as was noted earlier in our discussion, if science is to be a study of causes its subject-matter must be of a space-time character.

Ordinary usage on both everyday and scientific levels adopts such subject-matter for its *ontologic domain.* This means, it refers to sticks and stones and to their relations as "real or concrete" or "material", in a sense in which pure and other abstract entities are *not.* This employment of the term, 'real', seems to be quite natural; for, as it seems, it is only the concrete entities that make causal differences to us – not the least of which is man's own survival.

Science inherits the essential core of its subject-matter from day-to-day factual inquiry. Thus for both science and common sense space and time are not only distinguishing attributes of their common subject-matter but are also the earmarks of reality in a rather ordinary and quite familiar sense of this term. Space and time therefore assume a rather distinctive and special ontological significance, and questions about the nature of space and time take on a rather pressing character, preoccupying philosophers of science of both the analytical and the metaphysical sort. Both science and common sense tell us they are relations that are supposed measurable in certain ways. But just what is the nature of the assumptions underlying these measurements. Can we say anything more about what it is we are measuring? Furthermore how are space and time themselves related? Are there fundamental reasons other than those concerning causation for dubbing things "real" if and only if they turn up in space-time?

B. SCIENCE AND COMMON SENSE

1. Except possibly for some of the last few references to space and time, our attempt to frame a synoptic definition of science has taken us to the brink of involvement with ontological and epistemological matters as well as with questions regarding the nature of perception (or "experience") which lie much beyond our scope here and which we shall refrain from pursuing in all that follows. A more appropriate alternative will be to attempt a preliminary account of the nature of science by comparing it with that wide range of knowledge usually labeled *ordinary* or *common sense* knowledge. This makes a convenient point of departure for our task. Common sense knowledge is familiar to all; in some elemental respects its standards of cognitive validity are virtually undisputed; and from it science is in some major ways an outgrowth, sharing with it certain fundamental, observational methods and a basic subject-matter. We apply the labels 'ordinary' and 'common sense' here with some hesitation, lest they

be understood to refer only to the most commonplace levels of information and know-how. What we mean by these terms includes not only the plain facts of day-to-day experience but also vast stores of knowledge ranging from the practical arts to prescientific technologies that command highly specialized information, an impressive degree of organization and even considerable quantitative precision. Often, though not always, these arts and technologies antedate science and furnish it with major portions of its factual content. The term 'pre-science' might suggest itself here as a preferable alternative to 'common sense'. But it seems to take in too much bringing to mind certain philosophic speculations that common sense, as we construe it here, does not encompass. The term may also tend to suggest that what is refers to is something which historically comes only *before* science. Common sense knowledge does indeed antedate the sciences but it also coexists with them in a form ranging from everyday know-how to some phases of technology which, though employing many of the results of science, in certain basic respects fall short of it. One could of course institute three categories; common sense, technology and science. But for our purpose which is to distinguish science from other levels this would seem to be rather unnecessary and perhaps even misleading. For what is often labeled *contemporary technology*, has, except for its practical slants, much in it that is genuine (though applied) science. And as present-day technological and theoretic inquiry become increasingly interdependent, often living under one roof, it becomes more and more artificial to maintain fundamental distinctions between science and advanced technology.

That science is largely an outgrowth of common sense and shares its deepest core with it would be futile to deny. Not only do these two levels of knowledge both deal with the natural world, i.e., the world of space-time things and causal connections, and do so by ultimately similar observational methods, but, historically, the continuity of development from one to the other makes it in some cases difficult to say just where a prescientific technology ends and a genuine science begins. Indeed with respect to several (though not all) features, the basis for any convenient distinction at all is bound to be somewhat arbitrary and artificial, largely marking differences of degree rather than of kind.

2. Nevertheless, it would be grossly naive to regard science as simply a more elaborate version of common sense. In some highly crucial respects science owes much to logic and speculative philosophy both of which depart radically from common sense, yet provide characteristics that differentiate science decisively from it. Without doubt, the general state of knowledge prior to the sixteenth century is a very different affair from

what it is afterward. And today, as in the past, scientific knowledge exhibits certain features of content and structure that place it considerably beyond ordinary cognizance even when the latter includes large accumulations of the well-ordered information and substantial amounts of highly specialized technological know-how.

Quite obviously, science as a cognitive enterprise, generally surpasses common sense (though not always with respect to range).[11] To what characteristics of science may we attribute this superiority? Several features readily come to mind, and we attempt to label and list the most notable of these as follows : specificity of formulation, over-all logical consistency, perspicuity of language, systematic structure, and theoretical content. Systematic structure of a certain sort and theoretical content are beyond doubt the most dramatic and novel additions that science makes to common sense. The others are not absent from common sense but are characteristically well-developed in science. It will be instructive to give some brief attention to each of the features listed.

C. SOME DISTINGUISHING FEATURES OF SCIENCE

1. Specificity and Precision

High specificity, both qualitative and quantitative, is one of the most obvious features of scientific discourse. It is loosely referred to as "precision" especially where numerical quantities are involved,[12] and is so closely associated with science that being precise is often virtually equated to being scientific.

To illustrate, consider the throwing straight up of a compact dense object from a given initial place, with a certain initial speed, and on a

[11] Technology and everyday practical knowledge often ranges over far wider areas than can at the time be explained or otherwise systematized by science. Thus the technology of fermentation was well known millenniums before biochemistry could give a scientific account of it. At the present time we can in practice do much more, for example, with alloys, crystals, complex vibratory systems and various drugs than can be given a genuinely scientific account of.

[12] In philosophical discussions the term 'precise' is often used in a sense opposed to that of 'vague' or 'unclear'. Numerical precision, however, has to do with being *specific* which is opposed to being *general.* A concept may be general and either clear or vague, as the case may be. We regard a measurement as precise when it has been reliably made to a high degree of quantitative minuteness. Thus a determination of length that has been reliably made to the nearest tenth of a milimeter is, in this sense, more precise than one that has been reliably made to the nearest meter.

calm day. Also, neglect any effects due to air resistance. It is a time-honored rule of ordinary knowledge that, whatever is thrown straight up comes down, returning to the starting place. It would be rather difficult to discredit this old rule of thumb; not only because for all ordinary throwing it holds with good approximation, but also because it is itself, sufficiently non-specific as to allow for a somewhat indefinite range of variations in what actually happens. A more specific statement, and one that might be made in prescientific ballistics, would say that the object returns to the initial place and that the greater the speed with which it is thrown up, the longer it takes to return and the greater is its terminal speed. Now, compare this with the far more specific claim which one might find in a text on (scientific) mechanics : With close approximation, the object (1) returns to the initial place, but (2) with a terminal speed equal to the initial speed, and (3) after a time (in seconds) numerically equal to 1/16 the initial speed (in feet per second). This claim is far more specific than the proceeding one and, so much so, that it is comparatively vulnerable to disconfirmation. As a matter of fact, for initial speeds that are too large or for initial places that are too removed from sea level, (3) will become noticeably untrue, while, if the body is thrown up very high, it will tend to return somewhat displaced from its starting point, which is contrary to (1).[13] Moreover, beyond some height, the body does not return at all.

High specificity and numerical precision in science is not only typical, despite exceptions, but it is often crucial rather than optional. Virtually all of more recent physical theory has at one time or another to square with discrepancies barely outside the limits of experimental error often involving magnitudes of atomic and subatomic dimensions. Indeed, many of the laws of natural science are laws of microdimensions.

Nevertheless, it would be naive to suppose that even comparatively great amounts of specificity and numerical precision would be sufficient to make any discipline a science. The epochal technologies of Egypt, Babylon and Rome did achieve notable degrees of specificity and precision. But would they have been sciences if they had been even more specific and precise in the cataloguing of their alloy mixtures, in their architectural specifications, in their astronomical tables, or in their surveying? By the same token, would we call the genetics of Mendel *fundamentally* less scientific than the atomic physics of Sommerfeld because the latter turns on experimental determinations that are vastly more refined and numerically

[13] This is the Coriolis effect due to the rotation of the earth. For a basic discussion see : H. Goldstein, *Classical Mechanics*, Addison-Wesley : Reading, Massachusetts, 1959, p.p 135-140.

more specific by several orders of magnitude? Galileo is said to have used his pulse for some time measurements. If he did, his chronometry was cruder than some which today we may find being used by the ordinary man. But were his doings therefore unscientific? With the help of modern instruments certain present day crafts can achieve enormous precision as for example in tool and die making or even in certain phases of structural and road engineering. But despite any high precision and the use of scientifically refined implements, these and certain other branches of modern technology seem, in some important respects, to bear greater resemblance to prescientific technology and common sense than they do to genuine science.

2. Logical consistency

It has been pointed out by some that in science there is more concern for over-all logical consistency than in practical affairs, in which immediate results are of primary importance.[14] This is indeed a noteworthy feature of science, and it has lead to many theoretic advances. Common sense and even advanced technology may at times be heedless of it. We can, for example, easily imagine a producer or designer of optical instruments in the early twenties felicitously contemplating some device that incorporates both a photo-electric cell and a classical lens system without the slightest concern for the fact that, as he might well know, such devices are based, respectively, on corpuscular and wave properties of light – properties that seemed so flagrantly paradoxical to his scientific contemporaries as to disquiet them to the brink of intellectual rebellion.

But again, this is not to say that logical consistency is the mark of science alone. It is at the heart of all informative communcation and without it cognitive discourse would come to nothing. In principle, therefore, scientists are not the only ones that must be consistent. They are merely among the few who try very hard to be. And this is no odd compulsion. It rather stems fundamentally from a profoundly distinguishing feature of scientific knowledge, namely, the attempt to bring a wide range of subject-matter under a few principles. Inconsistency seriously threatens the desired unity; hence, every effort must be made to eliminate it wherever it arises. Indeed, in their attempts to expunge paradoxes, fastidious investigators have scored some great gains in the systematic structuring of scientific knowledge.

[14] Nagel, E., *The Structure of Science*. Harcourt Brace & World : New York, 1961, p. 6.

3. Systematic Structure

That scientific discourse is highly structured is beyond question. But it is also quite obvious that possessing any structure whatever is by no means a distinguishing characteristic of science, even the most common collection of assertions may be well-ordered reporting time-sequences, classifications, or even regularities expressed either as causal connections or as correlations. Thus, for example, a chronological account of developments leading to a given event would surely be well-ordered yet might not be science, even if the account were a reconstruction based on indirect evidence rather than a flat description of what had actually been observed. Similarly, a carefully reported classification of items, e.g., a categorization of occupations or of merchandise or even of living organisms, would, without further qualification, hardly be science. Nor, for that matter, would an experiential generalization or even a catalogued set of these *ipso facto*, pass for science, though they might be quantitatively formulated as, for example, some agronomic rules relating quantities of a given plant food to increment of yield under specified conditions. Nor, again, would a set of correlations between atmospheric factors, e.g., barometric pressure, relative humidity, wind direction, etc. and rain, together with any predictions based on them, by themselves constitute a piece of science, however specific or elaborate those correlations might be. Pre-scientific technology, the practical arts and crafts, even day-to-day knowledge illustrate all such modes of structure abundantly.

Nevertheless, it would be futile to deny that structure is one of the most distinctive features that differentiate science from common sense. But what sort of structure is this? A scientific account is more than a well-ordered description giving either the history of things or their place within some classification, or even their immediate dynamic and causal relations. Indeed it would not have to be a science that *explained* (or *predicted*) the brittle and sonorous property of a piece of alloyed copper on the grounds that nearly 20 percent of it was tin. The structure of science is more than this. It is, we might say, *profoundly* systematic. By this we mean that – (I) the key structural components are arguments (also called inferences).[15] That is, they consist of statements E and H, so related that, in some sense, H (the conclusion) is assertable on the grounds of E (the base or premises). (II)

[15] The term, 'inference', is sometimes used to refer to a mental process rather than to a certain sequence of statements. Our use of the term, 'inference', however, will have no such psychological connotations and will be considered interchangeable in all its occurrences with the term, 'argument'.

These arguments are linked to each other in such a way and are themselves so ramified that they logically integrate wide ranges of discourse sometimes encompassing highly diversified subject-matter. The metallurgical example just given is indeed an argument. From general facts of copper metallurgy and information about the composition of a certain metallic object, a conclusion is drawn (derived, inferred) giving certain other properties of the object. The example therefore satisfies (I), above, but, by itself, fails to satisfy (II). Mesopotamian culture which was pre-scientific but had an ongoing and productive technology no doubt cultivated a sizable stock of rather precise experimental principles on which to base arguments which took explanatory, predictive, or other forms. If, collectively, such arguments did not constitute a science, it is, at least in part, because they did not comprise a more comprehensive body of discourse resting logically on a rather limited number of broad principles – principles which in some cases can be surprisingly few and, in certain respects, remarkably simple.

A large portion of all that follows will be given to a discussion of the inferential structure of science. It will be convenient at this point, however, to give some preliminary attention to a traditional dichotomy which distinguishes *two* basic kinds of argument : (i) *deductive*, also referred to as *demonstrative* and (ii) *inductive* or *non-demonstrative*.

(i) A deductive argument is one in which the premises, E, and the conclusion, H, are so related that asserting E but denying H would be a contradiction. Logicians express this by means of such phrases as : E logically implies or entails H; H follows logically or necessarily from E; or 'If E then H' is *logically true* (also referred to as *logically necessary*). It is clear, of course, that E and H in our account of deductive arguments, need not themselves be simple statements but may each consist of combinations of other statements. In the more quantitative sciences, many deductive arguments are often rendered by long sequences of mathematical expressions. A quick thumbing through any text on theoretical physics or chemistry will easily convince one of this. In the less mathematical sciences such as certain branches of traditional biology, qualitative chemistry and non-mathematical geology, deductive arguments are usually of the more familiar verbal kind; that is, they consist of statement sequences expressed in ordinary language, more or less supplemented by specialized technical terminology.

Though some sequences of statements may have the superficial appearance of deductive form, their E and H do not satisfy the required relation. Such purported "arguments" are called *invalid* to distinguish them from genuine deductive ones which are referred to as *valid*. It is easily shown that if arguments have certain forms or proceed by certain steps then they cannot possibly be invalid. Logicians refer to such argument forms or steps as *valid argument forms* or *valid steps* and these constitute what is largely the concern of deductive logic.

Our example from the metallurgy of bronze is, if stated fully, a deductive argument. In explicit form, it says : (E) all metal consisting of four parts copper alloyed to one part zinc is sonorous and brittle. This metal is such an alloy. Therefore, (H) this metal is sonorous and brittle. It would be manifestly absurd, i.e., self-contradictory, to assert E and deny H.

The following is a second illustration :

(E) All unimpeded objects near the earth's surface fall with a velocity that constantly increases 32 feet per second for each second. We call this the acceleration, g, of the object. The object I now hold will be released near the earth's surface and will be allowed free fall for three seconds.

(H) At the end of this time it will have fallen 144 feet.[16]

(ii) An inductive argument is one in which the premises E and the conclusion H are so related that : (1) assering E and denying H would not be a contradiction, and (2) E purports to give a degree of support to H thus, in some sense, making it probable.[17] As an example, imagine a nineteenth-century chemist asserting that : (H) all metallic objects are conductors on the basis of the fact that : (E) of a very large number of diversified metallic objects tested, all have been conductors; or consider forecasting that : (H) it will rain, on the grounds that : (E) atmospheric conditions like the present have, in a very high proportion of cases, been followed by rain; or again someone judging that : (H) an odd-looking fruit is edible because (E) another very much like it was eaten sometime ago without ill effect. Each of these arguments presupposes, some supportive relation of E to H. Yet it is clear for all of them that the premises do not logically necessitate

[16] From E it necessarily follows that the object will accelerate at 32 feet per second^2 for 3 seconds; whence from what we mean by such an acceleration we get that the distance fallen is 144 feet. This may be a bit more explicitly shown as follows :

(1) By definition, $\bar{v} = s/t$ where s = distance fallen and t = time of fall.
(2) Hence, from (1), $s = \bar{v}\,t$
(3) But it may be shown by valid steps that $\bar{v} = \frac{1}{2} v_t$, where v_t is the velocity after time, t.
(4) $v_t = at$ by virtue of what we mean by 'a' (the constant acceleration) and by 'v_t'.
(5) But by (2), (3) and (4), $s = \frac{1}{2} at^2 = 144$ feet.

[17] Contemporary views are much divided on just what it means to say that E gives a degree of support to H. One even finds a recent attempt at developing a notion of inductive support which, contrary to traditional views is non-probabilistic in character. See : L. J. Cohen "A Logic for Evidential Support", *The British Journal for the Philosophy of Science* XVII (1966) 21-43, 105-126.

the conclusion; that is, it is possible for premises, E, to be true and for conclusion, H, to turn out false. And the reason for this is, very roughly speaking, that H says something more than E does; it is about objects, attributes, instances of these, etc., lying outside the scope of E. This is a feature characteristic of all inductive arguments. There are scientific arguments for which, on the occasion when they are made, it is hard to say whether they are deductive or inductive. For example, on the basis of having found that : Out of one thousand residents of a certain village in Sicily, three hundred have blue eyes, we might wish to infer that : (H) the probability that the next resident we encounter from that village will have blue eyes is 3/10. In this case, whether E logically necessitates (logically implies) H or, instead, supports it inductively depends upon what we mean by the term 'probability'.[18]

Apart from their common inductive character the above examples of inductive argument differ from one another in what may be some important respects. For example, the first argues from particulars to a universal, the second argues statistically, the third from one particular to another. But whether or not these differences reflect irreducible varieties of inductive argument need be of no concern for the present.

What does seem appropriate to indicate at this point, however, is that there is much disagreement not only on the nature of the inductive relation but also on its role in actual scientific practice. Thus, though some might concede that in science we do sometimes speak of the likelihood or probability of a particular event, they would nevertheless deny that there is any such thing as a "valid" inductive relation leading from limited evidence to a general hypothesis such as an experimental law or a theory. What are ordinarily labeled inductive conclusions are merely hypotheses suggested psychologically rather than "compelled" logically by evidence – in any intelligible sense of being so compelled. Such hypotheses are suggested sometimes by the data, sometimes by broad and even far-fetched analogies, sometimes by creative fancy, And often the choice of hypothesis is guided by considering such features as symmetry, simplicity, and familiarity or, perhaps, resemblance to another hypothesis already in use in the same or in another area of study. Once put to scientific use such hypotheses are logically cultivated for their systematizing power (explaining, predicting,

[18] On the so-called *logical* conception of probability, the probability of any hypothesis, H, on given evidence, E, follows *deductively* from E. On the so-called empirical conception, the probability of H is posited on the basis of inductive support by E. See R.Carnap, "The Two Concepts of Probability" H. Feigl and W. Sellars, *Readings in Philosophical Analysis*, Appleton-Century-Crofts : New York, 1949, pp. 330-350.

etc.) as long as they continue to yield true consequences. As soon as they are found to lead to falsehood they are revised or discarded altogether.[19]

This view has the considerable advantage of circumventing serious difficulties associated with assigning probabilities to universal laws or theories. It is in effect a simplification of the logical bearing of evidence on general hypotheses by allowing for *refutation* only – a matter governed solely by deductive logic.[20] Its major weakness, however, stems from the fact that, in typical instances, experimental evidence along with certain structural criteria (simplicity, analogy, etc.) do seem to figure somewhat systematically in the framing of any generalization or in the appraising of competing hypotheses. Those who reject induction refer to the suggestiveness of data. But however non-rational the overtones of this latter term, an aspect of reasonable choice or appraisal seems to be undeniably involved, and it is this sort of appraisal that inductionists are talking about.Accordingly, many students of scientific methods continue to concern themselves with induction.

But even among those who allow that inductive inference is a species of rational argument there has been sharp disagreement. Until relatively recent times, the term 'induction' derived from Aristotle's term, *epagoge*, had connotations which in some respects differed profoundly from those we give the term today. For Aristotle, the conclusion of an inductive inference, though based on the sense perception of particulars was not only "better known" than one gotten from "demonstrative" (deductive) inference, it was certain.[21] The basis of the inferences was not one of formal validity but one which we might call *epistemic*, in which an allegedly infallible faculty of intuition discovered and guaranteed the result. Even in Bacon and Galileo there persists the notion that empirical discovery has something of the ultimate in it. What is modern in these men, with respect to induction, is the crucial roles they give to controlled investigation and experimentation in the process of such discovery.

More recent views on induction, characterizing the predominant empiricist philosophies, are, in general, agreed that inductive arguments confer, at best, probability rather than unqualified truth on empirical statements.

[19] For a strong defense of this view see K. R. Popper, *The Logic of Scientific Discovery*, Basic Books : New York, 1959, pp. 53-56.

[20] By an elementary theorem of formal logic known as *Modus Tollens*, if theory, T, has the deductive consequence, O, and O is found to be false, then T must likewise be false.

[21] Aristotle, *Analytica Posteriora*, tr. W. D. Ross, Oxford University Press : London 1942, II, 99^b-100^b.

How to assign probability appraisals to the results of any inductive process and how to justify the chosen method is, no doubt, the central problem in the contemporary philosophy of induction. It is to be carefully noted that specific viewpoints regarding the solution to this problem are highly diversified and, in some cases, separated by considerable philosophical distance. As a result, there is at present no adequate philosophy of induction available, though a wide range of technical literature has appeared on the subject. Nevertheless, with respect to actual scientific practice the following, it seems, may be safely said of induction :

(a) It is hardly deniable that, as our examples show, inductive type inferences occur in scientific discourse. Indeed, such inferences, in a variety of forms, seem to figure centrally in expanding the factual content of scientific knowledge by providing the grounds for asserting both *singular* statements like :

The man walking toward me is my brother Jack.

and *general* ones like the universal statement :

All men are mortal.

or the statistical statement :

Nearly all animals that contract rabies die from it.

Inductive inferences often occur rather explicitly in the formulation of experimental laws, i.e., laws that relate quantities which are more or less directly observable, e.g., the gas laws, Ohm's law, Galileo's law of falling bodies, etc. The history of science provides us with a generous abundance of them. Very striking examples are Cannizzaro's method for assigning atomic weights and Milikan's measurement of the electronic charge. Indeed, every time an experimental physicist or engineer fits a smooth curve through a set of experimental points either by actually drawing it or by algebraic interpolation,[22] he has made a reasonable generalization from limited data – that is to say – an inductive inference. It seems to matter little in such cases whether we say that the data suggest the generalization or, instead, lead to it inductively.

Expressed either way, there seems to be a definite scientific strategy leading to the generalization. It follows explicit rules and is in some sense defensible.

(b) It seems unquestionably natural to say that as we continue to use an experimental law and find only favorable evidence for it, its acceptability

[22] We refer here to established mathematical procedure used in countless areas of experimental endeavor in which from a set of n experimental points on a graph an equation of degree n-1, of the representative curve may be developed. Standard mathematical devices such as Simpson's rule, Newton's fomula for linear interpolation, or

is enhanced. In this way, we speak of laws that have had long and successful careers, as highly confirmed or even as (practically) true, as for example, when we say that it is true that all metals are conductors or that light travels faster *in vacuo* than in any other medium. What is expressed here is an objective supportive relation between evidence and conclusion, suggestive of the logical rather than of the subjective or psychological sort. This relation seems, at least, to allow for comparing various stages of confirmation of one and the *same* hypothesis, in terms of "more" or "less"; so that, we can at any given time say, of a certain hypothesis, that it is more confirmed now than it was before. In this regard, it is of course clear that the confirmation of any universal hypothesis can never decrease with time except for out-and-out falsification, the obvious reason being that any amount of negative evidence suffices for such falsification.

(c) When scientists actually make generalizations leading to experimental laws, they do not explicitly qualify them with a numerical probability or other measure of reliability. The inductive process, as it occurs in science, though usually incorporating some degree of technique and systematic approach is, in actual practice, not pitched to yield such measures of reliability. Nor, it seems, is there any obvious need for metrics of this kind. Universal laws are generally understood to be unexceptionable and incompatible with any negative instance.[23] They are accepted therefore only as long as they continue to hold "true". Also, despite the possibility of speaking of a law as being more confirmed now than before, actual scientific practice seems to provide no instances in which differential probability appraisals have been called for to decide between *competing* experimental laws. The history of scientific procedure amply illustrates that if the scientist were ever to be faced with a choice between two

---

solving a set of simultaneous equations of degree n-1 for their coefficients, usually involve the making of a number of tacit assumptions about continuity, simplicity and other properties of the function to be calculated.

[23] Our remarks here do not apply to scientific statements reporting the value of certain natural constants, e.g., the velocity of light *in vacuo*, the universal constant of gravitation, the atomic weights of elements, etc. Though these may be regarded as experimental laws the reported value is ordinarily qualified by statistical measures of reliability reflecting on the maximum precision with which the constant has, thus far, been determined as, for example, in reporting the atomic weight of copper to be $63.546 \pm .003$. But such statements are essentially statistical in nature. They are therefore more appropriately regarded as statistical rather than universal laws. Deviations from the reported value, in future measurements, would be admitted as counter-instances but would change the statistical reliability. Our own concern at the moment is, however, with universal rather than with statistical laws.

alternative hypotheses, each of which correctly represented similar data and also enabled him to make correct predictions (not necessarily the same ones), he would decide his choice not on the basis of confirmation derived from external evidence but rather by considering the simplicity, compatibility with existing knowledge and perhaps other inherent features of the contending hypotheses. As a matter of fact, it is only with respect to such features as these that two non-refuted universal hypotheses have in practice ever competed. For, where distinct universal hypotheses are concerned, inductive logic has not yet provided the means for calculating any sort of positive confirmation metrics, or even the means for making rank order comparisons in terms of "more" or "less".[24]

(d) Where the formulation of theories is concerned, generalization seems to diverge even more from anything like standard logical process. Unlike experimental laws, which describe observed regularities in comparatively limited domains of experience, theories usually have vast range and generality. They typically make some reference to items that are not, in any ordinary sense, directly observable, and, by and large, they are logically more removed from immediate data. Accordingly, their formulation, or perhaps better, their *creation* is something that many students of the subject prefer to describe as sophisticated or *strategic* guesswork, rather than as any matter of logic. Thus, where invention of theory is concerned, serious doubts have been expressed regarding the possibility of any so-called "logic of discovery". This is a controversial question to which we shall return in the next chapter. As in the case of experimental laws, it seems natural to regard a theory as an hypothesis whose confirmation increases as it endures with continued success. But, as in the case of all universal hypotheses, no system of inductive logic is available for making specific appraisals of this kind. Nor, as in the case of experimental laws, does science seem to call for such appraisals. From the view-point of actual

[24] The prevalent inductive logic of confirmation yields the rather cheerless result that the confirmational probability of any universal statement – however successful it may have *actually* been – is zero. We cannot develop this point explicitly here except to mention the reason as stemming ultimately from the fact that, while universal laws talk about an unlimited number of instances, they are always based on finite evidence. It should of course be noted that – as seems agreed by students of induction – the confirmation of any hypothesis is not only a function of the "quantity" of available evidence but also on the extent or "range" of its content. Consequently even if it could be said of two distinctive hypotheses that there was more evidence for one than for the other this would not suffice for comparing their respective states of confirmation. In terms of current notions of the "range" of a hypothesis – it would seem that the range assignable to any universal hypothesis of science could *not* be finite.

scientific practice it would seem to be of little consequence if not altogether gratuitous to provide confirmation metrics for any scientific theory. As far as all future evidence goes, a theory is no less subject to revision after a long and successful career than it is the moment it is conceived. To say of a theory that it has a long history of successful prediction and is therefore highly confirmed may be of interest to anyone gambling on its continued success or to someone seeing it as a likely account of the sort of world we live in. But what such a confirmational or probability qualification implies about the theory's continued role in scientific inquiry is far from clear. Nor, as we have already noted, are confirmation coefficients presently available. Systems of inductive logic have been developed but they fail to provide any acceptable manner of computing adequate confirmation metrics for any universal hypothesis.

In actual practice, once a theory is formulated and offered for acceptance, it generally remains a highly tentative hypothesis until it turns out wrong. It bears mentioning at this point, however, that there is a very special sense in which theories may be said to become established or *entrenched* in scientific discourse merely because of their vast systematizing power. On this basis, such theories may form an important class of exceptions having a modified susceptibility to falsification. We must, however, put off consideration of this important phase of scientific practice to a much later part of our discussion.

It is noteworthy that where two scientific hypotheses have vied for survival as alternative theories for the same general domain of experience, they have never been so closely similar that they could not be ranked in order of preferability according to such criteria as : how logically "neat" or simple they are, how well they "fit in" with existing knowledge, how readily experiments can be set up for testing them in untried domains, how clear or experimentally significant their descriptive terms are, how extensive their scope and systematizing power are, and so on to include perhaps still other features. Indeed the history of scientific thought provides much reason to believe that persistent controversy over any two scientific hypotheses has often stemmed as much from philosophical, theological and other non-scientific considerations as not.

(e) Science clearly makes use of inductive methods in statistical inquiry, which provides both statistical laws, having some such form as : A certain proportion (but not all) of S's are P's, and statistical probability metrics which can attach to statements as in saying : The probability of rain today is 80 percent.

Statistical probability metrics are initially gotten by enumeration, or counting, in order to determine the relative frequency of some sort of event under consideration. But like all other inductive methods, this procedure fails to provide any coefficients of reliability (also referred to as extent of confirmation, likelihood, probability, etc.) for universal hypotheses which, moreover (as we have already noted) do not seem to call for such metrics. In the case of statistical laws, however, reliability metrics seem not to be inappropriate. Thus, the statement, 'Rabies is fatal to nearly all dogs', unlike a universal law, is not refutable by one or even several negative instances, i.e., instances of rabid dogs that recover. It seems entirely reasonable therefore to regard positive instances as favorable or, to *some extent confirmative*, negative instances as unfavorable or, to some extent, disconfirmative of the hypothesis. But this is essentially an inductive relation tying evidence to conclusion in what, broadly speaking, appears to be a logically supportive manner. Science, it seems, could do here with a defensible method for calculating confirmation metrics. The methods of statistical analysis can provide such metrics.

The inductive relation, as we shall see in another section, recurs very centrally in probabilistic, explanation, prediction and related types of what will be called *subsumptive* arguments, all of which make use of statistical premises. Probabilistic accounts and accompanying statistical techniques are of growing importance in the social sciences. Some areas of physical science incorporate statistical statements and probabilistic accounts in a central manner as, for example, in statistical mechanics, in the theory of measurements, and, very fundamentally, in quantum mechanics. It would be far beyond our present scope to attempt any detailed consideration of the connection between statistical notions in these areas of inquiry and inductive inference. What may be indicated, however, is that closely associated with statistical notions are probabilities construed as relative frequencies in the long run, Accordingly, probability statements are hypotheses about future frequencies, formulated by inductive generalization from the enumeration of a limited number of cases.

4. Theoretic Subject-matter

There can be little doubt that common sense, well-informed and technically developed though it may be, does not quite make the pages of science until it has invoked the *sub-sensory* in order to account for the *sensory*. We apply the terms 'sub-sensory' here to subject-matter that, in principle, is either (1) nowhere to be found (e.g., perfectly evacuated spaces, frictionless media, heatless substances, perfect economic behavior and other idealized states), or (2) not, in any ordinary sense, directly perceivable (e.g., elementary physical particles, sub-conscious states and other *theoretical* – also called – *hypothetical* entities). The component of science which introduces such subject-matter is called *theory.* It follows therefore that if a necessary difference between science and common sense is sub-sensory subject-matter then (roughly put) *science is theorized common sense*; or as one might also put it – *no theory, no science.* For theory, by introducing subject-matter not immediately tied to the senses, makes possible the comprehensive, logical integrity which characterizes the structure of scientific knowledge, to a more or less extent, depending on its state of development. A consideration of the manner in which theory gives structure to knowledge and provides scientific generality will have to await later discussion. We might, for the moment, merely note that ordinary sense experience which, after all, is the bed-rock subject-matter of all empirical knowledge – scientific as well as pre-scientific – presents us with a vast range of qualitatively diverse phenomena. To bring coherence and unity to such diversity would seem to require unifying principles whose content cannot be limited to some particular kind or range of familiar attributes. At least some great principles of science, it would seem, must be about entities, events, states, etc. which cannot themselves be specified in terms of the familiar atributes like color, shape, or taste but which account for a range of such attributes in other things. Our Cartesian preference for reductive analysis prompts us to try to break things down to *smallest* elements. Consequently our theoretical entities, states or events are often (but not always) imperceptibly minute rather than of ordinary size. Minuteness, however, need not be a characteristic of theoretical subject-matter. What is essential to it is its *abstractness*, in the sense of lacking some of the familiar attributes of sensory objects.[25] The historical development of science strongly suggests that the extent to which theory structures the content of any empirical discipline is a direct measure of the scientific maturity of that discipline. Thus the common sense astronomy of the Babylonians, charting only observable celestial regularities eventually gives way to the Ptolemaic science of concentric but unobservable spheres, eccentrics, equants, and epicycles, which in turn becomes Copernican science, dropping common-sense geocentrism for the more abstract and less immediately apparent sun-centered view. Aristotle's dynamics of motion which required the action of a sustained cause in order for there

[25] In applying the term, 'abstract', here, to theoretical subject-matter there is no intent to detract from any space-time existence which may be claimed for such subject-matter. The term is used merely to note a lack of certain sensory properties, with no intended implications for the controversial ontological status of theoretical entities.

to be motion is better characterized as *true* common-sense than as *false* science. It is a simple matter of ordinary experience that for inanimate objects, at least, unless we continue to push or pull them they eventually return to rest. Galileo initiates a scientific dynamics of motion, not, as the school books are fond of saying, because of any unprecedented passion for empirical check. Aristotle makes frequent reference to ordinary experience throughout his discussion of motion in Books II-VII of the *Physica*. What is most novel in Galileo's approach is *idealization*. On the basis of graduated experimental steps, he formulated generalizations about ideal or limiting cases rather than about subject-matter that is concretely given in sense perception. If Galileo is to be considered the founder of a genuine science of dynamics it is because it was he who first talked about an ideal or *theoretical* state like a frictionless region through which a moving body can go on indefinitely with unchanged velocity, and requiring no external force to keep it going. Frictionless regions are not to be found anywhere, though the concept of such a state – like many other limiting concepts – is experimentally applicable, *with close approximation*, to a vastly diversified range of contexts. The history of chemistry also illustrates development from common sense to science in several steps. Genuine quantitative measurement and analysis in the early studies on oxidation begin only after gases, especially Priestley's odorless, tasteless and colorless *dephlogistonated* air (more elusive and "spiritual" than any liquid or solid) were given full recognition and, with Pascal's pump, brought under manipulable control. Finally with the introduction of the idea of basic particles or atoms – not subject to direct inspection yet capable of accounting for a variety of experiential facts – chemical technology takes on the unmistakable features of a full-fledged science. Similarly biology remained largely common-sense taxonomy until, after microscopic evidence had accumulated for about two hundred years, Theodor Schwann and others at the start of the nineteenth century finally gave clear expression to the idea of an elemental (usually microscopic) unit of life or biologic cell of which all living things are composed. Further scientific maturity was achieved with the recognition of subcellular structure and eventually with the introduction of the elusive gene as the elemental determinant of heredity. The new molecular biology of the past decade brings biological study to nearly full parity in scientific sophistication with the physico-chemical science of which it is now virtually a part.

In calling science "theorized common-sense" one makes an essentially *systematic* distinction between scientific and common-sense knowledge both of which share an empirical heritage. What seems appropriate to add, however, is the *historical* note that not only is the practice of theorizing a radical departure from common sense, but that this departure occurs – long before the advent of empirical science – with the earliest philosophical thinking. Philosophy is man's first *reasoned* break with common sense. The break occurs in several ways and philosophical theorizing is one of them.[26] But such theorizing is, of course, not science. It does not become so until the missing observational criteria – already implicit in common sense – are provided. Hence, just as one would say that, roughly, science is "theorized" common sense, one might also say it is "sensified" theory. Such descriptions of science must of course be taken with a substantial grain of salt. Speculative entities like the "seeds" of Anaxagoras, along with the swarm of entelechies, vital substances and animistic forces which metaphysical speculation continued to spawn are not to be narrowly classed with the theoretical subject-matter of empirical science, e.g., atoms, electrons, electromagnetic and gravitational fields, photons, etc. A resemblance there is indeed, but only a very limited one. Though not *directly* observable (and sometimes misleadingly referred to as "unobservables") theoretic entities, states or events are, in a genuine sense, quite detectable; qualitative seeds, entelechies and vital substances on the other hand are not. A failure to make this distinction has led to the scientific malpractice of introducing such metaphysical notions into empirical science and of attempting to have them do the work of genuine theory. It seems, by now, to be a matter of rather general agreement among scientists and empirically minded methodologists that, in order to have scientific significance, theoretical subject-matter, must, in some specifiable manner, correspond to experience. Entelechies, vital forces and other postulated items which fail in this correspondence fall subject to serious question regarding their admissibility as scientific subject-matter.

D. DISTINCTIVE ASPECTS OF CONTROL ON SCIENTIFIC INFERENCES

It would be plainly false to suppose that science is the only level of factual knowledge where inferences, deductive or inductive, occur. Crop failures were explained and correct generalizations about what could be safely eaten were made long before there were agricultural and dietary

[26] Anaxagoras, in the fifth century B.C., is the first, it seems, to postulate theoretical entities in order to account for the familiar reality of sensory experience. He took as basic entities imperceptibly minute particles and mind, "the thinnest of all things", and therefore also beyond the senses. He also spoke of the infinite divisibility of matter and the infinity of mind, both of which are extra-sensory concepts.

sciences. But in science inferences are abundant enough and so connected as to produce an over-all logical structure that is unparalleled on any other level of factual knowledge. What is more, there are a number of distinctive methodological controls on such inferences, which contribute crucially to the structural profile of science. This is *not* to say that there is a unique logic of scientific inference with features that are totally absent in prescientific procedure. Intelligent experimentation and controlled investigation must have been familiar in those crafts and technologies where practical necessity called for hard facts rather than philosophical or theoretical bias. Students of intellectual history may be generally agreed that modern science does not make a decisive appearance until a century after the great Copernican dawn of the sixteenth century. But it would be less than plausible to regard the experimentalism which followed as something totally novel – a methodological revelation, as it were, coming straight out of the blue.

Nor does scientific discovery seem to be a uniquely standardized process. The mere *possibility* of any so-called logic of discovery remains a matter of some doubt and the efforts of Francis Bacon, J. S. Mill and, more recently, H. Reichenbach and R. Carnap to recommend rules for making good inductive inferences have led to no conclusive agreement on results.

Nevertheless the grounding and systematizing of scientific knowledge involve investigational procedures, both observational and intellectual, which are relatively controlled and attuned for isolating and manipulating the causes of phenomena. And it does make some definite sense to speak of a distinctively scientific approach to discovering and formulating the patterns of nature. We turn now briefly to some of the methodological refinements that constitute this approach.

1. Perspicuity of Scientific Language

Scientific discourse achieves considerable clarity and precision in formulating both the base and the conclusion of any inference. It is very unusual in science, and equally rare in common sense, to provide discriminatory definitions of key terms, thus establishing fundamental notions for subsequent discourse. The distinction between force and pressure, or between mixture and substance or, again, between acceleration, momentum and kinetic energy (as species of the prescientific concept of motion), have contributed vastly to fruitful generalization and prediction on the basis of the highly specific determinations which these refinements help make possible.

2. *Use of Numerical Quantities*

Closely paralleling the verbal perspicuity of scientific usage is its tendency to use numbers, and more generally, mathematical formulations to describe the nature of things. The use of numerical language is by no means new with science and is anticipated even on the most rudimentary levels of discourse. But in science it becomes the vehicle for the most specific and minute discriminations ever imagined possible for the human observer. The high place in cognitive discourse which Plato and the Pythagorians gave to number and to mathematical form, later echoed in Galileo's view of mathematics as the key to the universe,[27] could not have seen greater fulfillment than with what has actually taken place in the history of scientific knowledge. The change from qualitative to quantitative description parallels much of the development of human knowledge as it passes from common sense to prescientific technology and eventually to genuine science. Even within science itself the most characteristic mark of advanced development is the change from qualitative study to quantitative measurement. This change is occurring to a remarkable degree in the biological, psychological and social sciences.

Quantitative measurement equips science with the most effective means known not only for *achieving* high precision but also for assessing it, that is, for determining the extent to which one has achieved it in any experimental procedure. Let us dwell briefly on a few aspects of this crucial feature of scientific procedure.

3. *Precision and Precision Measures in Quantitative Measurement*

The precision of a measurement is a matter of essentially two features. The first of these is the degree of specificity, or as one might say, minuteness with which a measured quantity has been discriminated and reported. Thus, for example, if we measure the amount of mercury pollution in a nearby fresh water lake to the nearest gram of mercury per one hundred million cubic centimeters of water and report it as being 0.0000036*2* (rather than 0.0000036*1* or 0.0000036*3*) grams per cubic centimeter, we are said to have done it with greater precision than if we had made the measurement to the nearest gram per million cubic centimeters and reported it as 0.000003 (rather than as 0.000002 or 0.000004) grams per cubic centimeter.

[27] Galileo leads the Aristotelean character, Simplicio, to admit that "To want to study natural problems without mathematics is to attempt something that cannot be done". Cf. Galilei, Galileo, *Dialogo sopra i due Massimi Systemi del Mondo*, Opere, Ed. Naz. VII, 38, 256.

This aspect of precision is usually achieved with the help of highly refined measuring devices and techniques. The second aspect of precision has to do with the reliability of the measurement. This is, essentially, a matter of the stability and reproducibility of the measurement and constitutes much of the so-called "objectivity" of scientific results. To see how this is so we note that the scientist tries to assure this aspect of precision by usually reporting not a single measurement but a measure of central tendency or "average" of a series of measurements. From the viewpoint of precision the value reported, often referred to as the "best" value is the arithmetic mean. The arithmetic mean (or simply the *mean*) has certain useful properties which mark it as the "best" or most reliable value. In the first place, the laws of large numbers guarantee – as every student who has averaged his grades knows – that as the number of measurements increases their mean becomes less and less sensitive to the random variation of any individual measurement. But assuming, as is reasonable, that under approximately fixed circumstances these variations do not change very much in their distribution, the computed mean of repeated measurements must fluctuate less and less as more and more measurements are made and used in the computation. Thus the mean value of a set of data, though itself subject to some random variation, becomes, on the whole, increasingly more stable (predictable, reproducible), as the measurement sequence is extended, than single measurements which are subject to unmitigated random variations however minute these might be. (The pollution measurements on our lake, for example, may differ considerably from one another in the eighth decimal place.)

But more important than the stability of the mean – indeed most important, from the viewpoint of reliability – is the fact that for any random sequence of measurements the *mean* is that value from which the sum of squares of deviations (from the mean) of individual measurements is less than it would be for any other value we might have chosen to report. Roughly, this says that the mean is the most "central" value of the measurement sequence. That is to say, the mean is the value from which any single measurement is *likely* to vary less than from any other value. It is therefore, in this sense, the most reliable value to report as the result of measurement. Moreover, as can be shown, mathematically this reliability increases with greater repititions of the measurement.

We see then that in reporting the mean of several measurements the scientist enhances the reliability and hence the precision of his results. Prescientific knowledge, of course, had to achieve some degree of precision and reliability in its results in order to have achieved any lasting and useful technology at all. This came largely through the exercise of high craftsmanship and care in the use of a variety of tools and standards of measurement. No doubt, some highly reliable and reproduceable determinations were made in this way. But in science, particularly on the more mature and quantitative levels, the concern for precision is both much greater and far more deliberate under the demands of contemporary theory which call for experimental discriminations on the most extraordinary levels of minuteness.

But there is considerably more than this in the scientific concern for precision. For science is committed not only to being precise, to any extent required by the occasion, but also to assess its own precision. In typical instances, it must do this even when its measurements are not so refined. The measurement of a distance to the nearest foot by mere pacing, despite all its apparent crudeness, begins to take on scientific character as soon as we attempt to estimate the experimental error. This is an assessment of reliability which, in the absence of any better known value of the desired length, must be made statistically on the basis of the stability with which results may be reproduced by repeated measurement. Roughly speaking, the more closely individual determinations tend to cluster, randomly, around the reported value (usually an average such as the mean), the more reliable is that value.

The assessment and assignment of measures of precision to experimental results is one of the most distinguishing phases of scientific procedure. It is entirely absent in common sense inquiry and in early technology.[28] Much of contemporary physical theory, on the levels of both cosmic and micro-dimensions, has significant implications only on the highest levels of experimental precision. As scientific theory has reached out to the cosmic spaces and down to the fine and hyperfine structure of things, the random experimental error or "noise" in every determination has played increasingly crucial roles. It has had to be carefully reckoned with in the confirmation or disconfirmation of theory where minute discrepancies, barely exceeding experimental error, have called for drastic revision in the main body of knowledge. It is by now a commonplace of scientific procedure that the success or failure of the major part of contemporary physical theory is decidable only after the experimental error inherent in test situations has been carefully accounted for.[29]

[28] It is of course undeniable that contemporary technology, though its concerns may be largely practical, is in many areas heavily committed to all aspects of precision. But these are areas which are profoundly science-based and share many basic aspects of inquiry with science proper sometimes to the point where any attempt to distinguish such technology from so-called "pure" science becomes conspicuously artificial and at best only a matter of academic interest.

[29] The significance of random error in contemporary physical science now goes far

4. Systematic Structure and the Acceptance of Laws

We have already noted that logically integrated structure is a highly distinctive feature of scientific knowledge. Apart from the high degree of logical conciseness, intelligibility, and predictive power which this structure yields it has an important, though subtle, effect on the discovery and retention of scientific principles – that is to say, on generalization and confirmation. This is an effect which discussions on the nature of science rarely consider.

The systematic context in which scientific principles are couched tends to enhance their confirmational status in at least two ways :

(i) In the first place, it tends to *enhance* the weight of favorable evidence. To see this we return to a familiar contrast between ordinary knowledge and science. General statements belonging to the first, as for example, *a waning moon rises later than a waxing one*, are, for the most part, reportative descriptions directly grounded on some observed regularity. They are, so to speak, generalized reports of what has been observed to happen in some rather circumscribed domain of experience. Pre-scientific knowledge may be regarded as being, at best, a catalogued set of such statements. The principles of science, however, tend to form a logically integrated system, some of whose principles are subsumed under others of greater generality. Thus experimental evidence having a direct bearing on one part of the system also has some indirect bearing on another part. This is sometimes expressed by saying that, in general, there is indirect as well as direct evidence for any scientific law.[30] Thus scientific principles are in general subject to confirmation or disconfirmation from a broad range of subject-matter. That is, they have what might be called a relatively *wide empirical base*. Intimately related to this feature of scientific principles is the vast generality of the relatively small number of principles usually found at the heart of scientific systems – namely

---

beyond pragmatic requirements dictated by the experimental process. In the contemporary science of small dimensions, that is in quantum physics, the notion of random error occurs as part of the theory itself. This, then, makes error a matter of principle establishing it on the micro-level as an uneliminable ingredient in the measurement of any dynamical state. See : "The Uncertainty Principle" in any standard text on Quantum Theory.

[30] Thus, though Kepler's laws of planetary motion primarily express a generalized experimental regularity stemming from observations in a limited area of concern, they are also supported, though less directly, by the evidence which immediately bears on Newton's laws of motion. This is because, given certain *boundary* or *initial* conditions, Kepler's laws are immediately derivable from Newton's.

theories. In some cases these illustrate the wide empirical base dramatically. Such principles are often general enough to encompass a wide variety of apparently dissimilar subject-matter. Thus confirmation of the Newtonian laws of mechanics may come from such diversified phenomena as planetary motion, the oscillation of pendulums, freely falling bodies, the diffusion of gases, thermal phenomena, surface tension and capillary action, hydrostatics, dynamics of fluids, and more. What must be noted is that the "wide" empirical base of any scientific principle seems to enhance its acceptability considerably beyond the amount of confirmation to be gotten from sheer quantity of evidence iterated apart from scope (range or width) of evidence. And though no objective measure of this effect is available it seems rather natural. A prediction seems more crucial and its positive outcome more convincing, the more it reaches out to contexts that are removed from those which inspired the principle in the first place. There is little doubt that predictive power in this sense does much to promote the acceptance of any theory. Indeed this seems to be a contributing reason for the fact that some predictions are regarded as more decisive or "crucial" than others.[31]

It seems appropriate to note here that also contributing to the "cruciality" of an experimental test is the degree of high specificity or precision involved in the prediction to be tested. Again the judgment involved is inexplicit and informal, with no method available for assigning a numerical measure of cruciality or decisiveness to any experiment. In any case, the considerations involved seem based on the presumption that highly specific or precise predictions concerning particular phenomena would, in some sense, be antecedently improbable if the theory from which the prediction stems were not true. Thus, for example, it might be asked – if the theory of general relatively were not true, how probable would it be that light coming to us from a star like the companion of Serius would show a slight shift toward the red in the order of magnitude predicted by the theory? The sense of the question seems such as to suggest that the shift in point would be most unexpected in the absence of the relativistic effects of stellar gravity – that is, in case the General Theory were false. These considerations, however, leave much to be desired from the viewpoint of clarity. What determines such an expectation or antecedent probability – as it is sometimes called? And how does it determine a measure of cruciality for the prediction in point? Nevertheless judgments of cruciality with regard to scientific ex-

[31] There are other factors which enter into the logic of crucial experiments as, for example, the (antecedent) probability of the prediction being true in case the theory in question is false. But we can not concern ourselves with these here.

periments are rather familiar in science, and they seem to be based on factors which ultimately imply a highly developed theoretical context. By this we mean a context not only structured to an advanced degree of specificity but also held together by a logical webbing of theory that has consequences and therefore systematizing power for a *wide* range of subject-matter.

(ii) The second way in which highly systematic contexts enhance the confirmational status of scientific principles is by *undermining* the decisiveness of unfavorable evidence. The precise manner or logic of this is a subtle and involved affair on which one finds little settled opinion. Nevertheless, it is a matter of actual practice that there would be extreme hesitation in generalizing from a body of given data, however reliable it might seen, to any conclusion that was logically incompatible with any highly established *system* of principles.

The history of science is replete with instances where generalizations that seemed called for by the immediate evidence were in some way qualified or suspended pending fundamental review because they did violence to such principles as conservation of energy, conservation of charge, inseparability of magnetic poles, insuperability of the velocity of light, Galileo's law of inertia and even others of less consequence.

Thus, the systematic context exerts a measure of "control" over scientific principles by rendering them at least provisionally fortified against any unexpected rash of unfavorable evidence. Some philosophers of science have believed this control to be so complete that a theory, experimental law, or, for that matter, any scientific hypothesis whatever can be held true regardless of the evidence – provided we modify other parts of the system in the appropriate manner. This would make every scientific "truth" a matter of choice rather than one of genuine inference from evidence. The paradoxical and somewhat discomforting consequence of such a view is that scientific knowledge, despite its apparent relevance and efficacy for controlling natural forces, is after all just one way of looking at "reality". There are an unlimited number of other so-called "truths" of nature. And the one we adopt is ultimately a stipulation – a logically arbitrary choice governed by preference rather than compelled by evidence. If so, then scientific statements are more correctly regarded as ultimately vacuous conventions rather than as truth-bearing descriptions of actuality. Conventionalism, however, is not an established thesis. It remains a controversial philosophical issue to which we will return later for a bit closer look.

5. Experimentation

Experimentation is often singled out as the key feature of scientific inquiry. School texts never tire of contrasting the armchair speculation of pre-scientific, scholastic dialectic with the experimental methods of modern science. In similar fashion many commentators on the modern age loosely label the scientific approach or "method", as "experimentalism". So crucially do some regard experimental procedure for science that they would deny history and other social studies scientific status because in these areas experimenting is difficult or – in one sense of experimenting – even impossible as in the case of history, where particular past events can never again be observed.

Despite its frequent use the term 'experimentation' remains rather vague and ambigious throughout most discussions. In its broadest senses it is used to refer to any systematic investigation that is basically observational in character and ultimately dependent on the senses. In such a sense, the astronomer, paleontologist, archeologist and political historian, no less than any other investigator, are experimenting whenever they search out records, artifacts, traces of the past, or simply observe and record weather, celestial events, or other uncontrollable phenomena. What is more, understood in this way, experimentation is not peculiar to science. Ordinary factual inquiries on all levels make more or less use of it. On the other hand, the term 'experimentation' is sometimes reserved for those procedures that involve not only selective observation and systematic recording of data, but also some degree of controlled manipulation of the phenomena under investigation, in the sense that selected factors are either produced or suppressed at will. Experimentation in this sense is clearly not necessary for science. The absurd consequence of regarding it as such would be the denying of scientific status not only to social history but also to astronomy, paleontology, as well as much of geology and even much of embryology.

But, however we label the diverse observational procedures of science, it is obvious even to the most causal observer that they are conspicuously precise and highly directed affairs. In those procedures where the phenomena are manipulable at will, considerable sophistication and subtle strategy may be brought into play. What is too often overlooked in the usual school accounts of scientific method is that experimentation does not typically begin with a vast flood of measurements, roughly concerning the phenomenon under study. Such an approach could go on for ever without much profit. And, it is highly doubtful that one would do this even on the most ordinary levels of inquiry. The procedure best starts by making highly informed or "educated" assumptions as to what factors are relevant to the phenomenon under study. These factors are then varied, one at a time, keeping the others constant. In this way, the experimenter can observe the effects and possibly discover any regular dependence of the phenomena on one or several of the manipulated variables. In actual practice, the role of hypothesis in experimentation is apt to be much more extensive than this. Plausible guesses or hypotheses are made not only concerning which variables are to be considered relevant but also concerning the explicit relation (mathematical or other) between each variable and the phenomenon. The experimenter would then go on to make sharply focussed tests of each such explicit hypothesis.

Thus we might want to discover which features of a simple pendulum affect its period for small swings. Our general familiarity with pendulums suggests that the color and exact shape of the bob might not be relevant but that the mass of the bob, length of the pendulum and amplitude of the small swings might very well be. After experimentally eliminating the mass and amplitude as irrelevant we would go on to test the effect of length. We are sufficiently acquainted with pendulums to suspect that long pendulums take more time to complete one swing, i.e. have longer periods, than short ones. Accordingly we might suppose that the period increases directly with the length. This is easily tested, for if the length is increased by factors of 2, 3, 4, ..., then the period should increase by the same factors. Finding this not to occur the hypothesis is discarded and a new one made. This time we assume the period to vary as the square of the length. For the same length variations the period should now be increased by factors of 4, 9, 16, ... This hypothesis is easily tested and discarded. Trying another one we assume, this time, that the period varies as the square root of the length. The expected period should then increase by factors of $\sqrt{2}$, $\sqrt{3}$, $\sqrt{4}$, ... that is, approximately, 1.41, 1.73, 2, ... A few measurements will readily confirm this. The choice of hypotheses in such a procedure is generally governed by a variety of considerations, simplicity, symmetry, and analogy with other phenomena being only a few.

But the role of hypotheses in experimentation cuts to even deeper levels especially when the hypotheses are theories. Extreme inductivism such as that of Francis Bacon and to some extent of recent positivism makes the rather naive error of supposing the observational concepts, in terms of which experimental results are reported, to be theoretically neutral. The most cursory reflection on any set of basic experiments in mechanics, optics, electricity, molecular chemistry, biology and even the social sciences easily convinces one that not only the design and description of experiment but also the reporting of results is at the very outset considerably infused with theoretic concepts. Forces, waves, particles, rays, poles, charges, fields, atomic cross-sections, conscious states, group cohesion, open-societies, and much more are theoretical notions that govern experimental apparatus, strategies, and selective perception as well as clusters of collateral assumptions underlying any experiment. Theories rule not only on the relevance of any experiment (as do all general hypotheses) but they contribute crucially to a *language* of experimentation. Unlike the theory-neutral observational languages envisioned by naive inductivism the language of scientific experimentation and observation seems to be inextricably theory-laden. This important consideration has been a point of departure for some of the present-day reaction to the inductivistic philosophies of science of recent decades. We shall have occasion to revisit the point in the concluding portions of this work.

Where factors in the phenomenon under consideration cannot be manipulated, as for example in astronomy, investigations of a highly controlled character are nevertheless possible. In this case, aspects of the phenomena are observed as they are found to occur naturally. This, however, calls for a high degree of selectivity and care in isolating sequences of instances where variations in one or a few factors can be observed while all or nearly all others that might be relevant remain constant.

Inquiry of this kind – controlled by sophisticated judgments of relevance and by observing variational sequences that are either produced in laboratories or else selectively discriminated in the field – are highly characteristic of how science expands the inductive bases of generalized knowledge. In this respect the advance over common sense is no doubt only one of degree. But, with the help of a vastly expanding technology, which makes it possible to attain undreamed of levels of numerical precision and perceptual discrimination, this advance is so great as to bring scientific experimentation to a conspicuously distinctive level.

6. Scientific Guidelines on Deductive Structure

We have noted that the structure of science aims at comprehensiveness in which wide tracts of factual content are subsumed under a relatively small number of very general principles or theories. The logical form for this sort of systemization is deductive, and, indeed, it is the form which explicitly structures almost all of established scientific knowledge – inductions remaining largely implicit, transitory procedures throughout. In more qualitative portions of scientific discourse the mode of deductive exposition is verbal argument often of a rather discursive nature. Where the content is more quantitative, however, mathematics and related formal disciplines are the vehicles of deductive structure. But formal rigor cannot be the immediate concern of the practicing theoretical scientist who often forgoes it for the expediency of mathematical tools which he often – sometimes only provisionally – devises himself.[32] More typically, however, it is unnecessary for the theoretician to generate a brand new formal edifice for his science. Almost always it turns out that a piece of ready-made mathematics can – perhaps with the addition of an assumption or two and with an indication of how some of the mathematical terms are to be interpreted – provide the required deductive structure. Thus parts of Euclidean geometry, depending on the way they are interpreted, become geometric optics or kinematics, both, branches of physical science. In this way physicists have borrowed freely from some of the most abstract branches of pure mathematics. More recently, they have, for example, found several interesting, physical applications of group theory, a highly abstract branch of mathematics, formerly of interest only to pure mathematicians.[33]

But, whatever the mode of deductive exposition, the attempt is eventually made to reconstruct a piece of working science so as to make it, logically, more explicit and, therefore, clearer. The basic tenets of theory are abstractly and economically stated; the physical significance of key terms is appropriately brought into focus; and consequences are rigorously derived. The level of systemization which may result from such refinements is one that is sometimes found in the more orderly portions of our most highly advanced sciences.

The rigorous and explicit structuring of discourse is an entirely familiar matter in logic or mathematics. And the manner in which such formal exposition is best carried out is *axiomatization*. We shall give some consideration to this mode of exposition in later discussion. For the present, however, a few points may be briefly noted : The axiomatic mode of exposition has been used largely to restructure areas of science that have

[32] Entire branches of applied mathematics have been created by physicists themselves, throughout the history of science, to service their needs in the structuring of theory. Notable among these are the infinitesimal calculus, the calculus of finite differences, vector and tensor analysis and, to some extent, differential geometry.

[33] For an interesting account of applications of Group Theory in contemporary physics see : Freeman J. Dyson, "Application of Group Theory in Particle Physics", *SIAM Review* (1966) pp. 1-9.

already been established on rather high levels of sophistication. More recently, however, it has occasionally been used in the formulation of new scientific theory. It consists essentially of two major parts. In the *first* we state all the *formal* assumptions underlying some theory or body of knowledge and then derive formulas that follow from these assumptions. This provides the symbolic components, the rules for relating and manipulating these and the resulting formal structure of the system. In the *second* we give the *meaning* assumptions. That is we indicate what the symbols of the system mean, i.e., what sorts of things (objects, states or events) they refer to. Axiomatization seems to be the most satisfactory way known for the explicit and rigorous structuring of any scientific body of knowledge. Separating the purely formal aspects of a system from those of meaning and reference makes the system logically transparent on formal questions like rigor, economy of assumptions and, to some extent internal consistency. Such separation also permits a clearer focussing on all matters concerning the experiential significance or meaning content of the system. Indeed, the generality, succinctness and logical transparency which axiomatization gives to scientific theory result in the sort of systematic structure which most decisively distinguishes science from common sense or any other level of factual (empirical) knowledge.

E. EMPIRICIST BACKGROUND AND SIGNIFICANCE

Modern science, especially in its more contemporary dispositions, owes much to empiricism. This is the philosophy of knowledge which says that the ultimate base for any matter-of-fact claim is ordinary sense experience, and which therefore denies that any matter-of-fact may be known *a priori*. In effect this is tantamount to denying : (1) that about natural objects and events there can be any logically necessary truths since such truths are decidable *a priori*, which is to say, by purely formal procedures and hence independently of what actually happens in time and space; (2) that there is factual knowledge that we are born with (connate ideas) or arises within us (innate ideas) and (3) that there are factual statements whose truth may be known by some faculty of intuition, or by self-evidence, or by the inconceivability of the contrary, or as the consequence of only some convention or definition.[34]

[34] Points (1), (2) and (3) refer to various epistemic grounds that have been traditionally held for making factual claims about existential subject-matter. They are all *a priori* and have, at one time or another, been specifically rejected by empiricist writers.

From the viewpoint of the methodology of empirical science the basic tenets of the empiricist approach are nearly completely set forth by the great eighteenth century British empiricist, David Hume. Hume's empiricism is taken up in scientific context by the psychologist, J. S. Mill and by physicists and philosophers of science like Mach, Helmholtz, Poincaré, Duhem, Boltzmann and Einstein. Both Mach and, very notably, Einstein succeeded in making rigorous scientific applications of the empiricist approach. Einstein, in particular, questioned the physical (or empirical) significance of the classical notions of space and time and, also, of the common sense notion of simultaneity at a distance, that is, of two events in *different* places (knowably) happening at the very same time.

The Vienna Circle (1924-36) consisting of mathematicians, scientists and philosophers reformulated the empiricist philosophy, giving explicit expression to several themes already implicit in Hume's account.[35] Broadly speaking these may be summed up in the following *three* items :

(1) *Semantical criteria* : All significant discourse is constituted of terms that have either purely logical function or make some ultimate reference to sense experience.

(2) *Testability* : All statements either reflect matters of logic (logical statements) or express matters of fact (factual statements). Logical truth (or falsity) is entirely decidable by analyzing meaning relations and formal ones; and these are ultimately matters of convention or stipulation specifying – as it were – how we shall talk. Such truth or falsity therefore comes out independent of whatever happens in the world. Consequently, except for reflecting the conventions of language, logical statements say nothing at all about any material state of affairs. When logically true such statements are also said to be *logically necessary*. Factual statements on the other hand must be *verifiable*. This means that it must be logically possible to say, in experientially understandable terms, what sort of sense observational evidence would count for or against their truth. Failing in this respect, their significance and status as genuine statements becomes seriously questionable.

(3) *Unity of Science* : The basic concepts and criteria of science are essentially the same and belong to one coherent system. More specifically this can be understood to express the following :

[35] The Viennese empiricists were at first known as logical positivists and later as logical empiricists. When taken together with pragmatists, Cambridge philosophical analysts, and operationists they comprise a broader class of closely related philosophers sometimes labeled *scientific empiricists*.

(a) The principle that all descriptive concepts of science ultimately reduce in meaning to familiar experiential elements.

(b) The criterion that all factual truth is ultimately grounded in sense experience.

(c) The thesis that though the laws of science are not yet comprehensively systematized as deductive consequences of one basic set of empirical postulates, such "unification" is a real possibility and an underlying aim of all science. This thesis known as *reductionism* is a projective generalization or limiting form of what has in fact happened only within *parts* of scientific discourse. In its most usual form, reductionism envisages the essential vocabulary of the basic postulate set to be that of physics, thus "reducing" all sciences ultimately to physics. Conceived in this manner reductionism is also known as *physicalism*. Whether total theoretic unity will in fact ever be achieved or not (and its likelihood, though remote, seems less remote now, with the new biology, than a decade or so ago) it is noteworthy that the reductionist thesis can be clearly explicated as a significant possibility. Though the development of chemical biology is a great step in this direction, the question of whether psychological and sociological laws are so reducible raises involved complex philosophical issues on which opinion is still unsettled.

The principle of verifiability has, throughout its history, come under heavy philosophical criticism. To meet this challenge logical empiricists have had to subject it to a series of modifications with some resulting doubt cast on its ultimate defensibility as a general doctrine of meaning.[36] Nevertheless since the beginning of the present century there has been an unmistakable tendency within science proper to adopt criteria of "physical" or "operational" significance similar to the semantical and verifiability conditions (1) and (2) above. These criteria have functioned not only as a basis for clarifying the empirical content of both new and established scientific concepts but also to decide their *admissibility*. Most notable in this respect is operationism due largely to the work of physicist, P. W. Bridgman and, before him, Einstein, both of which were enormously effective in specifically adapting empiricist criteria of significance to actual scientific contexts in terms of *performable measurements*.

In psychology, operationism takes the form of behaviorism which requires psychological terms (even the subjective ones), and hence statements,

[36] For a survey of the career of the principle of verifiability see : C. G. Hempel, "Problems and Changes in the Empiricist Criterion of Meaning", *Revue Internationale de Philosophie*, Brussels, Vol. 4, No. 11 January 15 (1950) pp. 41-63. This article may also be found in : *Semantics and the Philosophy of Language*, ed. Leonard Linsky, Urbana, Illinois, 1952, pp. 163-185.

to have operational content. Despite the fact that behaviorism is only one of several competing schools of psychology, it has had an irresistible influence on all of them and on all of the social sciences as well. It seems now about as appropriate to ask about the operational significance of psychological and sociological terms regardless of their doctrinal origin as it is to ask about the physical significance of any term in the physical sciences.

But like the verifiability theory of meaning, the operational (or observational) analysis of scientific terms has had its own share of difficulties. It is implicit in all three of the positivistic tenets listed above that the content or "meaning" of any scientific term be ultimately reducible or translatable to an observation language whose vocabulary encompasses only familiar notions of ordinary sense experience, operations, etc. This, however, is only a *desideratum* which has not quite been achieved, theoretical vocabulary presenting the most serious obstacles to such translation. Admittedly, the positivistic program has made explicit stipulations on how to *coordinate* theoretical terms (e.g., 'electron') with observational ones (e.g., 'scintillation on a fluorescent phosphor screen'). Such coordinations, however, cannot serve as meaning equivalences or definitions of the terms in point. The term, 'electron', simply does not have the same meaning as the term, 'scintillation on a fluorescent phosphor screen'; for the two terms are not interchangeable in many contexts. We could not, for example, make the replacement in the following statement : 'The mass of an electron is 9.107×10^{-28} kg'. Substituting the term 'scintillation on ...' for 'electron' in this statement would result in non-sense. Nor are coordinations to theoretical terms univalent. Not only are scintillations coordinated with electrons but so are Geiger-Müller pulses, deflections of electrometers and galvonometers, reflection of light from polished metals, traces in cloud chambers, chemical affinities between substances and much, much more. To resolve such ambiguities, theoretical accounts would have to be provided showing how the *same* theoretical term could be interpreted in terms of differing phenomena for differing contexts. But even were this accomplished, some objections against regarding such interpretations as "definitions" of the theoretical vocabulary in point would still remain to be reckoned with. And though these interpretations have been liberally referred to as "coordinating" or "operational definitions", their definitional status is not established.[37]

[37] Contemporary reactions to some such difficulties in the positivistic account will be touched on in Chapter VIII.

III

AMPLIATIVE SCIENCE

I. DISCOVERY

A. PROPOSAL AND ACCEPTANCE

Our general characterization of science in the last chapter resulted in the observation that science is profoundly systematic, meaning by this that its structure is, at heart, inferential in character. But this is a very broad characterization, and one that is safely made only when science is viewed as a more or less finished product of inquiry. There are stages of science, however, other than that of the finished product; and we want now to ask whether inference, in some proper sense of this term, plays any role in them.

For purposes of study a fundamental distinction between *two* phases of all scientific inquiry suggests itself : an *ampliative* and a *subsumptive* phase. The *first* expands the content of science as, for example, in Mendel's theoretical proposals on the nature of heredity or in Spallanzani's substantiation of the biological law that every organism has a parent. The *second* invests this content with logical order as, for example, in showing how Galileo's law of falling bodies may be inferred from the known uniformity of acceleration of bodies falling near the surface of the earth and how this uniformity may, in turn, be shown to follow from the principles of Newtonian mechanics (and certain special considerations about relative masses, distances, negligible air resistance, etc.). As our chosen label, 'subsumptive', suggests, this second aspect of science incorporates scientific statements into the body of knowledge by logically subsuming them (as "special cases") under more general statements and systems of statements. In actual practice, these two phases of science occur not as sharply demarcated and disparate "parts", but rather, as blended aspects of a diversified and complex enterprise in which the establishing of both content and order have much to do with each other. The major part of our concern in all that follows will be with the systematic (subsumptive) phase of science or, as we might say, with the structure of science as a more or less established product of inquiry.[1] The ampliative phase of science, i.e.,the *growth of science* will be of much less concern to us. It seems desirable, however, before going on to our more central task, in the later portions of this chapter, to undertake, first, a cursory examination of the ampliative process, whose burden it is to beget rather than to elaborate the content of science.

The writings of C. S. Peirce on hypothesis formation strongly suggest dividing the ampliative phase of science into two distinguishable sub-phases : *conjecture* which results in the *proposing* of plausible hypotheses and *confirmation* (or refutation) which results in their acceptance (or rejection) on the basis of later evidence.[2] It has become customary to apply the term, *scientific discovery*, somewhat loosely : to the proposing of *plausible* scientific hypotheses. And we shall find it convenient to abide by this custom. It seems appropriate to remark, however, that though discovery and acceptance are for purposes of analysis separable aspects of the ampliative process, they are nevertheless actually admixed throughout. Thus, there is already some measure of acceptance in the plausibility of any initial proposal, while conversely, confirmation and more conclusive acceptance adds reason for looking back on any proposal as having been a "discovery". It is because of this somewhat blanketing sense of the term, 'discovery', that we have referred to its usage as loose. It seems desirable to make an additional remark at this point. Scientific discovery, particularly of the sort which results in the uncovering of observational facts, rather than in the proposing of theory, tends to raise a special crop of questions such as, for example : What sort of thing does one do when one observes a phenomenon, or a natural configuration, or system? How does one *selectively* perceive a pattern in a perceptual field that may be looked at in an indefinite number of possible ways? What is it to claim that one has discovered and consequently "knows" some observational fact? And so on. Questions of this sort belong to the general philosophy of knowledge and perception and will not detain us here. More specific to the methodology of the empirical sciences and basic to our concern in this section is what one might well call the fundamental question of the philosophy of discovery. As usually phrased the question is this : *Is there a methodology,*

[1] Methodologists have traditionally referred to all subsumptive systemization as *explanation.* But for reasons to come out in later discussion we shall find it more appropriate to regard explanation as only one kind of scientific subsumption.

[2] Peirce's distinction is between "hypothesis" and "induction". See C. S. Peirce, *Collected Papers,* Cambridge, Vol. II, Sec. 512.

or – as it has come to be put – *a "logic" of discovery?* Let us now confine our attention to *discovery* with this central question in mind. The second stage of ampliative science, namely, *acceptance* will be the subject of later discussion.

B. DISCOVERY : AN OLD QUESTION

How the understanding can arrive at basic knowledge about the nature of things is one of the oldest and most intriguing questions in the philosophy of inquiry. Plato was probably the first to attempt a sustained account of how we "grasp" underlying principles. He characterized the process as recollection (*anamnesis*), in which sense perception can be a suggestive factor in "seeing" a truth but in which ultimately philosophic reflection is what brings to full consciousness that which is implicitly in the mind already.[3] Plato had no specific rules or strategy of discovery except for a general method of reflective analysis or "dialectic" which when relentlessly pursued to the most abstractive levels could lead to ultimate and encompassing insights. The results of this process were supposed to be neither a matter of option nor of tentative conviction; they were final and compelling. Moreover, it seems fair to say that basic insights were not, for Plato, logically compelled by the dialectics which paved the way to them. The philosophical process, though in itself highly logical as a discipline, was propaedeutic, anticipatory, ultimately psychological, rather than logical in its bearing on scientific knowledge. And if this is correct, then Plato's account of discovery affords us little or nothing which might be characterized as a method or "logic" of discovery. And despite its underlying philosophy, it is, in effect, a thesis more for the psychology of teaching and learning than for logic.

Like Plato, Aristotle believed the discovery of basic or "first" principles to be an ultimate and incorrigible act of mind (intuition, insight). In contrast to Plato's view, however, this was not *recollection.* It was *new* knowledge coming in, grossly, *via* the senses, but distilled out by an intuition which finally "grasped" and guaranteed its truth. Aristitole referred to the process as *epagoge* which was much later translated as "induction".[4] According to Aristotle the intuition is able to sift out relevant evidence. That is, it can select those particulars in which, so to speak it will perceive the universal.[5] Such particulars, then, constitute a basis for arriving at

[3] For some brief but eloquent comments see W. Windelband, *A History of Philosophy* Macmillan : New York, 1923, pp. 116-121.

[4] Aristotle, *Analytica Posteriora* II, 99^b-100^b.

[5] *Analytica Posteriora* I, 72^b.

some *one* generalization rather than at some other. Thus, for Aristotle, not only does experience lead to conclusions, but it does so in a strictly determinate or, as we might say, *univalent* manner. An inductive conclusion is incontrovertibly *true* and therefore precludes the possibility of any alternative that might be incompatible with it.[6] In discussing this manner of discovery Aristotle makes explicit reference only to first principles – that is, to establishing the most basic principles of knowledge. Whether Aristotle would have wanted to extend his views on "inductive" discovery to the creation of scientific theory, in the modern sense of such theory, can, strictly speaking, be only a matter of conjecture.

At any rate, Aristotle's "epagogic" though based on the "perception of particulars" is guided by no set of explicit rules for arriving at conclusions. Hence, though it has the aura of an inference based on experience, one cannot confidently speak of it as following some specific logic without a considerable loosening of the term, 'logic', to include what many would today prefer to call *psychological* or at best *epistemic*. In Aristotle's account experience *is* necessary for the formulation of principles and for the "certainty" with which we grasp them. But validity (correctness, objectivity) in the whole process does not stem from any strategy (method, logic) relating evidence to conclusion in some specifically *formal* manner. Instead it rests solely on a special faculty of mind or intuition. Such a manner of validation would hardly do as a "logic" of discovery in any present-day sense of 'logic'. For, whatever else contemporary writers may have in mind when they speak of a "logic" of discovery, most surely they are thinking of rules or directives which are either already codified or capable of being codified. What is more, such rules would relate evidence to hypothesis *via structural*, or, one might say, *formal* features of statements and not on the basis of irreducible mental faculties for the fathoming of relevance. Aristotle prescribed no such rules. Nor did he ever address himself to the task of finding any, or else of presenting some examples of some. His *epagoge* offers no principles of relevancy and cogency on which to pattern discovery which could themselves be objectively held up for validation. Hence, as in Plato, where there is a similar absence of reference to justifiable rule-governed and structure-governed strategies, Aristotle's account can, in no modern sense, be regarded as constituting any significant step towards a so-called "logic" of discovery.

[6] In this respect Aristotle departs from virtually all contemporary conceptions of inductive inference.

Some writers attribute to C. S. Peirce a considerable depth of concern with the logic of discovery.[7] But, (and we refer here only to that portion of Peirce's work which has been codified) though he makes scattered references to discovery throughout his writings, nowhere does he give a sustained, systematic analysis of it. To his credit is the distinction between discovery (proposal) and later confirmation (acceptance). Also he hints that distinct logical processes are involved in each of these, labelling the first process "retroduction" to contrast it with the second which he calls "induction". Unfortunately, however, he offers no explicit exposition of the alleged difference. As a matter of fact, in an informal manner and adding somewhat to the perplexity, he even at one point attributes to retroduction what seem to be decidedly non-logical characteristics bordering on a guess-and-error "theory" of discovery.[8]

The philosophy of science of recent decades is for the most part negative on the question of any logic of discovery, though it is often unclear as to whether what is being denied is the possibility or merely the explicit existence of such a logic. The approved characterization of science has been chiefly in terms of the so-called *hypothetico-deductive* method, familiar to anyone who had the standard school lesson on scientific method. And, indeed, as a methodology of the positive sciences, particularly in their more matured stages, the hypothetico-deductive account is a welcome departure from the inductivisms of Bacon and Mill which take science to be a vast data-processing enterprise in which the sifting of evidence is expected to bring out the patterns of nature much like in the developing of latent fingerprints with powder. As a philosophy of discovery this sort of inductivism is wrong; for it is blind to the manner in which much of science actually happens particularly where theory creation is concerned. Nevertheless, the hypothetico-deductive reconstruction of science, though it is a decided advance over the Bacon-Mill account fails to provide any methodology at all of scientific discovery. Indeed, it views scientific discovery as a non-rational process governed at best by heuristics and not by anything like a logic.

Let us pause to say briefly what the hypothetico-deductive view of science is. In essence, it goes as follows : the growth of science stems largely from attempts to resolve perplexities – some practical, some theoretical. In

[7] As for example N. R. Hanson, in "Is There a Logic of Discovery?" *Current Issues in the Philosophy of Science* (Symposia of Scientists and Philosophers in Proceedings of the American Association for the Advancement of Science, 1959) ed. H. Feigl and G. Maxwell, Holt, Rinehart and Winston : New York, 1961, pp. 32-34.

[8] Peirce, C. S., *op. cit.*, vol. II, Sec. 755.

the course of ordinary scientific practice, a problematic situation may arise – a surprising phenomenon, or one which is not so surprising yet lacks explanation; a paradox or other kink in the current theory; or just not enough simplicity in some working hypothesis. A member (or members) of the scientific community then reflects on the problem. He formulates it as clearly and precisely as he can, fixes on the relevant parameters and on relevant features of the context in which the problem arises, and then makes a "shrewd" or "sophisticated" guess at the solution. The next step concerns the *testing* of this "guess-work". Such testing is not always a simple and straightforward affair. Typically, it first requires that the hypothesis be precisely formulated so that observational statements, i.e., statements having clear and immediate observational content may be deduced from it. The deduced statements may then be put to more or less direct experimental test. A favorable outcome of such testing "confirms" the original hypothesis, thus enhancing the scientific acceptability of the hypothesis. An unfavorable outcome, on the other hand, amounts to a falsification and disqualifies the hypothesis altogether. The deduction of consequences from hypotheses can be a very technical affair sometimes requiring highly specialized and sophisticated mathematical techniques. The resulting networks of explicit derivations (explanations, predictions and other systematic inferences) collectively comprise the logical structure of the "finished" science itself, particularly when the hypotheses involved are fundamental pieces of scientific theory. Another way of putting all this is to say that *deductive* elaboration within any science, besides paving the way to relevant experimentation and to the confirmation or falsification of its principles, sets out the systematic features or, as we might say, the logical skeleton of that science thus consolidating it as a theoretic *system*.

Admittedly the hypothetico-deductive view provides a generally correct and penetrating account of science as finished and systemized product. That is, it develops the logic of subsumptive systemization and points suggestively at the possibility of confirming or falsifying general hypotheses *via* the experimental testing of their observational consequences. The hypothetico-deductive account, however, remains remarkably silent on discovery. It may concede that scientific proposals are often matters of "shrewd" and "sophisticated" conjecture. But there is little follow-up along this line. In the final analysis the hypothetico-deductive account of the scientific process tends to discount scientific discovery as an ultimately non-logical process and one which is therefore outside the methodology of science.

But despite its distinguished achievements in clarity and philosophical penetration, the hypothetico-deductive account suffers from a remarkable lack of explicitness and univalence in its dismissal of the subject of scientific discovery.[9] Hans Reichenbach, for example, writes that the philosophy of science cannot be concerned with reasons for *suggesting* hypotheses but only with reasons for *accepting* hypotheses.[10] But he verges on inconsistency when elsewhere he argues for the fundamentality of induction by enumeration as a basis for suggesting (Reichbach's term is "positing") statistical hypotheses. Induction by enumeration is a logical process. It provides reasons based on formal considerations (i.e., the counting of instances), which, Reichenbach urges, are justifiable.[11] But, apart from the overall ambivalence in Reichenbach's philosophy of discovery, his claim that discovery is not a proper subject for the philosophy of science leaves us wondering as to whether he is denying the *possibility* or the *actual occurrence* of formally based reasons in discovery. This is a typical unclarity in the hypothetico-deductive dismissal of logicality in discovery.

Some hypothetico-deductive theorists rather than denying that logical strategies may be operative in discovery are more inclined to deny that discovery calls for any special logic. That is, they would maintain that whatever logical strategies may be operative in discovery are no different from those used for other scientific purposes, and that such strategies have been given due recognition within either deductive logic or the inductive logic of confirmation. As one writer puts it, discovery presents no "new and special type of reasoning which has not been sufficiently taken into account by formal logicians".[12] Stated in this manner the hypothetico-deductive thesis denies considerably less than do its other versions. But the contention that there is no strategy peculiar to discovery and different, let us say, from some strategies of acceptance is by no means obvious and cannot be granted gratuitously. At the very least, it calls for some examination of actual instances of discovery.

In summation, the hypothetico-deductive account either denies that discovery is (or can be) any sort of logical process at all, or it denies that discovery employs any special logic not already in use in other phases of scientific inquiry. In the first denial, it essentially maintains that at most one can ask only about the causal genesis of discovery. This would place

[9] For some concise historical references in this general connection, see : N. R. Hanson, "Is There a Logic of Discovery?" H. Feigl and G. Maxwell, *op. cit.*, pp. 20-21.

[10] Reichenbach, H., *Experience and Prediction*, Chicago, 1938, pp. 657, 382.

[11] Reichenbach, H., *The Theory of Probability*, pp. 432-434. Reichenbach's justification of positing on the basis of enumeration is ultimately his justification of the inductive "process" in general.

[12] Feyerabend, P. K., "Comments", Feigl and Maxwell, *op. cit.*, p. 38.

the subject of discovery entirely within the empirical science of psychology, sociology and (in one sense) history, where *causes* rather than *reasons* are the primary concern. The second denial makes any strategy of innovation merely another chapter within the standard (deductive-inductive) methodologies of consolidative science. But on neither of these claims is the hypothetico-deductive account conclusive and some reconsiderations therefore seem appropriate. We return therefore to the very notion of a logic of discovery and to the basic question of its possibility.

C. THE IDEA OF A LOGIC OF DISCOVERY

We ask now how does a scientist hit on something new like : an oral contraceptive; or the curative effect of the tuberculosis vaccins, BCG, on some forms of cancer; or the atomic theory; or the special theory of relativity? It takes only casual acquaintance with the scientific process to sense how much non-logic goes into the expansion of scientific knowledge. The history of science offers countless examples of how individual preferences, personal habits, special techniques, social and physical circumstances, and even mere chance can affect the direction and outcome of inquiry. Moreover, some of these factors could well be the basis of recommendable (i.e., rational) strategies on the grounds that they have proved to be fruitful in the past. But such strategies, justifiable (and hence "rational") though they may be, would properly fall under psychology, sociology or history or, perhaps, under an art or technology of discovery. They are not matters for any philosophy of science in the sense of a *logic* or *methodology* of inquiry. What, then, can one mean by a logical strategy or, more simply, a *logic* of discovery and does such a strategy figure at all in the process of discovery?

To begin with, a *logical* strategy, L, unlike any *psychological* or *social* strategy must be one based on what would ordinarily count as *logical* considerations, i.e., mathematical and linguistic form, conceptual content, the counting and distribution of instances, the range of information, etc. It would not be based on what one might call *material* considerations, which is to say on the psycho-social or physical determinants (causes, concomitants, etc.) of human action and decision. Thus, the rules (directives, recommendations, guidelines, etc.) which constitute L might direct us to examine the quantity or distribution of available evidence – perhaps to *count* up favorable instances or to compute relative frequencies on which to generalize. Or, it might call attention to the formal structure of some theory in a neighboring (or even a remote) discipline on which to draw some analogy. Or it might recommend that the hypothesis to be proposed have certain structural features such as symmetry, economy, continuity. Or, again, it might direct us to prefer an hypothesis which is logically compatible with already established principles. It may bear re-noting that the distinction between logical strategies and non-logical, or, as we might say, material ones, does not mean to suggest that the first are more rational than the second. Strategies of either kind might find some justification and hence provide a basis for rational decision. Moreover, for both kinds the justification might be no more than past successes. It is strategies of the first sort, however, which lend themselves to logical analysis and which would more naturally interest philosophers rather than historians, psychologists, or sociologists.

Secondly, L, as in the case of any rational strategy, must be *justifiable*. That is, it must be possible to show how the use of L leads to *plausible* hypotheses or, as one might also say, to hypotheses which are likely to succeed. This is not to suggest that for a given problematic context, our strategy, L, must guide us to a uniquely plausible hypothesis, H. L may only provide a basis for saying what general sort of hypothesis, a successful H must be. In such a case, any proposal governed by L would consist of a whole *class* of plausible hypotheses rather than of a particular one.

Finally, as an elementary point of logic, L cannot possibly be a strategy of the deductive sort. For, deduction can only unpack (i.e., analyze) what is already known or available, and what is desired, instead, is an H having *new* factual content. This might be put by saying that in order for H to be a discovery, its content must be novel relative to the general corpus of knowledge. This, then, means that it cannot be deducible from the conjunction of statements, E, expressing such knowledge. An H deducible from E may be "new" in some psychological sense; but it could not be new in actual content, surprising though it may be when first unshrouded by the deduction. Any claim therefore that deduction from known principles suffices for discovering new ones is either a logical blunder or the result of special senses assigned to terms like 'new', 'discovery' and, possibly, even 'deduction'.[13]

[13] Perhaps adding to some of the misunderstandings along these lines is the choice of language which one occasionally finds even in the greatest scientific writings. Isaac Newton, for example, speaks frequently of "deducing causes from effects" or of "deducing the properties of objects from phenomena" or "from experiments". See : I. Newton, *The Mathematical Principles of Natural Philosophy*, London, 1903, II, 314, and elsewhere. In many such passages, however, the reference is to inference, in general, even when the actual inference in point is not what we would now, more strictly, call deductive. That is, the term, 'deductive,' in such contexts is used as loosely synonymous to the term,

These consideration makes it apparent that a deductive manual for making *necessary* inferences from a given context of evidence to an hypothesis or even to a class of hypotheses is a chimera; and those who deny the possibility of any such manual are making a logically impeccable claim. But this is not to suggest that deductive inference plays no part in discovery. Indeed, without deduction much of scientific discovery might have never occurred. Although deduction is explicative rather than innovative, it can make explicit the pertinent implications of what is already accepted or assumed and thus raises – as it were – the launching platform for the innovative "leap" of discovery.[14]

D. ON THE POSSIBILITY OF A LOGIC OF DISCOVERY

1. Consistency and Intelligibility of the Idea

Having outlined a sense in which one might speak of a logic of discovery, we now return to the question of its possibility. To begin with, it seems fair to say that despite monumental practical difficulties that could beset anyone attempting to formulate such a logic as an explicit set of guidelines for discovery, the possibility cannot be gratuitously dimissed. To establish impossibility one must show either (i) that the notion of a set of rules governing proposals is inconsistent or otherwise incoherent or (ii) that no justification can be given for such rules, or, putting it another way, that no plausibility can be assigned to proposals governed by the rules.

It is difficult to see how anyone would go about establishing (i). Practically speaking it may be very unlikely (perhaps, hopelessly so) that any L can ever be made explicit in the context of natural scientific usage. A precise definition or explication of such a key notion as the *simplicity* of theories, for example, is not at present readily given except with respect to a very limited set of structural considerations.[15] But again this is no

---

'inference', in a sense general enough to include logical strategies of a non-deductive sort, some of which might, ordinarily, be classed under inductive inference.

[14] For a paradigm example of the extensive role which deduction can play in discovery, the reader is invited to follow, in detail, Newton's own account of how he arrived at his conviction that the force on a planet is of the universe square type. See *Lord Portsmouth Collection* (additional manuscript) 3968, No. 41, bundle 2, Cambridge, 1888.

[15] Nelson Goodman, for example, has outlined an approach for judging the simplicity of the basic extra-logical vocabulary of a theory. This, however, is far short of explicating the *overall* simplicity of theories and of dealing with related matters in

reason for doubting the consistency or the intelligibility of the idea of a rational strategy of discovery or, for ruling out even the practical possibility of such a strategy based, perhaps, on only *partially* explicated criteria.

Establishing (ii) would also suffice for the impossibility thesis. That is, even if there were no inconsistency or other incoherence in the idea of a strategy for governing discovery, the possibility of a logic of discovery could still be ruled out by showing the impossibility of justification for any such strategy. Without justification a set of guidelines would not be logically compelling in the slightest and would confer no measure of plausibility on any hypothesis. The decision to implement any unjustified strategy would, at most, reflect only the habits and preferences of individual scientists or of the scientific community at large. The strategy in point would be a logically indifferent recipe for arriving at formulations and as such would belong not to the context of validation and hence to the philosophy of science but rather to the context of the history, psychology or sociology of science.[16]

But as a matter of fact, asked about analogy, simplicity, economy, symmetry, compatibility with established context, simple enumeration, and perhaps, still others, a working scientist might well say that they are defensible strategies and point to an impressive historical record of past successes. Admittedly such justification by past record, known as *justification by induction*, has raised a number of traditional objections, some of these going as follows :

(a) Induction is no means of justification at all; that is, it is not itself a rational method or a logic of justification but rather a conditioned response to some experienced regularity.

(b) Where inductive methods have been precisely formulated, they are inapplicable to universal and other general hypotheses (e.g., statistical ones) because the latter cover an unending number of instances, whereas evidence is always finite in extent.

(c) Induction, viewed as a validating logic, must itself be validated. But how would this be done? Could we point again to past successes? If we

---

connection with curve-fitting and induction generally. For an illuminating discussion, the student is invited to read : N. Goodman, "The Test of Simplicity", *Science*, Vol. 128, (1958) pp. 1064-1068.

[16] As we have already noted earlier in discussion many philosophers of science in the analytical tradition have made this non-justificational claim for all techniques and approaches to scientific discovery. A strong version of this view is expressed by R. B. Braithwaite, *Scientific Explanation*, Cambridge, 1955, pp. 20-25.

did, would we then not be subject to the charge of circularity in using induction to justify induction?

But whether or not well-known indictments of induction, such as these can be made to stick is still a matter of strong philosophic contention. It may even be argued that the alleged circularity regarding the justification of induction does not invalidate induction. The important consideration for our immediate purpose, however, is this : Any attempt to rule out the possibility of a logic of discovery by claiming non-justifiability must reject past success as a basis of such justification. But past success does ordinarily count at least in part, as a basis for the rational acceptance of material principles. It seems entirely reasonable therfore to accept past success as a basis for the validation of any strategy whatever. And in the absence of conclusive demonstration against such a basis, this mode of justification remains an open possibility.

An additional point on the justification of strategies for discovery bears mentioning. It is an elementary matter of logic, and we have noted it earlier in this chapter, that no body of evidence can deductively necessitate any hypothesis and that an indefinite number of hypotheses are compatible with such evidence. Thus, much of scientific activity amounts to, as Nelson Goodman puts it :

"... the invention of and the choice among systems of hypotheses. One of the primary considerations *guiding* this process is that of simplicity. Nothing could be much more mistaken than the traditional idea that we first seek a true system and then, for the sake of elegance alone, seek a simple one ... Furthermore in the choice among alternative systems, truth and simplicity are not always clearly distinguishable factors. More often, simplicity is a test of truth. The martyr-bestrewn debate over what bodies in the universe are fixed has turned into the question : What choice of points or reference will give us the simplest description of their relative motions ...? Thus simplicity ... is not a consideration after truth is determined, but is one of the standards of validity that are applied in the effort to discover truth".[17] (My own italics)

Bolstered by present-day conventionalism which denies that any hypothesis can be saved from falsification by proper adjustments in the going system, the strong suggestion here is that much of what is intended by calling an hypothesis "true" or acceptable is something about its structure. But a logic of discovery is a set of structural criteria leading to what we regard as acceptable hypotheses, simplicity being high on the list. On this

[17] Goodman, N., "The Test of Simplicity", *op. cit.*, p. 1064.

line of thinking, therefore, the justification of such criteria would be an immediate consequence of what mean we by "valid" and "acceptable" when applied to our choices.

This last consideration takes its point of departure from relatively non-traditionalistic trends of thought in the methodology of the empirical sciences. And we shall take up some discussion of these trends in our concluding chapter. For the present, however, it merely sufficed to note that any claim that no logic of discovery it possible because it would be unjustifiable would fly in the face not only of justification *via* past success but also of justification in the conventionalistic sense, i.e., *via* simplicity as inseparable from validity ("truth", "acceptability", etc.). Indeed given this and our previous considerations, it is difficult to see how any argument could be launched to establish the impossibility of a logic of discovery *via* non-justifiability.

2. *Historical Indications*

But more concretely the possibility of specifying a set of decision guidelines based on what we have called *structural* or *logical* considerations is strongly suggested by the history of science itself. Indeed, there seems to be something in historical evidence to caution against the widespread claim that discovery is, throughout, a non-rational, non-systematic process ultimately beyond all logical (methodological) analysis.

An instructive example is a well-known episode in the history of modern science, namely, Newton's thought leading to his discovery of the law of gravitation. His considerations are a mixture of deductions and conjectures; and they may be summarized in the following steps :

(1) Newton first applied Huygens' laws for uniform circular motion to the known, *nearly* circular orbits of the planets (Kepler) and got that there is a force, F, on any planet which is proportional to the planet's mass, m, and inversely proportional to the square of the planet's distance, d, from the sun. That is :

$$F \alpha \frac{m}{d^2}$$

(2) Then Newton assumed that F is due to the sun. This *plausible* conjecture is an analogy drawn between the solar system and more familiar mechanical systems like whirling slingshots and revolving bodies inside hollow spheres, for which the centripetal force is provided by such visible means as a string or a container wall rather than by an impalpable gravit-

ational pull. Hence, that the sun is the cause of F, is not a deductively derived conclusion but only a proposal, though perhaps, a very plausible one.

There is also a note of *simplicity* in this conjecture. For if the forces are directed towards the center, where, (with approximation) the sun is, then if they are to be explained at all, why not simply say that they are due to the sun? One could of course resist making this supposition and maintain that the sun has nothing at all to do with it and then look for some other way of accounting for the centripetal force. But this alternative would seem to be less compact and straightforward. For, instead of postulating causal ties between known parts of the solar system, it would veer towards the less *simple* or less *economical* alternative of devising an account in terms of factors more extraneous and less obvious.

(3) Now, having designated the sun as the source of the force on each planet we can apply Newton's already known principle of two-body interaction.[18]

This gives that :

(4) *Any planet pulls on the sun with a force equal and opposite to the force of the sun on that planet.*

But here Newton makes two other proposals (hypotheses) :

(5) *The force of interaction between sun and any planet is proportional not only to the mass, m, of the planet but, also, to the mass, M, of the sun and hence to their product, mM*; that is :

$$F \propto \frac{mM}{d^2}$$

which by introducing a constant, G, can be written as :

$$F = G \frac{mM}{d^2}$$

This conjecture is essentially a *symmetry* argument suggesting that for two interacting bodies, not just *one* but *both* masses are causally relevant, there being no reason to believe that either the sun or the planet is cosmologically special in the general respects concerned.

Finally Newton *proposes* that :

(6) *The above expression for F gives the force of attraction between any two bodies of the universe and this force depends not on G, but only on the*

[18] This is generally known as Newton's third law and says that if body, A, exerts a force, F, on body, B, then body, B, exerts a force equal and opposite to F on body, A.

masses of the bodies and on their separation. This conjecture is a very bold one. It is a generalization – one of the most sweeping in the history of mathematical physics. Like all other such generalizations it is, at heart, *analogical* in form : In all relevant respects the sun and any planet seem to be like any pair of bodies in the universe; hence why not with respect to gravitation! This vast *analogical* generalization was a breath-taking proposal. And what was proposed seemed indeed to be a plausible hypothesis. It represents discovery *par excellence*!

Hypotheses (2), (5), and (6) are innovative proposals. And if our account is right they were not – or, at least, need not have been – merely blind guesses, irrationally compelled inspirations, or mysterious hunches. Indeed behind the proposals there seemed to be considerations guided by analogy, simplicity criteria, and more.

The development of Newton's thought leading to the law of gravitation is but one illustration of the play of logic-like reasons (structural considerations) in hypothesis formation. History offers a vast array of others in which generalization, analogy, concern for simplicity, and other methods are well illustrated. Note for example :

The sweeping generalizations of Galileo about swings of pendulums and about other terrestrial motion, of Robert Boyle on the compression and rarefaction of gases, of J. L. Proust on chemical proportions; or the more methodical and quasi-statistical *generalizations* of Cannizzaro on atomic weights and of Millikan on the electronic charge; or the brilliant and vastly *simplifying* hypothesis of Copernicus on the planets, of Avogadro on the distinction between atoms and molecules, of W. Pauli on electronic distributions around the atom, of Faraday explaining his own laws of electrolysis with *simplicity* by assuming that small integral multiples of some indivisible amount of electricity are characteristically attached to each atom of an element; or the *analogizing* of Bohr with regard to atomic structure, of Maxwell with regard to the electric field, and of Schrödinger to obtain the mathematical form of a basic equation of micro-physics. Note also the quest for *symmetry* of Einstein, or of De Broglie. And the list can be extended much, much more.

E. PATTERNS OF DISCOVERY

Though history reveals a variety of approaches to scientific innovation, the underlying rational strategies seem to have been rather few and might perhaps be grouped for the most part (if not exhaustively) under a list of guidelines of *roughly* the following sort :

1. Generalization from Instances

(*sometimes referred to as inductive generalization*)

This applies to the formulation of experimental laws either of the statistical sort (e.g., antibiotics are effective in a very large proportion of streptococcal infections), or of the universal sort (e.g., All metals expand when heated). The stratagem consists of generalizing by the enumeration of instances and attributing to an indefinitely general population the proportions which have been counted through for the set of known instances.[19] This sort of inference also takes the form of curve-fitting in which from observed instances of a relation between quantities – usually plotted as experimental points on a graph – we generalize to an unlimited set of points, expressing the generalized relation mathematically as a curve or functional dependence between variables. The rule would be: choose the smoothest average curve (mathematical function) which passes through the neighborhoods of known experimental points.[20]

It should be noted that as reasonable as this procedure may seem it is not the only way we could have done it. For, as a matter of fact, an unlimited number of generalizations would also be compatible with the same evidential base. For example, having observed every past crow to have been black, we might still posit that a large proportion (but not all) crows in all of space-time are black, or even that only a small fraction of all crows are black. It could just have happened that only black crows have turned up in the past, whatever the ultimate proportions of black to non-black for the whole population of all past, present and future crows. Similarly an unlimited number of curves may be drawn *through* any set of experimental points on a graph, each curve representing a logically possible generalization encompassing the points.[21] In practice, however, it is considered reasonable to take what some have called the "simplest" alternative. Our next guideline therefore is a concern for simplicity.

[19] Generalization from instances has its counterpart in the *acceptance* or *confirmation* of hypotheses. After a hypothesis that all A's are B's has been proposed by generalization over a number of instances, later instances of A's that are B's tend to promote the acceptance of the hypothesis. Such acceptance may be regarded as a subsequent increase in the initial plausibility of the generalization, which itself, to some extend, depended on the number of initial instances of A's that are B's.

[20] This rule has its algebraic form in the several known methods of interpolation by solving for coefficients, such as the method of simultaneous equations, Simpson's rule, Newton's methods of finite differences, etc., as may be found in any basic text in applied mathematics.

[21] Three illustrative curves for points a, b, c, and d would be as shown.

2. *Simplicity*

This consists of attempting to formulate the simplest hypothesis that seems to provide the desired systemization. Such simplicity of course may not be won at the price of sacrificing the desired scope or generality of the hypothesis to be. In the case of having observed only black crows, the simplest hypothesis is : 'All crows are black' rather than : 'A large proportion (but not all) crows are black', though both would be consistent with the evidence. In our example on curve-fitting, graph (I), shown in the footnote as a straight line, would be considered the simplest. An explication of the logic of simplicity is beyond the scope of the present discussion.[22] We can briefly note, however, that long recognized criteria like economy, continuity, and symmetry figure centrally in the notion of simplicity.[23] Simplicity with regard to scientific theory has several aspects each requiring detailed explication. There is the simplicity of a theory's terms or vocabulary, the simplicity of its postulates framed in these terms and the problem of combining these into an *overall* simplicity for the theory.

As has already been noted simplicity ranks high in guiding the proposal of theories. Moreover, in practice, it seems to enter constitutively in the very notion of acceptability and effectiveness of a scientific theory. Indeed,

---

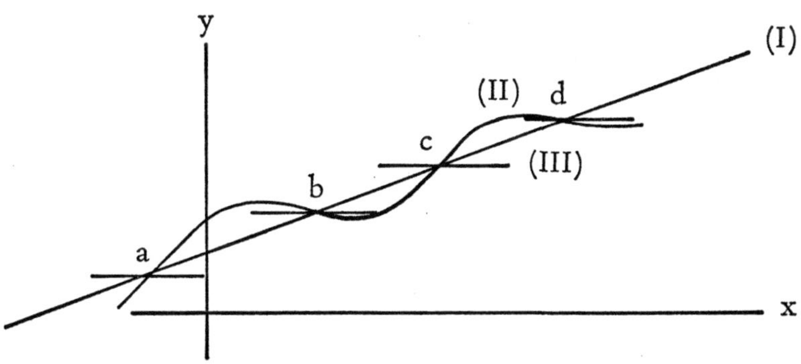

(I) is a straight line
(II) is a more complex continuous function
(III) is a step function and therefore discontinuous.

[22] For an introduction to the role of simplicity in generalization, see S. F. Barker, *Induction and Hypotheses*, Cornell University Press : Ithaca, 1957, pp. 99-95 and *passim*.

[23] Symmetry is a far-reaching aspect of simplicity. It touches not only on the symmetry of relational terms but also on such formal symmetry as invariance of postulates relative to some class of mathematical transformations. Symmetry criteria of the latter type have been the guiding principles in the development of relativistic physics.

in the absence of decisive corroboration or disconfirmation by data, simplicity may be the major basis of choice between competing theories. Historical evidence for the determinative role of simplicity considerations is not wanting. And the methodological import of simplicity criteria, though not systematically developed until very recent times, has been acknowledged in one way or another for centuries. Under the impetus of the conventionalists, P. Duhem and Poincaré, simplicity is now recognized by some as the most determinative single factor in the choice of hypotheses. Thus, old squabbles about which bodies in the universe are fixed have become questions about which choice of reference-frame will give us the simplest description of their relative motions.[24]

3. Analogy

Theorizing typically differs in some basic respects from experimental generalization. In the formulation of theory we do not simply report what appears to be some observational regularity. Theories introduce concepts which lie qualitatively beyond the pale of the initial data. As a result, the language of theories includes vocabulary far beyond anything needed for merely reporting experimental uniformities. The "leap" from data to theory therefore is something considerably different from what occurs when we pass from data to observational laws. In the latter no essential change in descriptive vocabulary is required; in the former there must be a linguistic expansion so as to include new theoretical content – content which is not, in general, exhaustively definable in terms of the concepts of the original or unexpanded language. No doubt, it is this conceptual "leap" in theory creation that has led many to argue convincingly against the possibility of a logic of discovery or of what might perhaps be better called a *logic of invention* or of *creation*. But analogy with some already known system whose characteristics seem pertinent is what often makes this radical expansion of scientific content possible. Analogizing permits making inferences about one object on the basis of what is known about another which seems similar to the first in relevant respects. When the analogy is extended to a whole class of objects it becomes an *analogical generalization*. Such generalization has figured significantly in the formulation of theoretical hypotheses. This strategy, perhaps better rendered by Cicero's phrase, *comparatio proportioque*, is not peculiar to science but is

[24] For a bit more on this see : N. Goodman, "The Test of Simplicity", *op. cit.*, pp. 1064-1065.

widespread wherever humans attempt to assimilate the new or strange, on the basis of some *presumed* similarity, with the old and familiar. The wealth of metaphor in everyday discourse attests well to the pervasiveness of the device. In this connection the history of theoretical science offers an abundance of examples. Heat was conceived as a fluid and its theory developed largely as an analogy with the flow of liquids whose laws were known, the point of resemblance was the fact that heat flows through solids with determinable speed, direction, and resistance, much like liquids. The kinetic molecular model for gases was essentially an analogy with elastically colliding billiard balls, and the Rutherford atom was an analogy with astro-planetary systems.[25] Electromagnetic wave theory drew much conceptual content from analogies with the stress-strain forces in elastic media.

Not only can analogy be suggestive of the conceptual content of a theory-to-be but it can also provide patterns for its formal structure. Thus the equations describing central force fields of the inverse square type and those of heat conduction bear a striking resemblance to each other so that formalisms in one domain can be fruitfully applied to the other. A most striking and far-reaching use of formal analogy occurs in the creation of quantum theory whose formal structure is an adaptation of the formal structure of its classical antecedent, Newtonian mechanics.[26] Not only the

[25] Rutherford shot α-particles at gold atoms. The α-scattering which resulted indicated a massive, positively charged nucleus, concentrated in a very small region of the atom. These results left open the question of just how electrons were distributed about the atom. Analogy with planetary systems strongly suggests orbiting electrons but other dynamical considerations strongly reinforce the credibility of this conjecture. A stable configuration of opposed charges at rest in free space is physically impossible and, in a central force field such as that exerted by the nucleus, smoothly circulating electrons provide the simplest stable dynamics.

[26] The Schrödinger equation which is the basic equation for the wave (rather than matrix) version of quantum theory was gotten by treating a mass particle moving through a conservative force field, in a novel but analogical fashion. The dynamical state of the particle was newly conceived in terms of the wave motion parameters, more particularly, in terms of a dynamically changing wave-length λ which the French physicist de Broglie for reasons of symmetry assumed to be inversely proportional to the classical momentum of the particle at any given time and place in its career. To invent a wave equation for these "de Broglie waves" Schrödinger then applied the classical formalisms of vibration theory. He appropriated a time invariant form of the familiar differential equation for a vibrating medium. One of the parameters of this equation is, of course, the classical wavelength λ which, in general, may vary from point to point depending on the local properties of the medium. Schrödinger then simply introduced the de Broglie wavelength expressed in terms of the particle's momentum. The latter is a quantity which for a conservative field can be expressed in terms of the total classical energy or the Hamiltonian

mathematical formalisms but even descriptive terms like 'particle' and 'wave' were introduced analogically into quantum mechanical discourse.[27]

4. *Maximization of Generality*

This consists is aiming for the most general (most encompassing, most fundamental) hypothesis that will do the job. Thus an hypothesis which explained transplant failures in terms of the mechanism of foreign tissue rejection would be less recommendable than one which (if it could be given) did so in terms of molecular phenomena. The latter approach is potentially capable of explaining not only transplant failures but also a vast range of biological phenomena such as healing, growth, regeneration heredity and, of course, the mechanism of tissue rejection itself.

5. *Compatibility with Existing Context*

The recommendation is to aim for hypotheses that will conflict least with the context of established principles. Admittedly, it would hardly be an exaggeration to say that the history of science is a graveyard of scientific theories. And this could raise some serious doubt regarding the actual role of this last guideline in scientific practice. But science, in its more typical or "normal" phases, does tend to preserve its content and any thought departing from this tendency will verge on revolution.[28] Indeed, compatibility with the existing context should not be a surprising criterion, given the overall scientific objective of systemizing with maximum compactness or simplicity. For this strongly suggests holding to some principles as long as is practicable with a view to integrating a maximum of content around them. In general it is only in those cases where clinging to established principles leads to intolerable complexity that the compatibility criterion is abandoned and radical novelties come under consideration.

---

for the system.

This purely formal transformation results in the famous Schrödinger wave equation. What Schrödinger did was draw an analogy from familiar physical theory, simply assuming that the new and radically conceived wave properties of matter would in some way conform to some of the formalisms of such theory.

For a mathematical exposition, see : Slater, John C., *Quantum Theory of Matter*, McGraw-Hill : New York, 1951, pp. 13-34.

[27] Failure to keep the *merely* analogical use of these terms in mind has lead to a number of notable misunderstandings of quantum theoretic notions.

[28] The most notable contemporary analysis of scientific conservativism and revolution has been given by T. S. Kuhn in his *The Structure of Scientific Revolutions*, University of Chicago Press : Chicago, 1970.

Several other guidelines may come to mind for adding to the strategies already mentioned. But, for present illustrative purposes, our list need not be exhaustive. The following general points, however, should be added to our considerations :

In the first place, by listing the criteria individually, we do not wish to suggest that they are entirely distinct. Some may be overlapping and mutually interpenetrating notions – generalization from instances, for example, incorporates aspects of simplicity since it generalizes *via* simple enumeration rather than in anyone of an indefinite number of other possible (more complex) ways. Analogy with known principles may, in a very general sense, be viewed as having some bearing on compatibility with existing contexts since the first of these will tend to promote the other by fostering certain formal uniformities.[29] Further, the close connection of all this with simplicity (uniformity, logical compactness, etc.) need hardly be elaborated.

Secondly, the problematic and complex nature of some of these criteria lend them an aspect of indefiniteness which can, perhaps, be too hastily dismissed as logical untenability. Simplicity, for example, which, in practice, underlies or, to some extent, governs all inductive generalization is still a highly unsettled notion and requires adequate explication in order to figure in any formal or even partially formal strategy for discovery.

Thirdly, reflection will show that criteria may at times conflict. Thus hypotheses of high simplicity and generality may not always be the ones which are compatible with established knowledge. Indeed, as history illustrates dramatically, the quest for simplicity can lead to hypotheses so divergent from established notions and principles as to precipitate a scientific rebellion. The change from Ptolemaic to Copernican astronomy, from Aristotelian to Galilean mechanics, and from Newtonian to Einsteinian physics are monumental examples.[30] As answers to the breakdowns and perplexities arising in the established context, they are remarkable for their great generality and simplicity. But as innovations they were radical

[29] Analogies, for example, may tend to perpetuate the widespread use of continuous mathematical functions throughout a whole cluster of sciences and thus promote compatibility in this broad sense of formal homogeneity.

[30] Very broadly speaking Galileo achieved a great simplification over Aristotle by *idealizations* like frictionless planes and resistance-free falls. Einstein's key to simplicity was a kind of *formal symmetry*. It was the requirement that the laws of nature be formulated in such a way that their structure would remain unchanged for certain changes in the underlying frame of reference. This possible feature of laws is technically referred to as *transformational invariance* of a kind depending on the class of reference frames over which the structure of laws is to remain unchanged.

and a rude shock at the foundations of the prevailing notions.[31] Similarly, the attempt to maximize generality may at times be in conflict with the strategy of analogizing with known phenomena. Thus an hypothesis attempting to explain heat transfer in terms of a special "heat fluid" would in certain important respects, be less general (or encompassing) than one attempting to do it in terms of colliding, fundamental particles. Yet, in first searching for something on which to model a theory of heat flow, the familiar phenomenon of flowing liquids and gases would seem to be much more suggestive (a stronger analogue) of heat flow than would be any system of colliding bodies. Again, simplicity and generality may sometimes conflict. Thus, one can no doubt entertain some appropriate sense of the term, simple, under which Mendel's genetics might be judged simpler than today's molecular genetics, though the first, which accounts for heritability in terms of genes and dominance-recessive mechanisms, is in very important respects less general than the second which does it in terms of molecular chemistry. Or – one might find certain aspects of simplicity undermined by an approach which is too blindly analogical. Some theories of the state, for example, have been based on analogies with biological organisms and have taken as a fundamental concept the notion of the state as a gigantic organism. But suggestive and picturesque though such theories may be, they could result in a loss of the simplicity which one might achieve from approaches that attempt to analyze the concept of the state in terms of simple relatable parts. Analogies may of course figure in the construction of more analytical social theory. Political and social theorists for example sometimes do speak of the "forces" or "dynamics" of the state or of society. But, in general, non-biological analogues in political theory seem to have been far less sweeping and less tightly drawn by their authors than have the biological ones. Indeed, organicistic analogies in some theories of state have been so sweeping as to detract not only from their simplicity but possibly also from their intelligibility and ultimately, perhaps, from their scientific utility.

At any rate the possibility of conflict between "criteria for discovery" strongly suggests the awesome complication of having to include directives on how, in certain instances, to weight or favor one criterion over another. Doing so would, of course, presuppose much clearer and far more settled

[31] The Copernican scheme, for example, conflicted not only with the prevailing cosmology but also with the current Aristotelean physics of falling bodies which required that the earth be fixed at the center of the universe towards which all bodies gravitate. For a discussion, see : E. J. Dijksterhuis, *The Mechanization of the World Picture*, Oxford University Press, : Oxford 1961, p. 290.

notions of each criterion than one finds today, perhaps even to the extent of associating quantitative (numerical) metrics with them.

Fourthly, it is important to note that there is nothing about our guidelines which might imply that though they are operative in the proposing (i.e., the discovering) of hypotheses, they cannot apply on the level of acceptance as well. In this respect, it should be recalled that such guidelines must function not only to help generate innovation in science, but also to assure the plausibility of the resulting hypotheses. That is, they also function in the context of validity or justification. This makes it natural to regard a logic of discovery as one that, together with a logic of confirmation or acceptance, can belong to a general logic of inductive inference. In such a logic, degree of simplicity, closeness of analogy to some successful and comparable theory, extent of compatibility with existing knowledge might – together with more conventional inductive reference to evidence – provide measures of acceptability for any scientific hypothesis.

F. REASONS VS. CAUSES

As has already been noted, scientific discovery is a human activity and one which no doubt has its decision-making aspects. But like all social phenomena it can be viewed as subject to causal determinants – physical, biological, psychological, etc. Discussions in this area, however, are often hampered by insufficient care in distinguishing between the *dynamics* (causes) of decision and whatever *rational bases* (reasons) such decisions may *also* have. These two aspects of human decision making, i.e., causes and reasons, are *not* mutually exclusive. Where innovative scientific thinking is concerned, the play of non-rational factors, e.g., preference, familiarity, unconscious habit, free association, inspiration, ideology, trial and error, random groping, and blind luck, is "visible" even to the "naked eye" of any historian. These are ultimately causal-descriptive factors and their dynamical relation to human action – in this case the proposing of hypotheses – is the proper concern of empirical science, in particular, psychology, sociology and, to some, extent, history. This, however, does not deny that there might also have been good reasons of the sort we have called "logical" (e.g. observed patterns, analogy, simplicity, etc.) for what any scientist, in fact, decided to propose. But the bearing of reasons on decision-making is very different from that of causes. Causes and not reasons are what determine or *explain* the proposing of an hypothesis, H; reasons are what *justify* the proposal. Causes, one might say, are connected "naturally" (a better term here might be 'materially') to the human *act* of proposing; reasons (of the sort under consideration) have only a logical bearing on the proposal, H. But the fact that good reasons may figure into any decision-making does not in any manner suspend the causal basis of such action even as directly concerns the reasons involved. And, the assumption that *all* human action like all natural events are determined by causes is not in jeopardy though reasons may be operative. Indeed what makes the having of good reasons for scientific proposals *historically* relevant is the tacit historical assumption that scientists usually act in accordance with such reasons. So that, even where it was known that good reasons were operative in the making of some decision, it would still make sense to ask the causal-genetic question; Why did X act in accordance with good reasons? And the answer might well be that X does in fact usually act in this manner. One might, of course, then even go on to explain that X is generally disposed to act in accordance with good reasons because he is intelligent, has been trained in a certain manner, etc. The point, however, is that the operation of reasons does not rule out the action of causes; moreover, among the causes of any propositive action by X is X's *being aware of* or his *entertaining* certain good reasons for the action and not the reasons, *simpliciter*.

Quite clearly, then, the possibility of there being reasons for proposing a scientific hypothesis is independent of the ultimately causal basis of such proposing. It is entirely pointless, therefore, in arguing against the possibility of reasons – and hence against the possibility of a *logic* of discovery – to insist as so many methodologists of the past have done, on the truism that the process of discovery is properly studied under psychology, sociology and history of science. This it certainly is, but not with the consequence that, therefore, discovery is not subjectable to logical analysis, apropos of reasons. Such a consequence can only stem from an improper distinction between the respective roles of reasons and causes in human action.

II. ACCEPTANCE

A. CONFIRMATION

It is true of both science and common sense that the acceptance of any scientific hypothesis depends in some important manner on its being supported or, as we shall say, *confirmed* by evidence. The process of marshalling evidence, i.e., experimentation and other forms of observational investigation, has in it much procedural technique that is peripheral to the philosophy of science. But the supportive relation between evidence and hypothesis, understood as a relation between sentences, and generally referred to as *confirmation*, seems to fall, decidedly, in the context of validity. Confirmation, therefore, especially as regards general (law-like) scientific hypotheses is of central and specific concern in the philosophy of science. Let us pause, therefore, for some brief considerations on this second aspect of ampliative inquiry.

To begin with, we note that a scientific hypothesis may relate to relevant evidence in one of two ways. The evidence may be either favorable or unfavorable to the hypothesis. Now, most would agree that once unfavorable evidence has been obtained for any universal hypothesis, its falsification is ultimately a matter of deductive inference. A universal hypothesis logically implies its evidence. Hence, the falsification of any such hypothesis is decisively accomplished by falsifying any one of its logical (deductive) consequences, O; for only a false premise can validly imply a false conclusion.[32] The falsification of any universal hypothesis thus requires no special kind of logical apparatus such as inductive inference, provided that we grant that the *obtaining* of negative evidence does not *itself* raise serious questions about any special logic. This last assumption may in principle be questioned.[33] But we have no need here to pursue the objection. It is sufficient to note that any useful universal hypothesis will in general have observational consequences that are so decisively defeatable that the conclusive falsification of such an hypothesis is, in effect, free of probability considerations. In this sense, falsification might be said to proceed without manifested and explicit recourse to inductive inference.

In contrast to falsification, *confirmation* is not so succinctly characterized. As has already been indicated, what it means to say that a set of statements

[32] The elementary rule of formal logic that sanctions such an inference is known as *modus tollens* conveniently schematized as follows :

$H \rightarrow O$	(If H then O)
$\sim O$	(not O)
$\sim H$	(therefore not H)

[33] For, as plausible as this assumption may seem, some would seriously object that no observational statement, however simple, is ever completely falsified. Its falsity can, at best, only be made probable; and so, falsification can never escape reference to inductive inference. There is another sort of objection to the very possibility of falsification, based on the doctrine of conventionalism. Some consideration will be given to this line of thought in our closing chapter. It might also be noted here that decisive falsification by *modus tollens* is, of course, not possible for hypotheses that do not logically imply their evidence, as, for example, hypotheses having an explicitly statistical form.

"supports" or "confirms" an hypothesis is a highly controversial matter, even among those who believe such confirmation to be a species of genuine logical relation.[34] Indeed the situation is far worse. There are those like K. Popper who deny logical status in any sense whatever to such a relation[35]. And there is much in Hume's account which similarly suggests a a denial that the inductive process corresponds to any sort of logical relation.[36]

Still, despite the unsettled nature of the complex issues concerning induction, there is an unmistakable sense – as most students of the subject seem to agree – in which it is reasonable to rest one's case for any factual claim on factual evidence. "Reasonable" here means justifiable in a sense in which whims or other psychological impulses are not – though, indeed, the only justification we might ever be able to give is the mere truism that if nature has any scrutable pattern at all, then the best possible way of fathoming this pattern is *via* some inductive method or other. Moreover, as we have already noted in earlier discussion, scientists *do* make what in a very ordinary sense seem to be reasonable, i.e., evidence-based, generalizations reporting some observed or experimental regularity. Also, it would seem similarly reasonable for these same scientists to bet, offering greater odds on the future of any generalization, the more extended is its history of uniform success.

Indeed, matter-of-fact hypotheses are, for the most part, never incorrigibly established by any body of evidence. This is particularly clear in the case of laws, theories and other statements having the form of unlimited generalizations. For, at any given time, such statements can be supported only by limited evidence, and, therefore, cannot be deductively implied by

[34] One may recall here the conflicting views of two inductivists like H. Reichenbach who allows as fundamental only a statistical notion of evidential support. See : Reichenbach, R., *The Theory of Probability*, Berkeley, 1950, and R. Carnap who admits of non-statistical conceptions. See R. Carnap, *Theological Foundations of Probability*, Chicago, 1962.

[35] Popper, for example, denies that science employs any modes of inference or "logics" other than deduction. Thus in ampliative science the only genuinely logical process is falsification which he takes to be deductive in character, and governed by *modus ponens* as indicated in a foregoing footnote. See Popper, K. R., *op. cit.*, pp. 27-35 and elsewhere.

[36] "... even after experience has informed us of their constant conjunction it is impossible for us to satisfy ourselves by our reason why we should extend that experience beyond those particular instances which have fallen under our observation ... Reasons can never show us the connection of one object with another, though aided by experience and the observation of their constant conjunction in all past instances. When the mind therefore passes from the idea or impression of one object, to the idea by belief of another, it is not determined by reason, but by certain principles which associate together the ideas of these objects, and unite them in the imagination". Hume, *op. cit.*, vol. 1, p. 94.

such evidence. This partial support or *confirmation* is – at least, according to prevailing views on the subject – the basis for the acceptance of any hypothesis of science. Its explication is therefore a major concern in scientific methodology.

Most students of methodology would generally agree that confirmation is to be regarded as an "inductive" relation between two sentences, E and H, the first expressing evidence, the second some hypothesis. There are, however, various ways of explicating the relation. In its most ordinary sense, confirmation is a qualitative concept on the basis of which one may say that some given sentence, E, does or does not support the hypothesis, H. Moreover, E is ordinarily said to support H if it instantiates H. Thus if H were : All A's are B's then the statement : (E) 'This A is a B' would ordinarily count as evidence in favor of H, just as the sentence 'this A is not a B' would not only count against it but would falsify it, outright. This, however, is a rather inexplicit notion of confirmation and one which is not very serviceable particularly where comparative judgments are needed for choosing between competing hypotheses. What is more, the task of explicating a satisfactory concept of qualitative confirmation has met with a number of formal difficulties largely raised by Hempel and Goodman which still await full resolution.[37]

With the aim of developing a more precise notion of confirmation inductive logicians have attempted to give quantitative definitions of confirmation so as to be able to assign numerical metrics as degrees of confirmation. The most natural course for a quantitative concept seems to be a statistical or frequentist analysis in which the confirmation of any hypothesis depends on frequencies of past occurrences. Thus it seems reasonable to expect that Mr. X will not be a nonagenarian since a very high proportion of past humans have not lived to be ninety. More generally, on the statistical conception, the degree of confirmation of any hypothesis that an A is a B will be equal to the proportion of known A's that have been B's.

In favor of this view, it may be said, first of all, that it accords very closely with the ordinary qualitative conception on which positive instances are regarded as supportive of an hypothesis. Secondly, wherever quantitatively measured expectations figure in actual scientific practice, they seem almost always to be based on relative frequencies. But the frequentist conception does raise questions. Thus, it might be objected that it admits

[37] For the so-called paradoxes of confirmation, see : C. G. Hempel, "Studies in the Logic of Confirmation", *Mind*, Vol. 54, 1945, pp. 1-16, 97-121. For Goodman's Paradox, see, *Fact Fiction and Forecast*, Harvard University Press : Cambridge, Massachusetts, 1955, pp. 63-126.

only statistical evidence for an hypothesis. Other considerations such as the *meaning* content of evidence or possibly even the *structure* of an hypothesis may seem relevant for purposes of scientific acceptance, even where statistical evidence is lacking or is so symmetrically distributed among alternatives as to be entirely non-determinative.[38] Another objection might be that the frequency conception seems to do violence to the notion of universal law – a notion according to which one firmly established counter-instance suffices for falsification. But on a statistical conception of confirmation, counter-instances of any hypothesis are not immediately decisive; they merely tend to reduce the degree of confirmation. To escape this objection, the frequentist may choose to regard all universal hypotheses as merely convenient simplifications of what are ultimately statistical generalizations. But then, the question of how universal laws actually function in science and what, in practice, the rationale for their acceptance is must be faced.[39]

Among the attempts to broaden the notion of confirmation, Carnap's is the most explicit and complete.[40] Essentially and very broadly described Carnap's account rests on the following :

(1) The notion of a system of discourse, L, having a basic or elemental vocabulary that makes reference to simple properties and to the individuals of some domain, S.

(2) The possibility of specifying in L, a set of descriptions each of which characterizes an alternative state of S and only *one* of which can be true. Carnap calls such descriptions Q-state descriptions.

(3) The possibility of assigning, in the absence of relevant knowledge, and hence *a priori*, a metric of initial likelihood to each of these alternative descriptions.

There is no need here to describe in any detail how Carnap proposes to assign *a priori* credibility metrics to descriptions. It suffices merely to mention that the method of assignment rests on an application of a refined version of the so-called *principle of indifference* or *insufficient reason* as it is also called. This is a principle which underlies the Classical Theory of

[38] The long history of this subject features the notion of equipossibilities, resulting from symmetrical distributions of evidence, as a basic notion in the assigning *a priori* confirmation metrics to hypotheses.

[39] For a statistical account of scientific hypotheses and a frequentist answer to this objection see H. Reichenbach, *The Theory of Probability*, pp. 344-442. A clearly presented non-technical account of the frequentist view is given by R. von Mises in his *Probability, Statistics aod Truth*, 2nd edit., New York, 1957.

[40] It is developed in his notable work : R. Carnap, *The Foundations of Probability*, University of Chicago Press : Chicago, 1962.

Probability of Laplace [41] and was emended by such writers as T. Nicod, J. von Kries and, later, by J. M. Keynes.[42] It is essentially the principle that when *all* fundamental alternatives are considered and there is no determinative knowledge on the basis of which to favor any one, it is *rational* to suppose that all alternatives are equally likely. The requirement that alternatives be "fundamental" is crucial for the resolution of certain paradoxes generated by cruder versions of the principle. Refined versions are, in effect, attempts to spell out what shall be intended by "fundamental". The calculation of a confirmation (or probability) metric for an outcome (hypothesis), H, is then, on the classical theory, a matter of taking the ratio of the number of possible ways of achieving H to the total number of possible basic outcomes. Roughly speaking, it is a matter of taking the ratio of the number of favorable to that of total possibilities.

By making some precise technical distinctions on how to set out basic possible alternatives for characterizing S ("Q-state descriptions" and "structure descriptions"), Carnap applies certain assumptions of *indifference* and arrives at a method of assigning an initial, *a priori* "likelihood" to any Q-state description. He then proposes that the numerical degree of confirmation of any hypothesis, H, on evidence, E, i.e., c(H,E) be given as the ratio of two quantities : the sum of likelihood metrics of each Q-state description of S compatible with both H and E; to the sum of the metrics of each Q-state description compatible with E, only. This definition of c(H,E) is, of course, strongly reminiscent of the classical ratio of the number of alternatives favorable to H given E to the number of *all* possible alternatives given E.

But despite the extraordinary explicitness with which Carnap's notion of confirmation is developed, it too faces difficulties. One objection, raised by some empiricists concerns Carnap's indifference assumptions. These are based on some notion of "rationality" which, in the absence of positive knowledge, is supposed to permit the *a priori* assignment of equal likelihood to some set of fundamental alternatives. The objection is that such *a priorism* is a vestige of rationalism and therefore cannot be justified in any empiricist philosophy of science. The counter-proposal from those expressing this objection would be that initial confirmation metrics can be assigned alternatives only on the basis of positive statistical evidence (i.e., relative frequencies) adduced in specific support of each alternative. Another

[41] Laplace, P. S., *Théorie Analytique des Probabilités*, 2nd Ed. (Paris, 1814).

[42] See : J. Nicod, *Le Problème logique de l'induction*, Paris, 1913; J. von Kries, *Der Prinzipien der Wahrscheinlichkeitsrechnung*, Freiburg, 1886; J. M. Keynes, *Treatise on Probability*, London, 1929.

objection has to do with unlimited general hypotheses. Carnap himself concedes that on his analysis, for the case of any large or unlimited system, the confirmation of any universal hypothesis dwindles to very small or zero. This may be seen in a very general way by nothing that a universal hypothesis of the form : 'All A's are B's' may be written as a *conjunction* of singular hypotheses of the form : '*This* A is B; and *This other* A is B; and *that* A is B; and so on'. But by a theorem of probability logic (which is quite applicable to Carnap's concept of confirmation), the probability of a conjunction of hypotheses is the product of the probabilities of each conjunct. And since, in general, the probability metrics are smaller than *one*, the probability of the conjunction in point becomes smaller and smaller as the number of conjuncts increases indefinitely. Scientific laws and theories, being unlimited in scope, are, then, subject to this consequence. And it remains a weakness not only of Carnap's system but also of a wide class of other analyses – despite Carnap's interesting suggestions for circumventing it.[43]

Difficulties of the above sort, especially those connected with the confirmation of general hypotheses, have encouraged some writers to urge that confirmation does not figure at all in the acceptance of scientific hypotheses. (One of these, as we have already noted, is K. Popper who regards falsification as the only form of inference in the logic of acceptance.) Such writers, however, have not demonstrated conclusively that the acceptance of any scientific hypothesis does not to some extend depend on supportive evidence. Indeed, any denial that supportive evidence for any hypothesis does in some proper sense increase its acceptability seems contrary to the most ordinary scientific practice. Does it not seem reasonable, for example, that a scientist should be more and more willing to *bet* on any hypothesis the greater has been its history of systematic success? What might be conceded is that, except in some of the more statistical phases of science, numerical *degrees* of confirmation, i.e., confirmation *metrics* play virutally no role in hypothesis acceptance. Thus, it seems fair to say that scientists simply do not calculate numerical degrees of confirmation in order to decide the acceptability of any hypothesis or even in order to choose between competing ones. If they could, perhaps they should and would; but no methods for computing such metrics are available. In actual practice, the acceptance of any universal hypotheses, H, seems to rest on the qualitative concept of confirmation with which we opened our brief survey of the subject and which, as we recall, bases the credibility

[43] See : R. Carnap, *op. cit.*, pp. 751-74.

of H on *confirming instances*. Admittedly, such a basis may not suffice to choose between alternative hypotheses. But, where only one hypothesis is in question, ordinary scientific practice leaves little doubt that it sanctions increasing credibility for that hypothesis as the number of instantiations for it grows.

One last sort of quantitative confirmation concept needs mentioning. This is the *subjective* concept, and it identifies the confirmation of H, given E, i.e., c(H,E), as the degree of coherent belief in H which E induces in some given individual or subject. Subjective confirmation differs from objective interpretations in that, for any given H and E, c(H,E), may differ from individual to individual. Another point of difference arises with the requirement of "coherence". Without this condition, subjective belief would have to be regarded as a purely psychological state subject only to empirical investigation and basically outside the context of validity or logic. In this connection some writers speak of the "justifiability" or "rationality" or "reasonableness" of belief.[44] Yet, such terms are not to be construed in any manner suggestive of criteria characterizing *objective* notions of confirmation. The logical restrictions imposed on subjective confirmation for which, following L. J. Savage, we prefer the term, 'coherence', include logical consistency as well as a notion of "proper" betting. This may be illustrated as follows : Assume a subject's willingness to bet on H given E to be a measure of his belief in H, given E. Then if his belief is to be coherent, he may not offer any sequence of odds which will make it logically necessary for him to lose the bet, given E; he may not, for example, offer five to one on H and ten to one on not-H. On such an arrangement if H comes out true he loses in proportion to 9; if H comes out false, he loses in proportion to 4.

The applicability and usefulness of any subjective notion of probability is doubtful. Since for any H and any E, it relativizes c(H,E) to individual subjects without providing a means for equalizing inter-subjective differences. This means that, if two scientists disagreed on their willingness to accept some H given some E, though each of their beliefs was (internally) coherent, they could not resolve their disagreement. Moreover, the subjectivist view in departing from general standards for assuring objective parity of confirmation estimates seems to be an intolerable departure from what we ordinarily mean by the term, 'confirmation', in the context of factual inquiry. This leaves it very questionable that such a line of analysis

[44] See F. P. Ramsey, *The Foundations of Mathematics and Other Logical Essays* (London, 1931), p. 192 and elsewhere. Also R. S. Lehman "On Confirmation and Rational Betting", *Journal of Symbolic Logic*, Vol. 70 (1955), pp. 251-262.

can count as a genuine explication of the general notion of scientific confirmation.

The above concepts of confirmation including the subjectivist one (provided it satisfies the condition of coherence) have one very important formal feature. They all satisfy the axioms of the probability calculus. This means that once initial measures of confirmation have been determined, on the basis of the favored concept, derived or complex probabilities can be mathematically computed *via* the calculus. In precisely this sense, the stated concepts of confirmation may be also regarded as alternative analyses of the notion of probability. Indeed, developments in the study of confirmation have generally been an integral part of far-reaching developments in the foundations of probability and induction.[45] A further consideration of these very important chapters in the philosophy of science is beyond the scope of our concern here.

B. DOES DISCOVERY DIFFER FROM ACCEPTANCE IN SOME IMPORTANT LOGICAL RESPECT?

Recall in this respect that one version of the argument against any "logic" of discovery is that as Feyerabend says : "The logic of discovery simply does not exist ..." for discovery involves "... no new and special type of reasoning that has not already been taken into account by formal logicians, and which therefore ought to be an object of special study".[46] But N. R. Hanson urges otherwise.

"It still seems to me that in some cases, one's reasons for entertaining an hypothesis and his reasons for accepting it are logically different".[47]

Reflection easily shows that some reasons for conjecture are of the same logical type as those for acceptance. In particular, we note *generalization from instances* in which one reasons from A's that are (uniformly) B's to the proposal that All A's are B's. This strategy figures typically in the discovery of experimental laws, i.e., generalizations of a non-theoretic sort. And it is, one might say, of the same logical type as the qualitative confirmation of general hypotheses in which particular instances of A's that

[45] But not all conceptions of evidential support, however, have been probabilistic. K. Popper, for example, develops a non-probabilistic notion of "corroboration" by evidence, see : K. Popper, *op. cit.*, Chapter 10. *Vide supra*, footnote 17, Chapter II.

[46] Feyerabend, P. K., "Is There a Logic of Discovery?" (Rejoinder) Feigl and Maxwell, *op. cit.*, p. 38.

[47] Hanson, N. R., "The Logic of Discovery", *The Journal of Philosophy* LV, (1958), 1073-1089.

are B's promote the further acceptance of the general hypothesis that all A's are B's. This means that both discovery and confirmation share at least one major type of inference, namely, confirmation by instances. The use of generalization from instances therefore provides no manner of differentiating any special logic of discovery. It is not one of the cases in which one's reasons for *proposing* an hypothesis are of logically different type from one's reasons for *accepting* it.

But where the proposed hypotheses contain novel concepts as in the case of theories, and hence are not merely generalized observational reports generalization from instances is of no avail. Other strategies must be employed, *analogy* perhaps being the most central of these. Now, is analogy to be typed differently, as a logical device, from inductive confirmation by favorable instances? The answer requires some brief attention to the form of analogical inference. We have already touched on analogy as a strategy of discovery and further discussion will occur in a later chapter on scientific theory. For present purposes it suffices to note that, in general, analogies are drawn between *two* domains of subject-matter, consisting respectively of, let us say, C's and A's which, though qualitatively distinct, are in some *relevant* respects *similar*. These domains are said to be *analogues*. The analogy may then proceed as follows : We may know that all C's are D's and from this, we conjecture that all A's are B's. In such a case we can conveniently label the domain of C's the *base analogue* and the domain of A's the *conjectural analogue*. The conjecture that all A's are B's is based on relevant aspects of similarity between the C's and A's, and the D's and B's. Thus, on the basis of certain similarities between hydrodynamic and electrical systems (e.g., requiring sources of energy, conductors, continuity, etc.); and a certain vague likeness between flow of water and electric current (e.g., interruptability, determinable rate of flow, dependence on energy, ability to transmit energy, etc.): and, knowing that increasing the hydrostatic pressure in a water system proportionately increases the rate of flow, one might conjecture that increasing the voltage across the terminals of any closed electric circuit will proportionately increase the electric current.

These considerations suffice to show that analogy does indeed differ from acceptance which, for general hypotheses, ultimately rests on confirmation by instances. Once a hypothesis is proposed it is the subsequent accumulation of such instances, i.e., future confirmation, that is required for later acceptance. In confirmation we go from A's that are (uniformly) B's to the *acceptance* of 'All A's are B's'. In analogy, however, we go from C's that are D's to the *proposing* of : 'All A's are B's'. C's that are D's are, of course, *not* instances of 'All A's are B's'. Hence there is here an obviously formal difference or, as one might say, a difference of *logical type* between the two sorts of inference.

Similar comparisons can be made in connection with other possible strategies for discovery such as employ criteria like simplicity, compatibility with established principles, high generality, and so on. These are all strategies which, though they are suitable to help arrive at proposals and though they may contribute considerably to the acceptability of hypotheses that have had some test and confirmation, seem, by themselves, insufficient to decide an acceptance. They can make crucial differences and precipitate an acceptance where very closely competing hypotheses are concerned and where extraordinary advantages (enormous simplifications, very high extent of generalization, etc.) are at stake. (Recall that the far simpler heliocentric astronomy replaced a geocentric one even long before the weight of instantial evidence had shifted in favor of the change.) Nevertheless, it seems fair to say that, in the absence of instantial confirmation, simplicity and other strategies underlying scientific discovery are, ordinarily, regarded as providing only a provisional plausibility which awaits critical confirmation by instances. Thus, as it seems, it would not take us amiss to say that, except for instantial generalization which can figure crucially in both proposal and acceptance, the sort of inferences governing the proposal of an hypothesis are of a different logical type from those which decide its acceptance.

C. SUMMARY

Our considerations in this chapter indicate that, in some important senses, *ampliative* science, that is, the content building phase of science (discovery and acceptance), has certain inferential aspects. By 'inferential' however, we do not mean *deductive*. The knowledge-content of science, particularly as expressed by general (law-like) statements, rests on limited evidence. And no amount of such evidence can deductively guarantee the truth of any law-like hypothesis.

Discovery, no doubt, is largely inspired by non-systematic factors. Nevertheless it seems to be profoundly and subtly subject to the support and interdicts of evidence, in accordance with defensible "logic-like" strategies for the formulation of plausible hypotheses. Moreover, there seems to be a genuine basis for urging that some of these strategies are of different logical type from others which govern the acceptance, i.e., confirmation, of hypotheses. Still, it is undeniable that there is no standard body of rules codifying such strategies. That is, one finds no explicit recipes for interpreting evidence so as to yield unique proposals in the scientific quest for plausible hypotheses. But though no such code of governance exists, its impossibility as a distinct body of general guidelines for discovery is by no means established, even for theorizing which is a far more creative and innovative process than is the formulation of observational laws, Indeed, if our discussion has not been largely misguided, there are considerations, both historical and systematic, which caution seriously against any impossibility thesis regarding a so-called "logic of discovery" – at least in one proper sense of such a *logic*.

Finally, in the acceptance of hypotheses, which is the second stage of scientific content-building, we find a more explicit manifestation of inferential processes. Thus, universal hypotheses are either falsified or confirmed. Their falsification proceeds deductively once unfavorable evidence has been established – perhaps by other inferential methods. Their confirmation, on the other hand, brings us squarely to the allegedly *supportive* or *inductive* relation between evidence and hypothesis. And though the status of this relation as a legitimate "thing" of logic is highly debatable, it nevertheless seems to have undeniably inferential aspects in a sense in which habits, whims, associations, inspirations, and psychological impulses do not.

IV

SUBSUMPTIVE SCIENCE : SYSTEMIZATION

A. A BROAD DISTINCTION : AMPLIATIVE AND SUBSUMPTIVE INFERENCE

We have given a brief consideration to the content-establishing or *ampliative* phases of scientific inquiry, consisting of *discovery* and *acceptance* with a view to characterizing their inferential aspects. If our findings are correct, inductive support and governance by appraisable criteria - in short. inductive-type inferences can figure significantly on these levels of investigation, though the precise nature and extent of this inferential aspect of creative knowledge-gathering remains deeply problematic. We turn now to view the established content of science for a parallel consideration, that is, a consideration of its inferential structure. But first, we note a precaution. The distinction between the content-establishing and content-structuring phases of inquiry may bring to mind the familiar and fundamental distinction between inductive and deductive modes of inference. Indeed, the traditional tendency has been to *equate* knowledge - establishing inferences with induction, and to *characterize* knowledge-structuring inferences as deductive. This, however, would be an error. For, inductive type inferences, besides figuring in the establishment of knowledge, can also provide explanatory and other systemizing structure to established content, as in the case of what we shall call probabilistic nomic inference, to be discussed in a later section. Following terminology introduced by J. S. Mill, we may characterize knowledge-establishing inferences as *ampliative*, and systematizing inferences as *subsumptive*. These terms have at best only rough and suggestive nuances and it would not be profitable, at this point, to consider any precise explication of their meanings. Nevertheless the connotations they do convey come close to rendering the sort of differentiation we have in mind.

A rough characterization of *ampliative inferences* would be to say that they are inductive arguments with premises that are no more general than their conclusions.[1] Thus characterized, they include not only inductive generalizations but also analogical inferences of a non-generalizing character.[2]

Since ampliative inferences are inductive in character, our general comments on inductive inference in Section 3 of the last chapter immediately apply. In this connection we recall indicating that, though most would agree that such inferences confer probability or likelihood on their conclusions, there is at present no general agreement regarding their more specific character or regarding the sort of truth-status connoted by terms like 'probability' and 'likelihood'. Moreover, there are some who would deny that such modes of discourse are inferences at all, claiming that they are at best only guesses subject to withdrawal whenever they fail to hold up.

In actual scientific practice – except possibly for certain statistical procedures whose logical foundations are also a matter of philosophical dispute – ampliative inferences are not generally expressed as explicitly structured arguments. At the present time therefore it seems rather fruitless to look to written science for any general and decisive model of ampliative scientific inference. Any philosophical account of it, therefore, must be largely stipulative and reconstructive in character.

But logical pattern is far more explicit in the established content of science than either in the search for such content (discovery) or in its subsequent confirmation. Thus in sharp contrast to ampliative inferences, *subsumptive* inferences are a matter of far more settled opinion. Taking the form either of deductive or of inductive arguments (the latter type will be referred to as *probabilistic* inferences) they provide the logical web that structures the working content of science. Whether the form of subsumptive inferences is deductive or probabilistic their distinctive feature is the one suggested by their label. Their conclusions are logically encompassed by, or, as it is also put, logically *subsumed* under their premises. This may also be roughly expressed by saying that their conclusions are less general than some of their premises.[3] Thus from Newton's laws of

[1] For rough purposes here it seems permissible not to concern ourselves with aspects of vagueness in the ordinary use of the term 'general'.

[2] An example of the latter would be an inference of the form :

(1) X_1 has properties $P_1, P_2, \ldots P_n, P_{n+1}$

(2) X_2 has properties $P_1, P_2, \ldots P_n$

---

(3) X_2 has property P_{n+1}

In which (1) and (2) inductively support (3) or, as we might say make (3) probable.

[3] Any attempt at a comprehensive and precise explication of the comparative generality or scope of statements would take us too far a field at this point. It seems sufficient for

motion and gravitation together with some initial conditions fixing limits on planetary energies we can infer (derive) the less general Keplerian laws of planetary motion. Similarly on the basis of the facts that Henry was an extreme manic depressive in circumstances X and that a high proportion of such manic depressives in circumstances X commit suicide, we can *probabilistically* infer that Henry committed suicide, the subsumptive inference in this case being of the probabilistic rather than deductive kind.

A system or network of subsumptive inferences can give logical unity to sizable domains of experience. In the most highly developed, theoretical sciences the resulting structure may take the form of comprehensive axiomatic systems in which a vast portion of the descriptive statements of a science are deductively subsumed under few and relatively simple assumptions (axioms).[4] The axiomatic mode, however, is a level of systemization rarely achieved in actual scientific practice. Instead, broad networks of subsumptive inferences theoretically encompassing and logically compact, yet, not quite developed to the level of axiomatization, systemize large portions of the more advanced sciences and, no doubt, constitute a feature that decisively differentiates them from pre-scientific levels of knowledge. Our concern for the remainder of the present chapter and the one that follows will be with the sort of scientific inference that constitutes the basic unit of such systemization.

B. NOMIC INFERENCES : INTRODUCTORY BACKGROUND

Most of the considerations taken up in this chapter are usually discussed under the topic of explanatory inference or, simply, *explanation*. It has been rather typical of such discussions to regard explanation as the basic and pervasive form of scientific inference. Prediction has gotten only brief attention and, at that, only as a pragmatic variation of explanation. Other forms of inferences such as postdiction, retrodiction and several for which there are no standard labels have hardly ever been mentioned. With a few notable exceptions, the prevailing view has been that there is formal parity between explanation and prediction or other varieties of scientific inference,

---

present purposes to indicate merely in a rough way that two statements may be compared as to their generality by comparing the inclusiveness of the classes of things about which they make assertions. Thus Newton's laws are more general than Kepler's because they make claims about *all* bodies irrespective of whether they are planets or not. For a detailed explication of the *generality* of laws see E. Nagel, *op. cit.*, pp. 37-42.

[4] These will come up for closer attention in later discussion.

the differences arising merely with respect to the time when the scientist makes his inference. Thus, for example, whether a scientist predicts or explains an event, the form of inference is the same. It is an explanation if the event is in the scientist's past, and a prediction if the event is in the scientist's future. Other variations may be distinguished by introducing time relations between the inferred event and the events on which the inference is partially based – usually referred to as *initial conditions*. Thus the case in which the inferred event comes earlier than the initial conditions has sometimes been labelled "retrodiction". But we leave distinctions of this sort to some later pages of our chapter.

The view that explanation is central to scientific discourse is an old and deep-rooted one that has remained largely unchallenged until very recent times. Moreover it is traditionally coupled with the belief that explanation is fundamentally a *deductive* form of scientific inference. Aristotle equated knowledge with a knowledge of causes – a knowledge that he thought essential to all scientific understanding. But to know the *cause* of an event is to know *why* it happens, hence to know what we would today call its *explanation*. In Book I of the Posterior Analytics, he says, "we possess scientific knowledge of a thing only when we know its cause". Also, "we do know by demonstration. By demonstration I mean a syllogism productive of scientific knowledge ..." Demonstration here is synonymous with deduction. Aristotle further stipulates that "the premises must be true : for that which is non-existent cannot be known ..." And he refers to such premises as "primary" or "appropriate basic truths". In some major respects this strongly anticipates what we now call *deductive scientific explanation*, though few of us would go along with Aristotle's further insistence that "demonstration", i.e., deductive explanation, must stem from "premises prior to and better known than the conclusion", and, "must rest on necessary basic truths".[5]

There was a departure from views such as these during the time of Bacon. It took hold largely with the more popular British commentators on science and focused attention on the inductive base of scientific knowledge. Except for this brief hiatus, however, deductive inference has been regarded as the ideal form of scientific systemization virtually throughout the entire history of thought on the subject, right up to very recent decades. With the onset of seventeenth-century science, deductive structure became the unchallenged logic of the mechanism and determinism that spanned the next two hundred years. The deductivist view received renewed

[5] *Analytica Posteriora*, I, 71^b-74^b, *passim.*

philosophical support from most major nineteenth-century methodologists such as W. Wheewell,[6] J. S. Mill [7] and W. S. Jevons.[8] The major portion of more recent writers – to mention only a few – K. Popper.[9] C. J. Ducasse,[10] H. Feigl [11] and J. Hospers [12] – have given undiminished support to the deductive model. The most comprehensive and influential contemporary account is that of C. G. Hempel and P. Oppenheim whose well-known article has become the starting point for virtually all discussions on the subject.[13] In summary, the major viewpoints have been as follows :

(1) Deductive explanation is the fundamental inference systematizing the cognitive results of science. It is fundamental because it yields the deepest scientific understanding and because it represents the form which, for the greater part, structures our most highly developed sciences.

(2) Prediction, a crucial function of science, is also an inference but is *structurally* identical with deductive explanation differing from it only pragmatically, that is to say, with respect to the particular circumstances of the explaining scientist, namely, the time at which he asserts the explanation and his state of knowledge at that time.

(3) Non-deductive explanations are only an incomplete form of deductive ones and in some cases can be put into explicit deductive form.[14]

But very recent philosophical discussion on the nature of scientific inference has challenged nearly all of these claims. Objections have come from an increasing group of contemporary writers who have argued that :

(1') Deductive explanation is neither fundamental nor pervasive in scientific discourse.[15]

– or that :

[6] W. Whewell, *The Philosophy of the Inductive Sciences*, 1840.

[7] J. S. Mill, *A System of Logic*, 1843, Bk. III, Ch. XII.

[8] W. S. Jevons, *The Principles of Science*, London, 1924.

[9] K. Popper, *The Open Society and Its Enemies*, Ch. 25, London, 1945.

[10] C. J. Ducasse, "Explanation, Mechanism and Teleology", *The Journal of Philosophy*, XXII (1925), 150-155.

[11] H. Feigl, "Operationism and Scientific Method", *Psychological Review*, vol. 52 (1945), pp. 250-9.

[12] H. Hospers, "On Explanation", *The Journal of Philosophy*, vol. 43 (1946), pp. 337-356.

[13] C. G. Hempel and P. Oppenheim, "Studies in the Logic of Explanation", *Philosophy of Science*, Vol. 15 (1948), pp. 135-178.

[14] For an example of such a transformation see : Scheffler, I., *op. cit.*, p. 35.

[15] I. Scheffler, for example, attempts to show that there are a variety of systematizing scientific inferences other than explanation or prediction neither of which is clearly central to science, *op. cit.*, pp. 46-57.

(2′) There are fundamental differences between explanation and predictions;[16]

– or that :

(3′) The deductive model of explanation does not account for the increased complexities and diversifications of contemporary scientific discourse which requires a fundamental place for non-deductive forms.[17]

But though contemporary discussion has added much to the traditional conceptions of the structure of science, it has been hampered by unclarity on the connection between explanation and causation and by the resulting tendency to conflate explanatory and non-explanatory subsumptive inference. This has in turn forestalled suitable categorizations of the various possible forms of systemizing inference and has also been largely responsible for the dubious view that explanation is the unitary and central form of scientific inference. Explanations as we shall attempt to represent them, may be more generally viewed as merely a species of that much broader class of inferences which – so to speak – "bring" matters of fact under more general matters of fact, as "special cases" or "instances", and which, for lack of more familiar labels, we shall refer to as *subsumptive* or *nomic* inferences.[18]

But though it seems plain enough that there are arguments which everyone would call explanations and others which virtually no one would so entitle, it seems far from clear as to just what makes a nomic inference an explanation. As a result some nomic inferences are not as readily decided, in this respect. The criterion sometimes given is that a scientific explanation is an answer to the question "why?" But no explication seems to be available to help say whether a given nomic inference is or is not a genuine answer to this question. Undoubtedly the notion of explanation is closely bound to that of causation. For it seems most appropriate to say that an event has been explained whenever a cause has been accurately assigned to it or a causal law has been appropriately brought into the account. Where the account given does not involve causes there may be good reason to refrain from calling it an explanation, though it may indeed be a genuine nomic

[16] I. Scheffler, *op. cit.*, pp. 43-46. See also : N. Rescher, "On Prediction and Explanation", *British Journal for the Philosophy of Science*, VIII (1958), 281; N. R. Hanson, "On the Symmetry Between Explanation and Prediction", *The Philosophical Review*, LXVIII (1959), 349; M. Scriven, "Explanation and Prediction in Evolutionary Theory", *Science*, Vol. 130 (1959), p. 477.

[17] N. Rescher, "Fundamental Problems in the Theory of Scientific Explanation", *Philosophy of Science* (the Delaware Seminar) ed. B. Baumrin, vol. 2 (1962-3), pp. 41-60.

[18] The general premises governing nomic inferences are either laws of nature or general statements in certain respects resembling laws. And the Greek term for law is *nomos.*

inference and one of a kind that structures much of scientific knowledge. The Hempelian account, for example, blurs the distinction between explanatory and non-explanatory nomic inferences and seems to attribute explanatory status to deductive nomic inferences irrespective of the causal nature of their premises. In their noted article, Hempel and Oppenheim occasionally make reference to "causal" explanations as though to suggest that these might be a particular *species* of explanation. But no such specification is actually ever made.[19] The Hempelian account of probabilistic explanation is similarly unmindful of the possible distinction between explanatory and non-explanatory statistical inferences.[20] In an article written considerably later than the original, basic one, Hempel explicitly refers to causal explanation as "... a special type of deductive nomological explanation".[21] But the distinction is invoked only to separate those nomic inferences that account for particular events from those that account for general facts or laws. A few examples later in our discussion will easily show that this manner of specification is insufficient. For there are nomic inferences which provide genuine scientific accounts *of* – but which do not account *for* – events and are not, therefore, reasonably called explanations.

C. NOMIC INFERENCES WITH SINGULAR CONCLUSIONS

1. Deductive and Probabilistic Inference

Nomic inferences may be conveniently divided into two types : (i) Those whose conclusions are singular statements [22] (sometimes called particular statements) and (ii) those whose conclusions are laws or general statements that in some important respects resemble laws. It will be convenient to confine our attention first to nomic inferences of type (i). We shall refer to these simply as *nomic inferences*, without qualification. Nomic inferences of type ii will be given some consideration in a later section where they will also be referred to as *derivations*.

Another important division among nomic inferences is that between those nomic inferences in which the premises (deductively) entail the conclusion (i.e., deductive nomic inferences) and those in which the premises make the conclusion merely probable. Except for the examples that

[19] Hempel C. G. and Oppenheim, P., *op. cit.*, p. 324.

[20] Hempel, C. G., "Aspects of Scientific Explanation", collected in *Aspects of Scientific Explanations*. The Free Press : New York, 1965, pp. 376-410.

[21] Hempel, C. G., "The Logic of Functionalism", collected in *loc. cit.*, p. 301.

[22] Statements about individual events or objects.

follow, in which both types of inference are explicitly and separately illustrated, this distinction will not be needed for our immediate purposes. The considerations which follow therefore apply with generality to nomic inferences of both the probabilistic and deductive kind.

2. Examples of Nomic Inferences

The following illustrations, D_1-D_4, are nomic inferences of the deductive type :

D_1 Any animal absorbing X amount of ionizing radiation will die from it. The animals in this pen were exposed to such radiation.

Hence they died from it.

D_2 Wherever there are living animals there is chemically free oxygen. There are living animals at the bottom of our pond.

So there is chemically free oxygen there.

D_3 The acceleration of the bob of a simple pendulum oscillating with small amplitude is proportional to its displacement from the center of the swing.

When the displacement of the bob was 2 centimeters its acceleration was 5 centimeters per second per second. The displacement is now 4 centimeters.

∴ Its acceleration is now 10 centimeters per second per second.

D_4 Jack usually takes the necessary steps to achieve what he wants.
Jack wants to pass his Latin quizzes.
It is necessary for him to study in order to do so.

Hence Jack usually studies for his Latin quizzes.

The following illustrations P_1-P_4 resemble D_1-D_4 in some obvious respects but are probabilistic instead of deductive.

P_1 Any animal absorbing Y amounts of ionizing radiation usually gets radiation sickness.

The animals in this pen have absorbed such radiation.

Hence, they have radiation sickness.

P_2 Wherever there are living organisms there is almost always [23] chemically free oxygen.

[23] Some bacteria are active in the absenee of free oxygen.

There are living organisms in this jar.

Accordingly, we find chemically free oxygen in it.

P_3 There is a high correlation between a student's intelligence quotient and success in school.

Jack has been very successful in school.

Hence his high intelligence quotient.

P_4 Jack is one who usually takes the necessary precautions to obtain what he wants.

Jack wanted to pass his Latin quizzes.

It was necessary for him to study in order to do so.

And so it was that Jack did study for his Latin quizzes.[24]

3. Structure Common to All Nomic Inferences

Though the above examples of nomic inferences differ in a number of important respects, they all share the following features :

(I) A set of premises containing :

(a) At least one statement, L, which expresses natural uniformities either invariable (universal) or statistical in character.[25] L may be a law of nature or a much less encompassing statement as, for example, a statement about some individual's dispositions as in D_4 or P_4. Nomic inferences which stem from premises of this latter sort have very low systemizing power. Nevertheless, the account would seem incomplete unless they were mentioned and illustrated explicitly.

[24] It should be noted that despite the apparent resemblance of P_4 to D_4 they differ in an important respect. The conclusion of D_4 is a statistical statement that can be shown to follow deductively from statistical premises. The conclusion of P_4 is not a statistical statement at all, and it does not follow deductively from the premises.

[25] L must be a *relevant* premise. It is an elementary truth of formal logic that to the premises of any inference may be *conjoined* any statement, S, without changing the validity of the inference. But S would then be logically superfluous or irrelevant to the conclusion and could be dropped without consequence. The L required of nomic inferences may not be such an S. Thus the following is not a nomic inference because its only general premise is irrelevant and therefore quite dispensable.

Jupiter has twelve moons.

All planets have elliptical orbits.

Jupiter has more than eleven moons.

In all that follows, we shall be concerned with nomic inferences in which L either is a law or has features closely resembling those of a law. All such statements will, for convenience, be loosely referred to as *nomic* or *law-like* statements or, simply, as *laws.* What entitles some matter-of-fact statements expressing natural uniformities to be called laws and others not is an unsettled question to which we shall turn later. For present purposes it seems sufficient to mention merely one feature of laws, namely that they all have some form of conditional or hypothetical structure. For simplicity this may be represented by the schema :

If ———— then ————

the terms *antecedent* and *consequent* are generally used for what fill the first and second blanks, respectively.

Thus, the universal statement in D_2 :

Wherever there are living animals there is chemically free oxygen. may be paraphrased :

If there are living animals at a place then there is oxygen at that place.[26]

The statistical statement in P_2 :

Wherever there are living organisms there is, almost always, chemically free oxygen

may be paraphrased :

If there are living organisms at a place then (almost always) there is chemically free oxygen at that place.[27]

(b) At least one singular statement, C, citing particular circumstances, c (states of affairs, events, etc.), which instantiate the attributes mentioned in the antecedent (i.e., the if-clause) of the nomic statements, L. The circumstances, c, are sometimes referred to as *initial* or *antecedent* conditions for the application of the nomic premises.[28]

[26] More symbolically :

$(x)\ (y)\ (Ax\ Py\ Lxy \rightarrow (Ez)\ Oz\ Lzy)$

where Ax = x is an animal
Py = y is a place
Oz = z is chemically free oxygen
Lxy = x is located at y
'---------- → ---------' is read :
"if ---------- then ----------".

[27] There are crucial differences between the 'if ... then' of universal statements and the statistical 'if ... then ...' but they need not concern us here.

[28] It is important to mention that C must instantiate the antecedent or if-clause of L in order to rule out what are sometimes said to be probabilistic inferences in which the singular premise instantiates the consequent or then-clause of some law, L, in order to conclude, *probabilistically*, with an instance of the antecedent clause of L. In just this form such an alleged inference would not fit the nomic model. This pattern of

(II) A conclusion, E, citing particular events, e, instantiating the attributes mentioned in the consequent of L.

(III) The conclusion, E, either follows from the premises *deductively* or it is made *probable* by them. Schematically we may represent this as follows :

Deductively	↓	L (= nomic statements, universal or statistical)
or		C (= statement of initial conditions)
Probabilistically	——	
		E (= statement asserting the occurrence of an event, e)

In precisely what sense a set of statistical premises can make the conclusion of a nomic inference probable is a matter of some philosophic contention. We shall return to this question in the next chapter.

4. *Must the Premises of Nomic Inferences Be True?*

Aristotle required that "... we must know the prior premises from which ... demonstration is drawn, ... the regress must end in immediate truths ..." [29] Similarly C. Hempel and P. Oppenheim are of the view that the premises of explanations must be true.[30] Others like E. Nagel recommend a weaker alternative, namely, that the premises be highly confirmed.[31] Our own concern is with nomic inferences in general. What truth-restrictions do the premises of such inferences call for?

Consider, first, that part of the premises consisting of singular statements, C, expressing initial conditions. It is basic to all subsumptive inference in science that C must be either *factually* true or else very highly confirmed. The initial conditions expressed by C, as we have noted, instantiate the antecedent clause of the covering nomic premise (law), L. This means that they are sufficient conditions for realizing the effects mentioned in the nomic premise. If C is false then the covering law has simply been *misapplied.* Thus, not knowing about the truth of C is tantamount to not knowing whether or not L has been correctly applied for deriving the events or states mentioned in the conclusion of the inference. And this is, in effect, not knowing whether or not the inference in question is admissible. It would, for example, be highly unsound to predict

inference will be considered in the next chapter for possible assimilation to the nomic form.

[29] What Aristotle calls "demonstration" corresponds, in essential structural respects, to what we call nomic or subsumptive inference. At any rate it is a species of such inference. See : Aristotle, *op. cit.*, 72^b, 21-23.

[30] C. G. Hempel and P. Oppenheim, *op. cit.*, pp. 138-140.

[31] E. Nagel, *op. cit.*, pp. 43-44.

that a certain lustrous object will conduct electricity on the basis of the law that all metallic objects are conductors unless one first made sure that the lustrous object was indeed metallic. In this case luster and general outward appearance would not suffice as, for example in the case of the non-conducting mineral, fool's gold. Thus even in day-to-day affairs we say that someone has been "unscientific" whenever he has not carefully checked the truth or reliability of C in his nomic inference. Experimentation and measurement in science are crucial not only because they test the conclusions of nomic inferences but also because they ascertain the truth or reliability of statements expressing the initial conditions which such inferences presuppose.

Consider, now, the nomic statement, L, in the premises of a nomic inference. Can we require it to be true or even highly confirmed? It seems not. Such a requirement would be too strong for nomic inferences with conclusions about future events, that is for predictive nomic inferences. It hardly bears mentioning that predictions from nomic premises of unknown truth-status constitute a typical and, in fact, crucial phase of scientific inquiry. Such predictions are never regarded as inadmissible, though their reliability may be subject to question. As a matter of fact, it is the possibility of making predictions of this sort that provides the very means for testing any nomic statement whose status in a science has not yet been established. Indeed, not only may an appropriate nomic prediction *with true singular premises* turn out true and thus help establish a nomic hypothesis but it may on the other hand turn out false. In this latter case : If the inference is probabilistic, the credibility of the nomic premise is to some extent undermined. If the inference is deductive the nomic premise is decidedly *false*. This is essentially the actual method of testing theories and other nomic hypotheses whose truth or degree of confirmation is at the time either low or virtually unassessed, in particular, because the hypothesis is new. Any blanket requirement of high confirmation for all nomic premises therefore, seems to fail scientific practice.

It is of course obvious that where nothing is known about the truth-status of nomic premises, the ensuing conclusions may well be regarded as unreliable in the absence of other independent evidence. But to say of such inferences that we do not know about their scientific admissibility would be very odd, indeed, from the viewpoint of actual science and would seem to reflect confusion between the admissibility of an inference with the reliability of its conclusion. Undeniably, true or highly confirmed premises would make for highly reliable predictions. But for some scientific purposes, particularly testing, the reliability which a predictive inference bestows on its conclusion does not figure at all in the scientific role of the inference itself. If so, then it seems appropriate to require at most that the premises of nomic inferences *not* be known to be false. This much must be imposed on nomic premises. For it is a basic canon of scientific practice to exclude statements from the working body of scientific knowledge once their falsity has been established.[32]

5. Some Further Requirements on the Premises of Nomic Inferences

We recall that the premises of nomic inferences must, in general, include at least one nomic (law-like) statement and a singular statement of initial conditions. Requiring that the general statement be a nomic one guarantees that it will have features such as, having factual content and certain aspects of scope or generality, which we shall discuss in a later section under scientific laws. Reflection, however, brings to mind other requirements having little to do with law-likeness. Thus consider the following:

1. Relevance

The general (i.e., nomic) premise must be relevant, that is, it must be logically indispensable for drawing the conclusion. Thus, the following inference is not nomic because its nomic premise is "dead lumber", that is, may be dropped without effect on the conclusion.

All men are mortal
Socrates is wiser than
Euclid, but Euclid is wiser
than Thrasymachus.

Therefore, Socrates is wiser than Thrasymachus.

2. Independence

The following probabilistic inference is trivial and impotent from the viewpoint of systemization : Jack has abdominal discomfort and offers to

[32] It is of course easily recalled here, as an elementary fact of formal logic, that false premises can guarantee nothing about the truth or falsity of any conclusion validly inferred from them. Similarly in probabilistic (i.e., inductive) inferences, false statements can have no bearing on any hypothesis. The state of confirmation of an hypothesis is always relative to the available evidence for it, and evidence is expressed only by true or highly confirmed statements. Thus purported nomic inferences, be they deductive or probabilistic, whose premises are *known* to be false are inferences *known* to be without systematic import and therefore not admissible in science.

explain it by telling us that he ate something bad. Roughly, but more explicitly the argument he offers is :

Eating very badly spoiled food almost always produces
abdominal discomfort.
Jack ate some badly spoiled food.
Hence Jack has abdominal discomfort.

Asked how he knows he ate something bad, he replies that he *must* have, because he *does*, after all, have a stomachache. To exclude such inference we require that the premises, P, of a nomic inference be supported by evidence, F, which is *independent* of the conclusion, E. More precisely, F must not be deductively implied by E or by any evidence on which E is based.

Subsumptive inferences for whose premises there is no independent evidence have, in general, no systemizing force. Such force derives from showing some matter of fact (expressed by E) to be a special case of more encompassing claims (P) established, as part of the corpus of knowledge, on grounds apart from the matter of fact in point. Thus where there is no such independent support for P, we have what some might see fit to regard as circularity. Such circularity, however, is not formal or deductive but – so to speak – *cognitive* or *epistemic* in character. The premises, as it were, do nothing to logically incorporate the conclusion into a more or less established piece of knowledge. To break this circularity Jack must provide external evidence for his contention that he ate bad food. He may, for example, remember that it was left overnight without refrigeration and that its flavor was not quite right. Or if there was some still around, it could be carefully tested for spoilage. Circularity of this sort, while perhaps not uncommon in day-to-day matters, is rather untypical of science.[33] A more subtle defect of rather similar nature but much better documented in the history of science is *ad hoc-ness.*

[33] It may be objected here that in scientific practice we may tentatively assume a general relationship between two quantities in order to *account* for some observed phenomenon, even where there are no other observed instances whatever of the phenomenon, This, it may be maintained, seems to be perfectly sound hypothetical procedure though in such a case there is no independent evidence for the proposed explanatory hypothesis. One might answer here with two points : (1) In such an extreme case of "no other known instances" there would, typically, be other considerations supporting the hypotheses – e.g. simplicity, similarity to other known laws, etc., such as we already noted in our discussion of discovery and which may initially guide the choice of hypothesis. (2) Any inference governed by so meagerly supported an hypothesis could be regarded as a degenerate extreme – having for the moment no systemizing import until more evidence comes in.

3. Ad hoc-ness

In the early decades of this century it was noted that electrons were emitted from radioactive nuclei with varying amounts of energy. On the basis of what was known about nuclear decay and some calorimetric measurements made to rule out a number of possibilities, it was concluded that either energy was being carried away to some undetected particle or energy was being annihilated in some unprecedented manner to the detriment of the basic principle of energy conservation. The puzzle became a crisis when all attempts at detecting the energy-thieving particle failed. Suppose at this point that a physicist had suggested as a peculiar feature of beta-decay [34] that the missing energy disappeared permanently, into an energy "black hole" catastrophically and momentarily created on the occasion of each disintegration. As for the "black holes" – they had only two physical characteristics. They were accompaniments of beta disintegrations only, and on each occasion they existed just long enough to annihilate the missing electron energies. Let us further suppose that, asked how such "energy holes" could be detected, the physicist answered that the beta-decay phenomenon was itself the evidence for their existence. Needless to say, this would be a most distasteful hypothesis. It would be condemned as uselessly *ad hoc* or, in some sense, circular. And one reason for its inadmissibility would be the impossibility of detecting "energy holes" independently of the discrepancies in beta-decay for which the "holes" were invented in the first place.

What was in fact postulated by Pauli and Fermi (1930-34) was the neutrino, a very small particle with negligible mass and no electric charge and capable of going off with the missing energies. Having these features the particle would be exceedingly difficult to stop and to detect. This hypothesis was itself not very palatable. But independent detection, though difficult and unlikely, was not ruled out in principle. Independent evidence for the neutrino turned up about twenty years later in connection with other related phenomena such as the recoil of the remaining nucleus and positron decay.[35]

For convenience let us speak of an hypothesis, H, formulated to give an account of some class of phenomena, D, in terms of a class of factors, F. We may say that H is *ad hoc* when its content is so defined as to imply

[34] Electrons emitted from atomic nuclei are called *beta-particles* and the process is referred to as *beta-decay*.

[35] For relatively non-technical discussions of beta-decay the reader is referred to : R. S. Shankland, *Atomic and Nuclear Physics*, The Macmillan Co., 1960, pp. 355-8.

that the presence or absence of F's can be asserted only on the basis oj the occurrence or non-occurrence of D's. (Our energy black-hole hypothesis is of this sort.) Obviously such an hypothesis is so framed that neither the occurrence nor non-occurrence of D's can disconfirm it. Nor, by hypothesis, can any other happening serve as counter-evidence. Thus, the falsifiability of H is ruled out essentially by stipulation. But if falsification is logically impossible then the hypothesis is compatible with all evidence, that is, it has been designed or contrived to come out "true" no matter what happens in nature. As such, it can say nothing about any matter of fact. Or as we might say, it is factually vacuous and scientifically dispensable.

Theoretic discourse is most vulnerable to the defect of *ad hoc-ness*. The framing of theoretic hypotheses, we recall, generally involves innovative additions to the conceptual framework. This usually takes the form of postulating subject-matter (e.g., elementary particles, fields, etc.) which because of the way it is conceived cannot be directly observed. It is in the postulation of such subject-matter that provision must be made to assure the avoidance of *ad hoc-ness*. In effect this amounts to conceiving the theoretical subject-matter in such a way that its detection is possible by means other than the class of phenomena which was to be explained or otherwise systemized at the outset. Thus in studying the masses of atomic nuclei, amounts in excess of what was due to the presence of protons were observed and had to be explained. For this purpose it was assumed that atomic nuclei contain non-protonic particles capable of providing the requisite nuclear mass. But although only a mass discrepancy was to be corrected, attributes other than simply mass were assigned to the "new" particles. They were understood to be electrically neutral (hence called neutrons), to have about the same mass as ordinary hydrogen atoms, but to be much more compact than them. From these postulated properties the possibility of their indirect detection was established. Such particles could be expected to have very high penetrating power, to be undeflectable by electric or magnetic fields and to produce quantitatively determinable effects on colliding with particles of known masses as, for example, nitrogen nuclei. On the basis of these deducible traits the postulated particles were experimentally "detected" twelve years later.

On the other hand, vitalism popular with many biologists up to about two generations ago, is an *ad hoc* account of organic life. Vital forces were postulated as having no determinate spatial attributes such as shape or specific location in the organism and none of the other material properties – mechanical, electromagnetic, thermal, optical, etc. Thus, the presumed non-materiality of such vitalistic "substances" ruled out the possibility of any empirical confirmation of their existence by anything other than the life processes which they were invented to account for in the first place. In similar *ad hoc* fashion, Rutherford might have attempted to account for the excess nuclear mass by postulating, *in lieu* of neutrons, a special power (or disposition) of nuclear protons to grow just enough only when part of an atomic nucleus, to produce the observed excess. Set up in this manner the hypothesis would be *ad hoc*, for a disposition restricted in this manner could not be measured outside of the phenomena it was devised to explain.

Some postulations are conceived with insufficient richness to permit their application beyond the original class of phenomena. In such a case the defect can be corrected by enriching the conceptual content of what has been postulated. Thus for a long time observers had noted a minute but regular perturbation (disturbance) of the orbit of Mercury that could not be explained by the gravitational attraction of all the other planets.[36] A theoretician attempting to explain this might have simply, though improperly, assumed that the planet's mass varied as a certain function of time. But, at the time, there was no technical possibility of testing for the mass variations, as for example, by sending instruments to the neighborhood of the planet. Thus, though the mass variational hypothesis did not *logically* exclude the possibility of external confirmation, it nevertheless did not explicitly provide any practical means for such confirmation. Though any hypothesis of this sort would have drawbacks very similar to those of an *ad hoc* formulation, it would not seem quite appropriate to simply label it *ad hoc*. For it could be corrected by enriching the conceptual content of what has been postulated. In our example, the mass variational effects were not in fact postulated simply but couched in a general theory which links mass variations to other parameters as, for example, speed. This enrichment of the account then provided the basis on which to seek external (independent) evidence. The same speed-linked mass variations could then be looked for in other sorts of events. The postulations of vital forces to explain physical life processes, however, is genuinely *ad hoc*. For to speak of physically detecting non-material forces (except by the biological effects they were invented to explain) would be a contradiction in terms. *Ex hypothesis*, vital forces do not have the physical attributes required for such detection. Caloric was, though with some qualification, another *ad hoc* postulation in past science. It was an invisible, massless substance

[36] In the 19th-century this was determined to be a precession of about 44 seconds of arc per century.

invoked to account for heat flow and heat retention.[37] Traditional metaphysics, when considered as a source of explanatory premises or "first principles", offers an abundance of examples, as for example : "real substances" stripped of all attributes; and causal "links", substantively distinct from the events they are thought to connect. When employed to "explain" the stability of objects and the regularity of phenomena, such postulated "realities" are, by the very manner in which they are conceived, beyond the reach of any evidence external to the general phenomena they were constructed to explain. It may be of interest to note that any hypothesis postulating some overarching reality conceived to explain *all* natural phenomena is necessarily *ad hoc*. For in that case no evidence could ever be external evidence and the postulated substance could never be detected independently. By sheer stipulation *every* phenomenon would be designated as "evidence" for the postulated "reality". Such stipulation, by the same token, rules out the logical possibility of all counter-evidence or falsification. Again we may look to traditional metaphysics for postulations of this sort as, for example, such universal causes as a "world spirit", an "Absolute", or a God.

D. EXPLANATION, A SPECIES OF NOMIC INFERENCE

1. Explanation not the Only Form of Nomic Inference

Nomic inferences as a class of entities can, like virtually any other class, be subdivided in a variety of ways. We have already distinguished between nomic inferences with singular conclusions and those with nomic conclusions (derivations), and between deductive nomic inferences and probabilistic ones. A third important categorization is between those nomic inferences that may be properly called *explanations* and those that may not. This last distinction has been largely neglected in traditional accounts where it has been customary to regard all nomic inferences as essentially explanatory in form. Predictions are in a secondary sense differentiated from explanations. But this distinction is generally taken to have merely pragmatic rather than structural import.[38] Accordingly, explanation and

[37] Unlike vital forces, however, it did have measurable flow, and hence some determinate space-time attributes. These permitted considerable quantitative and mathematically precise treatment of a wide range of thermal phenomena. Nevertheless the masslessness of caloric logically precluded what would seem to be important modes of detection for anything called a "substance".

[38] The distinction has been taken to rest only on the time location of the scientist with respect to the events involved.

prediction have been held to be structurally identical with the added implicit suggestion that therefore they are fundamentally equivalent in all essential methodological respects.[39]

More recently the structural identity thesis has been vigorously challenged. But in the attempt to differentiate prediction from explanation writers have tended to overstate the case. They have construed these inferences as mutually exclusive thus neglecting the fact that while some predictions are, in a distinguishable sense, non-explanatory, others are decidedly explanatory in character and may therefore be reasonably classed as explanations. Moreover, it appears that the distinction between explanatory and non-explanatory is applicable to all nomic inferences, regardless of whether they are predictive, postdictive or other. If so, we can non-trivially group all nomic inferences under two exhaustive headings : those which seem to count as explanations and those which do not.

2. *Explanatory and Non-explanatory Nomic Inferences*

Allegedly guided by ordinary scientific usage, some writers feel it necessary to maintain that one can speak of explaining only events *known* to have happened.[40] There seems, however, to be little point to this practice. For though it may, at first, appear a bit odd to talk of explaining what has not yet happened, we nevertheless do in fact ask such questions as : *Why* will the patient die before long; or *why* will the moon not rise until the morning hours during most of next week. These are essentially requests for *explanatory* accounts though what is to be accounted for has not yet occurred. And if these examples represent the matter appropriately, whatever entitles a nomic inference to be called an answer to the question *why* has nothing to do with the time of the event to be explained. It seems, rather, to depend on the sort of nomic premises supporting the inferences.

3. *What is an explanation?*

In science as well as on other levels of discourse, explanations are answers to the question 'why?' But answers that offer to meet this question can, it seems, vary in certain important respects. They may be about conscious purposes as in giving the motives for a crime; or they may outline a divine

[39] C. Hempel and P. Oppenheim, *op. cit.*, Part I. E. Nagel, *op. cit.*, p. 4.

[40] As for example, Scheffler, I. (*op. cit.*, p. 48) who applies the term 'posit' (as opposed to 'explanation') for events about whose occurrence there is no knowledge at the time of the inference.

plan as in some accounts of the general course of events. They may single out distinctive natural causes such as a pathogenic virus; or they may give a causal-genetic mechanism for biological development as, for example, in talking about natural selection; or, again, they may fix on some crucial function such as photosynthesis. Even proofs in pure mathematics have been called explanations. And the list could be made much longer.

Of all the accounts that claim to be explanations only a fraction are properly regarded as belonging to empirical science. Some of the accounts that have been called "explanations" are purely mathematical in content; others may be said to be metaphysical; and so on, depending on the sort of subject-matter and principles which they stem from. We might also say that whatever it is that would entitle one to call any such accounts explanations, regardless of what they are about, is a matter of some contention and need not keep us here, our concern being specifically confined to what science offers as explanations. But even with this narrower class we meet with serious obscurity regarding just what it is that is supposed to make them explanations. Nor does this discredit science in any way, since strictly speaking, the term 'explanation' is not a part of the language of science. It rather belongs to talk *about* science.[41] Scientists provide systematic accounts some of which would seem to be appropriately labeled explanations and others not. On the other hand, it is primarily philosophers who, in talking about science, have, it seems, been seriously interested in telling us something about how this label might be precisely applied. But agreement on this question has been settled only with regard to the general features of explanation leaving much room for divergence of opinion in a significant class of cases.

Following Aristotle (who used the term "demonstration") it has been largely agreed that explanation is some sort of inference in which a conclusion is drawn from more fundamental premises. The notion comes very close to what we have called a nomic inference, and contemporary philosophers have given extensive analyses of it, C. G. Hempel's work forming a point of departure for most writings on the subject. Except for some of the finer details what Hempel calls an "explanation" would, on our account, be a nomic inference. But as an account of explanation his model, which we need not repeat here because of its similarity to our own pattern of nomic inference, fails for D_2, one of our examples in the last chapter. D_2 fits the Hempelian model well, though in no established sense

[41] Correspondingly, the question 'Why?' is not, strictly speaking, a part of science. Rather, it is a request for a piece of science, that is, for an explanation.

would it pass muster for an explanation. Recalling the example, it is an argument which concludes *that* oxygen is present, the basis or premises of the argument being the presence of animal life and the principle that oxygen is necessary for such life. But this argument does not in the slightest sense tell us *why* oxygen is present. It concludes with, but does not explain the presence of oxygen.

Counter-examples of this sort can be furnished abundantly and they require us to conclude that the Hempelian model is not sufficiently specific as an explication of explanation. It fails to rule out cases which are clearly not explanations in any familiar sense whatever of what it is to be an explanation.

Later writings have extended the analysis of explanation to include a probabilistic variant which in essential respects corresponds to our probabilistic nomic inference. As a model for probabilistic explanation however it too fails. P_2 which, in all essential respects, fits the model is not an explanation for the very same reasons that D_2 is not. Again, like D_2 it *substantiates* [42] or *supports* its conclusion but does not *explain* it. D_3 and D_4 also fit the Hempelian deductive model, but whether or not they are explanations is not so readily decided. P_3 and P_4 which fit the probabilistic model are correspondingly indeterminate. In the case of D_3, for example, where we infer the observed, and therefore known, value of the acceleration of the bob of a simple pendulum from its position and other facts, one might reasonably hesitate to say that the bob has an acceleration of 10 centimeters per second per second *because* (given all the other conditions) its displacement is 4 centimeters from the equilibrium position. Or, if one did consent to accounting for the acceleration in this way, he would do well to remind us that he was not using the term 'because' in quite the same sense as when saying that the acceleration was 4 centimeters per second per second *because* the net restoring force on the bob was about 1/8 of the earth's gravitational pull at sea level. That P_3 is an explanation seems even more easily denied than that D_3 is. We recall the argument as one in which, on the strength of a high correlation between I.Q. and success in school, it is inferred that Jack must have a high intelligence quotient, from the fact that he has been very successful in school. Though the argument does *support* its conclusion in the manner of a nomic inference, one would hardly want to say that Jack has a high I.Q. because he has been a good student. Similar and corresponding difficulties arise in attempting to characterize

[42] A term introduced by I. Scheffler in a similar sense and context. See : Scheffler, I., *op. cit.*, p. 49.

D_4 and P_4 as explanatory or not. The reader may wish to reexamine these cases on his own. In contrast to all the others in our list D_1 and P_1 are – clearly enough – explanations. D_1, as we recall, explains the death of certain laboratory animals in terms of intense ionizing radiation and is deductive in character. P_1 explains the illness of given animals in terms of certain dosages of ionizing radiation, though its inferential structure is probabilistic rather than deductive.

Why, then, are nomic arguments like D_1 and P_1 explanations while others, e.g., those like D_2 and P_2 are not? Our examples clearly establish that the differentiating feature is not the inferential structure, be this deductive or probabilistic. Well, then what about the premises? Is there something about these which makes the essential difference? The premises of both D_1 and D_2 (also P_1 and P_2) involve causal connections in some manner or other. In D_2, for example, the presence of oxygen is assured wherever there is animal life because it is causally necessary for (though by itself not sufficient for) such life. But the initial conditions mentioned in the premises of D_2 are not what we would ordinarily call the *cause* of the event mentioned in the conclusion. The reason for this is to be found in the underlying law. To see this compare the nomic premise (i.e., the underlying law) of D_1 with that of D_2. Expressing these with greater logical explicitness, though much less idiomatically, we have the following :

For D_1 :

L_1 : For any x, if x is intense-ionizing radiation then x kills animals.

For D_2 :

L_2 : For any x, if x is an occasion on which there are living animals then x is an occasion on which there is free oxygen.

Both premises, L_1 and L_2, are universal conditionals having the form :

For any x, if ... x ..., then ... x ... .

In any such conditional the expression, '... x ...', may be referred to as the *antecedent* and '... x ...' as the *consequent*. But though L_1 and L_2 are both conditionals, there is an important contrast between them. In L_1 the antecedent describes the sort of state-of-affairs which, in any particular case, would be a cause of the sort of thing described by the consequent. This, however, is not true of L_2 where it is clear that any instance of the antecedent would be, not a *cause*, but a *symptom* or *indication* of some instance of the consequent.

In any nomic inference it is only when the antecedent and the consequent of the covering law (more generally, nomic premise) are related in the causal manner suggested – antecedent corresponding to cause and consequent to effect – that the initial conditions of the inference can be said to be causes of what is described by the conclusion. This of course follows from the logical nature of any nomic inference whereby, as was noted earlier, the statement of initial conditions and the conclusion instantiate, i.e., describe instances of the antecedent and the consequent, respectively. What is more, reflection on usage easily suggests that it is this causal relation between initial conditions and "inferred" event,[43] present in D_1 but not in D_2 which makes the former an explanation and the latter not.[44] The answer to the question 'why?' seems most appropriate when it mentions the cause (perhaps, better, *a* cause) of the event to be explained. And this, in informal discourse, often amounts to merely mentioning the initial conditions leaving the covering nomic premise unstated but understood.

Generally speaking, nomic inferences having underlying laws like L_1, in which the antecedent of the law corresponds to causes and the consequent to effects – let us call these *cause-effect* laws [45] – fall indisputably under the rubric of explanation, the initial conditions counting as causes.

Schematically explanations of this cause-effect type may be represented as follows :

	L (= nomic statements of the cause-effect type, universal or statistical)
Deductively or Probabilistically ↓	C (= statements asserting initial conditions, c, assignable as causes)
	―――――――――――――――――――――
	E (= statements describing the explained effect, e)

This schema shows cause-effect explanations to be a species of nomic inferences, their differentiating features being that the nomic premise is of the cause-effect type and (as a result) that the initial conditions, c, are causes of the effect, e. (It will be convenient to use the well-established terms, '*explanans*' and '*explanandum*', respectively, for the premises and conclusion of any explanation).

[43] We take here idiomatic license. Strictly speaking, one infers not events but *statements* describing events.

[44] The inferences P_1 and P_2 may be compared in quite the same manner, though they are probabilistic rather than deductive.

[45] We prefer here the phrase 'cause-effect' laws rather than 'causal laws', the latter term having connotations which are, at the same time, much too vague and too broad for the distinction in point. The underlying law, L_2, of the inference, D_2, may be said to be *causal* in the sense that oxygen has a causal bearing on, and in fact is necessary for, the existence of animal life; but it is not a law of the cause-effect type because the antecedent does not relate to the consequent in a cause-to-effect manner.

But does this schema represent all explanations? Are there explanations whose nomic base is not of the cause-effect type? A definitive answer cannot be given, simply because prevalent usage for the term 'explanation' is insufficiently clear to decide the question. D_1 clearly has the features of an explanation, and we readily call it one. D_2, on the other hand, whose nomic premise, L_2, is not of the cause-effect sort, does not pass for an explanation. As we have already noted, its premises do not tell us that there is oxygen present *because* of the presence of living animals. They merely support without explaining the conclusion about oxygen. Indeed, it is – at least in part – because of the oxygen that there are animals, and not because of the animals that there is oxygen. An entirely parallel comparison may be made between the probabilistic inferences P_1 and P_2, P_1 being explanatory and P_2 not.

But, though the presence of cause-effect type premises establishes the explanatory character of any genuine nomic inference, the absence of such premises does *not* seem, decisively, to preclude labeling some nomic inferences as explanations. Asked *why* a given event occurs, we do not always respond by singling out its causes. We may offer a derivation from premises that one would hesitate to count as causes. Thus to *explain* why the absolute temperature of a gas doubled while the pressure stayed constant, a physicist might say that it happened *because* : the volume of the gas doubled, and at constant pressure the volume and temperature of a fixed quantity of gas remain proportional to each other.[46] He would, however, most likely balk at saying that the volume increase *caused* the temperature rise! Examples of this sort can be multiplied endlessly by using for nomic premises functional relations (and statistical correlations) of the kind that merely report concomitant variations in two (or more) quantities with no one causing the other, though both may hang on some common cause.

What such indeterminacy regarding the explanatory status of a wide class of nomic accounts indicates is that the term, 'explanation', is generally used vaguely. Various authors on the subject who, since Aristotle, have relentlessly claimed that a scientific explanation gives us the "causes" of things, have generally made little or no explicit acknowledgment of cases like the one just mentioned. It may at this point be suggested that despite all the looseness of usage, it would be futile to deny the basically causal connotations of the term 'explanation'. One would, for example, hardly demur in saying that the temperature of a gas rose *because* of

[46] This experimental law is known both as the law of Charles and the law of Gay-Lussac, being due to the work of both French physicists.

contact with a hotter object, but might easily object to any claim that the rise in temperature was due to a concomitant increase of volume. Accordingly, a stricter explication of what it is to be an explanation would rule against so entitling nomic inferences in which the cause-effect relation was not appropriately involved. But even so, this restriction would not guarantee that the notion of explanation is an adequately understood one. For then it inherits all the difficulties which afflict the concept of cause.

To note some of these difficulties, we pause for a brief review of what one might mean in saying that one thing (event, state-of-affairs, etc.) causes another.

E. A DETOUR : THE CAUSAL RELATION

The question of what it is for something to be the cause of something else has been one of the most preoccupying in the history of philosophy. Throughout the major part of philosophic tradition the relation between cause and effect has been taken to be that of a *necessary* connection between objects, states, or events. This connection has been construed as having a reality of its own distinct from the entities, states or events which it relates and as having an "ultimate" character which cannot be exhaustively described in terms of space-time observables. Such a metaphysical conception of the causal relation makes way for the rather questionable belief that discovering causal ties rests on faculties of cognition, e.g., intuitions, rational insights, etc., fundamentally distinct from sense perception. It is also generally coupled with the time honored view that in discovering *necessary* connections science brings to light a body of eternally true statements whose truth, once known, forms the unassailable core of scientific knowledge. The dubious character of this view becomes more and more apparent as science continues to review even some of its most established principles for possible revision when these begin to prove fruitless in the solving of new puzzles. The history of science says little for depicting science as a set of eternal truths, and much for seeing it as a body of perpetually tentative though more or less established statements.

David Hume attempted to free the notion of a causal connection from metaphysical content by offering an empiricist account of it, that is by explicating it in terms, *only*, of space-time relations, hence in terms, *only*, of observable attributes. More specifically, he construed cause-effect statements about events exclusively in terms of the events themselves and their constant concurrence, i.e., their immediate proximity (contiguity) in space and time.[47] Hume's explication forms the background for essentially all contemporary accounts of causation that claim to be in the empiricist tradition.

But even contemporary empiricist accounts leave the question of what it is to be causally related unsettled and raise serious questions regarding the applicability of causal language to a science that finds it increasingly appropriate to speak of functional relations (i.e., mathematical dependencies) and correlations rather than of causes. We pause to consider, for example, a representative account given by a contemporary writer, which following Hume, lists four features of the causal relation.[48]

The first of these is that the causal relation must be invariable in the sense that whenever the alleged cause occurs so does the alleged effect.

Such a condition, however, seems much too strong. It excludes all statistical, or, as it is sometimes expressed, *stochastic*, causes. Certain degrees of exposure to ionizing radiation will *cause* radiation sickness in *nearly* (though not) all exposed mammals, as in our example P_1. There is now little doubt that smoking is a cause of certain physiological disorders but not in all smokers, perhaps not even in most. Nor will it do to reply that these are cases of insufficient knowledge where accompanying factors, if known, would complete the list of sufficient causal conditions. Such an objection is quite beside the point, for we do in fact explain the occurrence of such effects by their probabilistic *causes*. But much more than this, contemporary physical science, in particular quantum physics, makes use of probabilistic causes and does this in the context of a theory which says that the knowledge of such causes is, in a profoundly fundamental sense, never sufficient to guarantee invariability in the outcome. One speaks, in such contexts of exciting (i.e., energizing) an atomic nucleus which, as a result, responds with a calculable and uneliminable probability of decay (i.e., disintegration). In such cases there surely seems to be little impropriety in speaking of the excitation, or of the exciting factor, as the cause of or *a* cause of the decay.

Thus not only does ordinary scientific and common sense usage recommend against restricting causes to universal connections, but such narrowing seems to strike a sharp dissonance with some of the most significant developments in modern physical theory.

The second feature listed is that causally related events are spatially contiguous. This is perhaps a less objectionable restriction than the one

[47] Hume, *op. cit.*, vol. I, pp. 79-80.
[48] E. Nagel, *op. cit.*, pp. 74-75.

about invariability. The history of science offers ample illustration of how tenaciously we hold to this notion of causation. Thus when Newton's law of gravitation and Coulomb's inverse square laws of electro-statics and magnetostatics threatened to introduce action (causes) at a distance, intervening physical fields (gravitational, electric and magnetic) and, more recently, even particles (gravitons) were postulated to restore spatial contiguity.

Ultimately, however, this requirement seems to be no less arbitrary and restrictive than does the requirement of invariability, though the attempt to satisfy it has stimulated vast theoretic advances. Causal efficacy at a distance is not only logically possible, i.e., the notion is free of contradiction, but it is imagined with little more effort than is contiguous action, quite besides any intellectual preference we may have for the latter. This is amply attested to by the sciences of mechanics and electro-magnetism which often formulate their laws so as to express interactions between spatially separated entities. However incomplete or counter-intuitive some may be inclined to regard such accounts, scientific theory based on them is still an open possibility for future advances in the systemization of knowledge.

The third and fourth features require, respectively, that causally related events be temporally ordered so that the cause always precedes the effect and that the causal relation is asymmetric in the sense that, for a given occurrence, if S is the cause of P then for the same occurrence P is not the cause of S. These two conditions are obviously not independent. The first logically implies the second since it is impossible for an event both to precede and follow another.

The requirement that the cause precede the effect in time, accords with much of familiar usage in both science and everyday matters. Causes are tied to later effects quite naturally and one sometimes speaks of *remote* and *proximate* causes depending on whether the alleged cause and effect are separated by large or small time intervals. Our ordinary understanding of the temporal aspects of the causal relation, however, seems to require that whatever the time separation between a given cause and effect there is always a string of causally connected states-of-affairs (more simply, states) which are *contiguous* in time and fill the gap completely. The time interval between any two related events is, so to speak, a *causally connected* one. Such back-to-back states or events are sometimes said to be related as *immediate* causes or effects. But again, as in the case of spatial contiguity, there is no inconsistency in the notion of causal action between events that are separated in time and not mediated by intervening causes, unnatural and unfamiliar though such action may seem. And accounts of nature in terms of causal strings broken by finite time intervals that are devoid of connecting causes remain, as in the case of spatially separated action, an entirely open possibility. On such an account an electron, for example, might be in state, S_1, at time, t_1 in state, S_2, at a later time, t_2, and unaccounted for (and therefore in no causally relevant state) during interval t_2-t_1.[49]

But whether or not we allow causal strings to be temporally disconnected the time asymmetry which they feature permits a distinction of prime practical significance, namely the distinction between causes and symptoms. To illustrate, consider the law taken from one of our previous examples, namely, (L) that wherever there is animal life there is chemically free oxygen. As in the example, the presence of animal life is taken to be the *logically* antecedent condition for concluding, on the basis of L, that free oxygen is present. The presence of animal life, however, for the case in point, is not the *cause* of the free oxygen but only a symptom of it. The reason for this is that, in the nature of the case, it is not animals which *temporally* precede oxygen but quite the other way around. A law of the type we have just illustrated might well be called an *indicator law*. It is a law which, in contrast to those of the cause-effect type mentions symptoms or indications instead of causes in its antecedent or if-clause.

Indeed, on pain of contradiction no cause can come later than its effect. But this does not in the least imply temporal ordering, i.e., temporal asymmetry, for the causal relation. And though we often have such ordering in mind throughout discourse in both science and everyday affairs, it fails as a blanketing requirement of causally connected events or states. There is a wide class of laws constituting major portions of knowledge in the basic sciences which express relationships between quantities that vary together, i.e., co-temporally, rather than one before the other, and for *some* of which we quite naturally single out a particular one as causing the other, though not *vice versa*. Such laws are usually formulated as mathematical equations expressing functional relationships between variables with no time-lag or transient state indicated; and they are highly abundant in the more quantitatively developed sciences. As an example, consider a simple pendulum of fixed length and small oscillations. For such a pendulum the frequency of the oscillation and the gravitational field are co-temporally related to each other.[50] Now, if the pendulum is to swing freely, neither the frequency

[49] Quantum physics which departs radically from classical and common sense notions provides a theoretical basis for such a case.

[50] For a simple pendulum of fixed dimensions the frequency, f, and the square-root

nor the gravitational field may change without the other changing also. Hence from strictly this point of view it would seem arbitrary or, at best, a matter of pragmatic orientation as to which variable we took to be manipulable in order to bring about changes in the other. We could say either that we were taking the pendulum to the top of a mountain and thus reducing the frequency by reducing the gravitational field or that we were taking it there and thus reducing the gravitational field by reducing the frequency! Odd as this might seem, there is, by means of the pendulum alone, no experimental way of distinguishing between those two alternatives. Nevertheless, while it would seem entirely appropriate in actual scientific discourse to say the first, it would be strange, indeed, if not utterly bizarre, to say the second. Despite the undeniable symmetry in the functional relation between frequency and gravitational field, the language ordinarily used by scientists continues to reflect an aspect of asymmetry, expressed by singling out the gravitational field rather than the frequency as the causally determinative factor.

Can this be accounted for in terms of the structure of scientific knowledge? A difference that seems to suggest itself between gravitational fields and the frequencies of pendulums is theoretic fundamentality. Gravitational forces, unlike the frequencies of pendulums, are theoretically basic quantities, mathematically tied to other key quantities like masses and distances which directly account for a wide range of phenomena like the falling of bodies, the weights of objects, and very much more. In order to pursue this point a trifle further let us take up the more general case of the relation between force and change of motion. These two quantities are functionally related in a basic principle of Newtonian mechanics, namely, Newton's second law of motion according to which, the rate of change of momentum of a body and the net force on that body are numerically proportional to each other.[51] As in the case of the pendulum we have, here, co-temporal variables, symmetrically related by a mathematically expressed law. But again despite this, scientific usage attributes an aspect of causal asymmetry to the relation. In Newtonian mechanics it seems unquestionably natural to say that forces affect the associated rate of change of momentum but hardly *vice versa*.[52] Other examples are easily brought to mind, such as the

---

of the gravitational field, g, are directly proportional to each other. This would be given mathematically by the expression :

$$f \propto \sqrt{g}$$

[51] The law also says that these quantities are always in the same direction, but this need not concern us here.

[52] There are certain seeming exceptions to this in some mechanical formulations

relationship between mass and the *resulting* gravitational field or between the electromotive "force" and electric current which it is said to *cause*.

In cases of this sort, designating some of these quantities as the *causes* of others seems to reflect something more than mere idiomatic preference. Indeed, it has to do with the roles which the designated quantities are made to play in the scientific account of phenomena. Consider, for example, the quantities in Newton's second law which relates net force to the rate of change of momentum. (The momentum of a body is the product of its mass and velocity.) Assuming for simplicity that the mass is constant, changes of momentum can be measured by directly noting changes of motion. Force on the other hand does not lend itself to such direct and independent measurement. Obviously, if a force were something which could be detected only as a change of momentum then, for scientific purposes, the only role of the term, 'force', would seem to be that of a shorthand for the expression, 'rate of change of momentum', and the second law would collapse to a mere terminological convention – i.e., a definition of the term, 'force', and one essentially devoid of empirical content. As a matter of fact, however, a physicist does, in a real sense, look for "forces". That is, he looks for the factors that he takes to be causes of the change of momentum. And he does this in ways other than by looking for changes of momentum. He looks for such things as massive bodies, their spatial configurations, elastic distortions or some of their other physical characteristics. Moreover, the basis for doing so is nothing other than Newton's second law. For this law may be reasonably taken to say that for any given body in motion a *simple* mathematical function of such quantities as neighboring masses, their positions, deformations, etc., can be found so that when this function is substituted for the term force in Newton's law, the result is a mathematical equation whose solutions describe the motion.[53] Thus, one might say, we look for the existence of neighboring bodies (among other things) on the basis of which to account for changes of motion and not the other way around. Moreover this way

which introduce special constructions such as "centrifugal" and "inertial" forces as though they arise from accelerations. These, however, are generally viewed asdispensable devices used merely for formal convenience.

[53] Thus Newton's second law for a particle may be written :

$$\vec{F} = \frac{d\vec{p}}{dt}$$

where the vector, $\vec{p}$, is the momentum of the particle. The vector, $\vec{F}$ is, in general, the sought-for function of other quantities. When this function is inserted explicitly in the expression, the result is a differential equation whose solutions lead to a description of motion of the particle.

of doing it is not gratuitous, but favored by established physical principles which largely provide the force-functions needed for use in Newton's second law of motion. The law of gravitation, for example, furnishes an inverse-square function of distance between bodies, and Hooke's law expresses a linear function of deformation or elastic strain.[54] We have here a sense on which one might say that bodies along with some of their attributes are invested with a theoretic fundamentality which scientific discourse reflects by marking them as *determinative of* rather than *determined by* certain dynamical effects.[55] This *causal* asymmetry persists despite other aspects of symmetry, more particularly, despite the absence of temporal order between what is called *cause* and what is called *effect*.[56]

[54] The law of gravitation may be stated : $F = G\frac{m_A m_B}{d^2}$
where F is the force of attraction between masses, m_A and m_B, separated by distance, d. G is a universal constant.

Hooke's law is given by :

$$F = -kx$$

where F is the force and x the attending deformation or *strain* of an elastic body. The constant, k, depends on the sort of body involved.

[55] This is strongly suggestive to the author of what one might want to characterize as a theory-based notion of *physical substance*. Such a consideration, however, will not keep us here.

[56] Newton did not hesitate to speak of forces as "causes" of motion. And one need only glance at physics texts to be convinced that physicists – perhaps in their less guarded moments – ordinarily manifest no hesitation whatever in referring to forces as causes of motion. Thus Georg Joos in a widely adopted text on theoretical physics states : "... from common experience we know that different amounts of force are necessary to *produce* a given acceleration, depending on the nature of the body acted upon, ..." (my talics). G. Joos, *Theoretical Physics*, G. E. Stechert : New York, 1934, p. 78. Similarly, another well known, contemporary writer, in discussing Newton's second law of motion as applied to a single particle, refers to "... the total force *acting* on the particle ..." (my italics). H. Goldstein, *op. cit.*, p. 1. There is no question that all references to causes throughout these contexts are references to an asymmetric relation. At the close of the 19th-century, H. Hertz, with the idea of giving expression only to that which can be "actually" observed (i.e., mass, space and time) eliminated the concept of force from his reformulation of mechanics. But then, change of motion of a mass is, in his system, due (if not to a force) to a "connection" with other masses, with no suggestion whatever that one could just as well speak (conversely) of change of motion as causing "connections with other masses". (A concise, non-technical synopsis of Hertz's thesis may be found in E. Mach's *The Science of Mechanics*, tr. T. J. McCormack, Open Court : La Salle, Ill., 1960, pp. 317-324.) E. Mach, in his own account of mechanics, defines mass in terms of acceleration, again without explicit reference to forces (*ibid.*, p. 266). Nonetheless, causal reference to "bodies" persists throughout his discussion (*ibid.*, 302-305, *passim*). In particular, he speaks of bodies inducing in each other contrary accelerations and, of the action of body, A, on body, B, and of body, B,

There are many scientific laws relating co-temporal quantities for which the causal asymmetry illustrated in our previous examples cannot be claimed. To mention only a few we note : the experimental law of classical mechanics which expresses the proportionality between the gravitational and inertial mass of an object; the law relating the displacement of a freely vibrating mass with its acceleration; and the general gas law. In such laws the variables relate symmetrically. And neither time nor fundamentality tip the balance of symmetry. Can these variables be said to be causally related? Some would extend the notion of a causal relation so as to allow it to be symmetric. Thus, according to R. B. Braithwaite, two symmetrically related variables can be considered both as cause and effect of each other.[57]

To illustrate we return to one of our examples. For an ideal gas expanding at constant pressure, Charles' law tells us that the volume and the temperature are directly proportional to each other. But does the increase in volume cause an increase in temperature or does the increase in temperature cause an increase in volume? Or should we follow Braithwaite and say it is both? Indeed, given the law of proportionality, one can from known changes in either variable deduce changes in the other. But to call each of these changes both cause and effect of the other seems to be a sufficient violation of ordinary usage to recommend against it. On such a view one would be allowed to say that, at constant pressure, the change in volume of a gas caused the change in temperature simultaneously with the change in temperature causing the change in volume. But in so doing the sense in which the term 'cause' was being applied was surely different from the sense in which one could say, for example, also that *both* variations had a common cause, namely, changes in the average motion of the constituent molecules. In this latter case the causal relation is not symmetric. For, according to the pertinent theory, molecular motion accounts for both the temperature of a gas, and the amount of space it must take up at a given pressure, but not, by any means, the other way around.[58] On the other hand, as regards

---

on body, A. But again there is no reason to believe that the causal relation is construed in any sense other than an asymmetric one. Thus, a two-body interaction consists of two simultaneous and distinct actions in which the causal relation remains asymmetric for each action. That is, the presence of body, A, causes the acceleration of body, B – but not conversely. The acceleration of B does not, of course, cause the presence of A.

[57] Braithwaite, R. B., *Scientific Explanation*, Cambridge University Press : Cambridge, 1953, p. 123.

[58] The kinetic molecular model of gases was devised to account for the characteristics of gases and other thermodynamic effects. The dynamics of gas molecules therefore might, in this sense, be said to be causally prior to such experimentally relatable effects as temperature, pressure and volume.

the relation between the temperature and volume of a gas at constant pressure, it is difficult to see what saying that each is the "cause" of the other adds to the fact that both quantities vary together in time much in accordance with a certain mathematical relation.

Occasionally, one encounters attempts to explicate causation simply as cases of *necessity*, or *sufficiency*, or *both* in the senses associated with conditional statements of the form, *If C then E.* These may be equivalently expressed either by *C is sufficient for E* or by *E is necessary for C*, where E and C are themselves statements. Thus, according to one such view, when we say 'C is the cause of E' we sometimes mean that C is a sufficient condition for E, sometimes that it is necessary condition for E and sometimes that it is both a necessary and sufficient condition for E.[59] Accordingly, an organism may be called the cause of a certain disease if its presence is necessary (though not sufficient) for the disease. On the other hand, one sometimes says that decapitation is the cause of an animal's death if it is sufficient for such death. Finally, we sometimes say that mass is the cause of certain gravitational effects. In this case (on Newtonian theory) mass is not only sufficient but also necessary for gravitational effects.

But though it is often the case that causal statements are, for certain purposes, expressible as conditionals, i.e., in terms of necessity, sufficiency or both, any claim that this is in general the case fails to square with important aspects of established usage. We say quite naturally that smoking causes lung disorders or that absorbed neutrons cause atomic nuclei to emit electrons. Yet, strictly speaking, none of the mentioned causes is either sufficient or necessary for the effect associated with it. There are smokers who do not get lung disorders and atomic nuclei which have absorbed neutrons without emitting electrons. Similarly there are non-smokers with lung disorders and nuclei which emit electrons without having absorbed neutrons. Some will want to reply that such examples involve statistical laws whose causal nature is debatable. What is more, explication in terms of the conditional applies only where the underlying laws are not statistical ones.

But even on the dubious restriction of calling a relation causal only if it is non-statistical, any account of causation simply in terms of necessity and sufficiency is by no means free of difficulties. For such an account requires us to specify how to decide whether to translate 'C causes E'as 'C is necessary for E', or as 'C is sufficient for E' or as 'C is both necessary

[59] For one version of this sort of view see : I. Copi, *Introductictn to Logic*, Macmillan : New York, 2nd edition, 1961, pp. 356-360.

and sufficient for E'. One answer sometimes given is that we equate cause with necessary condition whenever we wish to eliminate some phenomenon by eliminating the necessary condition. On the other hand we title a sufficient condition a cause whenever we wish to produce an effect by providing the sufficient condition.[60] There are, to be sure, a variety of instances in which conditions – some necessary, some sufficient, are properly referred to as causes. Take as an example the presence of a pathogenic organism, *necessary* for a disease or a quantity of botulin, *sufficient* to kill. But the claim that we call conditions *necessary* for x "causes of x" whenever we are interested in *preventing* x and conditions *sufficient* for x "causes of x" whenever we are interested in *producing* x is in general false. It decidedly fails for the following case : Death is a necessary consequence of decapitation. Moreover though we might in every sense want to avoid decapitation – and, of course, avoiding death would necessitate avoiding decapitation – it would be not only untrue, but utterly preposterous to say that death is the cause of decapitation. Similarly, as in our example, D_2, though the presence of animal life is a sufficient condition for the presence of free oxygen, it would be plainly wrong to equate 'live animals are a sufficient condition for the presence of free oxygen' with 'live animals cause or produce free oxygen'. And to do so would be to confuse causes with symptoms or indications.

Those holding the necessity or sufficiency view we have been considering might also want to maintain that any condition which is both necessary and sufficient for x is a cause of x.[61] But there again the account fails. This time take, for example, a pendulum of given dimensions oscillating freely in a gravitational field. For such a system the frequency of oscillation is necessary and sufficient for the gravitational field. But as we have already noted it would be strange indeed to say that the frequency causes the field.

A concern of another sort arises with any attempt at literal translation of causal statements into conditional statements. How – it may be asked – does one understand a statement of the form : *If C then E?* Precisely what are the ordinary everyday and scientific senses of such a locution? In formal logic we ordinarily render the conditional as a material implication for which precise semantical rules (i.e., rules for assigning truth or falsity to a statement) are provided.[62] But does material implication correctly

[60] *Ibid.*, pp. 356-360.

[61] As for example in the account given by I. Copi, *Ibid.*, 356-360.

[62] In formal deductive logic a material implication is a statement made up of two statements, an *antecedent* statement, p, and a consequent statement, q, arranged as follows : $p \rightarrow q$ and read as the conditional, *if p then q*. By convention, this conditional

translate the relevant senses of the conditional? Thus, consider the conditional : 'If a red-hot object X is quenched in water then X cracks'. Ordinarily this statement would pass for true if the object were a piece of window glass and false if it were a chunk of quartz. But construed formally as a material implication, the statement remains true no matter what the X is as long as it is *not* quenched in water, for by a definition of the logic of sentences any material implication is true whenever the antecedent (in this case : 'X is quenched in water') is false.[63] Similarly the statement : 'If Jack is a bachelor, then Mt. Everest is the tallest mountain', makes perfectly good sense as a material implication in the sentential calculus, and, by the rules of this branch of formal logic is true whether Jack is a bachelor or not. (Any material implication is true if its consequent, in this case, Mt. Everest is the tallest mountain, is true.) But granting that there is no significant natural connection between Jack's marital status and the height of Mt. Everest it would be ludicrous to make such a statement in any serious geographical account about the height of mountains.

Thus, equating the notion of causal connection with that of necessity or sufficiency in the sense of material implication is a gross oversimplification which fails under obvious counter-examples.[64] The view that such an account is possible is no doubt closely connected with what is now the widely established custom of representing universal nomic statements as material

---

is false for the case that 'p' is true and 'q' is false but true for all other possible truth-assignment to 'p' and 'q'.

[63] As has been often argued, statements of this kind, which express natural laws and dispositions are more aptly phrased in the subjunctive mood and as such therefore could not be considered ordinary conditionals. Our conditional would then be phrased : If the red-hot object, X, *were* quenched in water then it *would* crack. The logic of subjunctive or counterfactual conditionals, as they are also called, is still a highly unsettled topic. The mention of such statements here merely serves to underline the fact that the conditional connective 'if ... then' has a varied usage. We shall return to the question of counter-factual conditionals in a later section.

[64] To meet the challenge of devising a formal explication of the causal relation some logicians have turned to alternative systems of logic generally labeled *modal logics* in which the conditional or implicative relation between sentences is formally construed in a manner fundamentally different from that of material implication. The introduction of a special logic of modalities for the analysis of scientific discourse is, however, a measure which many Anglo-American philosophers of science have been reluctant to go along with. Some have raised serious questions regarding the sort of ontological or metaphysical commitments that such a modal logic might entail. (See Quine, W. O. V., *Word and Object*, M.I.T., 1960, pp. 191-211). Moreover, it is seriously debatable as to whether, in a purely formal reconstruction of science, a logic of modalities is needed. This question, however, is a rather technical one extending considerably beyond our present scope and need not be pursued here.

implications.[65] But this practice makes no distinction between universal laws that express causal connections and those that do not. Material implication is used to express both, and obviously, adherence to such usage must rule out equating the causal relation with material implication.

Having considered some prevalent explications of the concept of a causal relation and having found difficulties with these, we turn to a somewhat different question, namely : What sort of things constitute the domain of the causal relation. That is what sort of "things" can be related to each other as cause to effect? The long history of writings on the subject of causality offers a vast range of answers to this question. Thus physical objects, events, and states; minds, ideas and manifestations of will or purpose both conscious and unconscious; deities and other transnatural entities; mathematically expressible forms, numbers; and more have at some time or other figured as belonging to the *domain* of the causal relation.[66]

Causal language in contemporary science continues to remain rather loose in this respect. Even the *lack* or *absence* of something is in certain cases cited as a cause of certain events, e.g., the lack of iron in the blood as a "cause" of anemia or the lack of available oxygen in the atmosphere as a cause of death.

[65] The practice of representing all universal nomic statements as material implications generates additional puzzles in connection with the subject of confirmation, which cannot concern us here. Such representation, however, does have high utility from the viewpoint of constructing valid inferences. We refer here to the fact that despite the demonstrable differences between material implication and what we might call *nomic* or *natural implication*, the rules of logic which govern inferences involving the former, can be shown to remain entirely valid when applied to references involving the latter. In particular, the inference rule known as *modus ponens* provides that given '$p \rightarrow q$' and 'p' we can infer 'q' or, in argumentative form :

$$\begin{array}{l} p \rightarrow q \\ \underline{p \qquad} \\ \therefore \; q \end{array}$$

But this rule of inference would yield a valid argument even when the premise was not simply a material implication but a universal law based on a natural or, as we have said, *nomic* connection. The reason for this validity is that, whatever else universal laws are understood to say, they most certainly mean to guarantee that if the antecedent is true then the consequent *must* also be. And obviously this is all that *modus ponens* requires for its validity.

[66] We have noted that, in general, the causal relation seems to be construed as *asymmetric*, i.e., $(x)(y)\,(Cxy \rightarrow \sim Cyx)$, where $Cxy = x$ causes y. Ordinarily, it is also invested with another formal property, namely that of *irreflexivity*, i.e., $(x) \sim Cxx$. In some cases where God or certain faculties of mind (i.e., *will*) have been given causal roles irreflexibity has been abandoned in talk about autonomous or self-caused *will* or a self-caused God. Whether such usage is, in some sense, viable or not will not concern us here. It is easily noted, however, that in its reflexive instances the causal relation cannot be asymmetric.

Nevertheless, scientific discourse does in a vague sort of way tend to reflect what might be regarded as a somewhat mechanistic view on the nature of causes. When seeking a *causal* account of some phenomenon, first preference seems to be for dynamical swirls of time-space objects sharing common neighborhoods (i.e., touching, colliding) or, in the absence of bodily contiguity, acting upon each other through intervening force fields. It is when causal efficacy has been depicted in such terms that many inquirers feel that a satisfying explanation has been achieved. This is one reason why some of the abstract and formal phases of contemporary scientific theory which seem to leave out causal language altogether and, instead express quantitative, functional relations between certain classes of physical states or events, though directly applicable in their algorithmic and symbolic forms to natural phenomena, are often supplemented with *models* that can be intuited or visualized in concrete geometrical and temporal terms.[67]

It is undeniable that we do speak of relatively fixed states, properties, or dispositions as causing certain events as for example in saying that the high density of the object caused it to sink or the cat's curiosity caused its destruction, or that the sugar's solubility caused it to disappear in the water. But we nevertheless seem to regard such accounts as incomplete until they have also made place for explicit interactions between concrete space-time entities.[68] We might also add that in singling out the causes of phenomena, we opt for even greater concreteness than this. In mechanics, for example, we sometimes talk of energy exchanges between particles and fields or even between fields and fields. But fields are, so to speak, non-substantial in the sense of 'substantial' that we attach to discrete entities like stars, billiard balls, elementary matter-energy particles, etc. Consequently it is not unusual to regard such talk as only a part of a more complete account in which the so-called "real" causes, i.e., the more substantial sources of forces and force fields are included, namely bodies, charges, energy packets and other "entities". Once again, however, what we choose to regard as our causal bedrock for natural phenomena is often a loose reflection of fundamentality in some theoretic sense. In such a case this fundamentality is generally a matter of how we construe our theories and how we go about applying them. Newton's second law of motion

[67] We shall return to this topic in the more systematic context of theories and their interpretation.

[68] The so-called mind body problem and some of its proposed solutions, are, in part, about whether or not mental states can belong to the domain of the causal relation.

which bids us to look for forces in order to account for changes in motion is an example of this. But sometimes the causal priorities which we maintain are not explicitly reflected in the actual formulations of theory. Thus on the basis of Newton's law of gravitation it is theoretically indifferent as to whether we choose force or massive bodies as causally more basic. We in fact choose masses, but the fundamentality here is more implicit and underlying. It is to be traced to the common sense notion that material bodies seem to be in certain obvious respects more concrete and more stable than forces.

In summary our reflections regarding the causal relation come to this :

(1) General usage both in science and everyday affairs does indeed characterize some natural occurrences as cause-effect sequences, others as not. We say for example that a certain amount of ionizing radiation causes death, but that the presence of live animals *indicates* (rather than causes) the occurrence of chemically free oxygen. This difference, however, is not reflected in the structure of the natural laws expressing the relationships. Thus both our examples, above, can be expressed as universal laws of the form – whenever S then P – though in the first example S causes P while in the second it does not.

(2) There is, however, a vast portion of established science in which the term, 'cause', has very dubious applicability. We refer here to that scientifically important class of laws typically expressed as functional (usually mathematical) relations between temporally concurrent variables. In the case of some of these laws, it is undeniable that those regularly engaged in scientific discourse do single out certain quantities as causes of others as, for example, in saying that masses cause gravitational effects (and not *vice versa*) or that forces cause accelerations (and not *vice versa*). But for a wide range of relations representing major portions of contemporary scientific knowledge, the term 'cause' seems not to figure in any familiar sense at all. Referring again to one of our examples, it makes no more sense to say that at constant temperature a change of pressure "causes" a change of volume, than it does to say that a change of volume causes a change of pressure. And it makes little sense, indeed, to say that each causes the other.

(3) The notion of causal relation is itself problematic. Contemporary accounts of what is to be understood by it are subject to serious question, regarding their appropriateness and applicability when reviewed from the standpoint of *actual* science. Not only do some of the alleged properties of the causal relation seem to fail in a number of important respects but it is not clear from these accounts and from actual usage, itself, just what sorts of things are causally relatable.

Returning now to explanation we recall that the general run of contemporary discourse in the philosophy of science provides no clear specification as to precisely when a nomic inference is or is not an explanation. On the level of ordinary usage, however, some cases are clearly decidable. Thus, on the basis of a law connecting a particular kind of pathogenic organism with a given disease, we may say that (granting certain conditions) someone has the disease *because* of that pathogenic organism. On the other hand, though, every time the skin yellows a certain kind of disorder is present, we deny that the disorder occurs *because* of the yellowing. What is clear is that nomic inferences are explanations when their underlying laws are of the *cause-effect* type. On the other hand we readily deny explanatory status to nomic inferences which are governed by what we have called *indicator* laws, that is, laws in which the antecedent clause makes reference to symptoms rather than to causes. But explanatory status seems unassignable for another major portion of nomic inferences in science. These, as we have noted, are inferences based on laws expressing mathematical relations of so symmetric a nature that no one of the related quantities could be singled out as the cause of the other. Application of the term explanation to such cases is a matter of loose usage and somewhat subject to the objection that explanations are answers to the question 'why?' which in turn is most appropriately answered by giving a causal account. At any rate, it might be added, indiscriminate application of the term' explanation' tends to blur the distinction between two sorts of nomic inferences – both providing systematic structure, both capable of yielding predictions, but one singling out causes and the other *not*.

In our considerations we have distinguished between inferences governed by cause-effect type laws, which are clearly explanations, others governed by indicator laws, which are clearly *not* explanations; and still others constituting a major portion of science, which are governed by relations whose causal content is problematic and which are sometimes labeled explanations, though not without drawing reasonable objections. We have also reflected, as indicated in items (1)-(3), on the problematic nature of the notions of causation to which the concept of explanation is intimately bound. These considerations raise serious doubt as to whether it can be reasonably maintained, as has been traditional done, that explanation is the primary or central mode of scientific systemization. Let us stop for some brief reflections on this question.

F. BACK TO EXPLANATION AGAIN

1. Is explanation the Primary Form of Nomic Inference?

On first consideration one might well be inclined to regard explanations as epistemically superior to non-explanatory inferences, in the sense of providing greater depth of knowledge or understanding. Such a view seems quite natural and is well documented historically. Aristotle regarded science as the search for causes on the basis of which to construct "demonstrations". And this has been the dominant tradition right to the present, though contemporary writers now speak of "explanation" rather than "demonstration", claiming it to be central to science and a major distinguishing characteristic of it. A prime reason often given for the high importance which we continue to assign to explanation and to the cause-effect premises on which it rests is the control which a knowledge of causes makes possible. And no doubt such knowledge is a major step towards the control of phenomena. Thus, among our examples, the nomic premise of explanation, D_1, tells us that X-amount of radiation suffices to kill animals. Such information is obviously applicable for achieving certain practical results. In sharp contrast to this, the content of D_2 though it allows for prediction, by itself provides no such basis of control. The removal or introduction of animal life at the bottom of the pond will in general have no immediate effect on the presence or absence of oxygen there. Similarly, for a given simple pendulum swinging freely, a knowledge of the dynamical relation between the displacement, x, and the acceleration, a, of its bob offers no possibility of controlling either parameter without bringing in causally determinative ones, as, for example, the gravitational field which may be varied in order to control both x and a.[69] Undoubtedly, where practical control is of great importance, the search for causes remains an immediate and often even underlying motive of systematic inquiry. The practical phases of science and, especially, such contemporary technologies as medicine, pharmacology and engineering remain largely oriented in this way.

But this does not mean that laws which are not of the cause-effect type – e.g., indicator laws or laws expressing symmetric functional relations –

[69] The relation between x and a is given by :

$$a = -kx$$

where k depends upon the length of the simple pendulum and the gravitational field. If the pendulum is to swing freely neither x nor a can be manipulated as in *forced* vibrations.

are not relevant to practical application and control of phenomena. Major parts of our going sciences are expressed as mathematical, a-causal relations, yet have found vast applications in technologic control. What such applications require is the introduction of auxiliary cause-effect laws which tie large tracts of mathematically expressed science to manipulable parameters.[70] Thus as we have noted, for a free pendulum of fixed dimensions, nothing can be done to vary the acceleration and displacement patterns on the sole basis of the law relating these two variables, although of course from a knowledge of one, measurable values of the other can be predicted. But if we introduce laws relating the acceleration to the gravitational field and this field to the altitude then we have a basis for manipulating the altitude so as to control both acceleration-displacement variations. It is therefore highly inappropriate to associate causes with control in such a way as to suggest that formulations which are not of the cause-effect type play no part in the application of science for control purposes. It would be highly implausible, for example, to maintain that a major portion of the mathematical formulations of quantum theory (in the two versions in which it is prevalent) are cause-effect type laws. This, however, does not preclude the possibility of controlling quantum phenomena by introducing auxiliary cause-effect laws (e.g., laws telling how to excite an atom) which would provide suitable conditions for producing calculable quantum phenomena.

It is also sometimes claimed that explanatory inference provides a mode of understanding, singularly distinct from what we get from nomic inferences of the non-explanatory kind and this may well be true. But it must also be remembered that depth of scientific understanding seems to be not only a matter of causes but also a matter of the scope and generality of successful theory. Compare, for example, the depth of understanding which might accompany each of the following cases :

(1) the knowledge that an electrical discharge across a small air gap in an electrical circuit will cause another discharge across a similar air gap in an independent circuit placed at some distance from the first one.

(2) knowledge and application of theoretical electrodynamics for the calculation of this result as a special case of energy propagation by means of transverse electromagnetic waves traveling with the speed of light.

[70] We are not saying here that abstract theory must be coordinated to empirical subject-matter in order to be a part of science. This is true enough. What we are saying is that even when experientially interpreted such theory cannot be used to control concrete happenings unless there are laws available which tie the functional relations in the theory to manipulable variables.

Maxwell's equations which are the basis of such a calculation are not cause-effect type laws. They are formulations which imply symmetric co-temporal relations that cannot be read as cause-effect type laws. Heinrich Hertz's experiments which established the experimental regularity referred to in (1) came some twenty years after Maxwell's electromagnetic theory. It was therefore hailed as a great historical event, a confirmation of Maxwell's predictions. Viewed, however, as the discovery of a cause-effect type experimental law it virtually vanishes in significance when compared to Maxwell's formulation with respect to generality, systemizing and predictive power, and depth of the scientific understanding which it affords.

What is more, it would be similarly implausible to claim that explanation, in the specific sense in which we understand the term here, makes up the bulk or, in some related sense, the primary portions of systematic scientific discourse. We have offered examples D_2 and P_2 of *bona fide* nomic inferences that are clearly non-explanatory. The experimental regularities that serve as premises for these inferences are not in any sense odd or unusual in science. Physical or chemical tests for the presence of specific substances or objects, the significance of historical indicators (i.e., fossils, artifacts, etc.) assuring us of past occurrences are, essentially, experimental regularities, universal or statistical, particular applications of which are properly expressed as D_2 or P_2 type nomic inferences with initial conditions which in general do not give the causes of the inferred event.[71]

But more generally than this, our considerations, thus far, must have already suggested that mature science expresses vast portions of its nomic content, both theoretic and experimental, as functional (usually mathematical) relations that resist characterization in terms of ordinary causal language not only because such functional relations are equations that would violate the asymmetry usually required of the causal relation but also because scientific theory is often framed in formal language that makes no simple reference to any sort of space-time entities or events. Causal language, on the other hand, is closely tied to concretely imaginable (usually mechan-

[71] As a highly simplified example we may offer the following :

(L) A series of fossilized footprints occurring in strata at place P indicates the past existence of animals at P.

(C) There are such indicators at P

(E) Therefore there once were animals at P.

In this nomic inference the initial conditions C permit us to infer E, given L, but do not give the cause of the event mentioned in E. The inference is therefore not an explanation. This inference may also be expressed (perhaps more plausibly) as a probabilistic inference in a form quite similar to that of P_2.

istically conceived) space-time objects, states, events, etc. that can be singled out as the "causes" of the predicted phenomena. It has already been noted in past discussion that scientists do tend to append models to theoretic formalisms sometimes (though not always) with much apparent benefit to productive scientific thinking. It is perhaps with reference to these heuristic advantages that causal language is said to afford understanding. (We will return to this question in a later section under the topic of the structure of theories.) But whether or not explanation provides a stronger measure of "understanding" in some psychological sense of this term, it seems rather *not* to be central to modern scientific discourse. The search for explanation as ordinarily construed is a search for causes and these, as ordinarily understood, are not the only and, it seems, not the main forms of systemization and even prediction in mature science.[72] What is more, the dethroning of causation (and therefore of explanation) – so central to prescientific or, at best, proto-scientific knowledge – calls for no regrets. As has already been noted the notion of a causal relation is somewhat vague and rather gross in application. Its replacement by the language of variables and their functional dependencies measures a vast and decisive gain of refinement and perspicuousness not only in virtually all of the *natural* sciences but in sizable portions of the *social* sciences as well.

2. Explanation and Prediction

We have separated explanations from non-explanatory nomic inferences by indicating that the former, unlike the latter, have covering laws of the cause-effect type. Conceived in this manner the distinction between an explanation and a non-explanation is applicable to the entire class of nomic inferences and this takes in a variety of distinguishable types of which *prediction* is only one, a few others having been occasionally tagged by some writers with such labels as *post-diction* and *retrodiction*.

In the Hempelean account, explanation and prediction of an event, e, were taken to be structurally identical differing only in what might be called a pragmatic respect, that is, in the relative time at which the inference is offered by the scientist. If the inference is offered before e occurs then it is a prediction. Otherwise it is an explanation.[73] Recent writers, however,

[72] It is to be noted (and this will be discussed further in the next section) that prediction can be quite independent of any knowledge of causes. The result of a rabbit test may well permit the prediction that Mrs. Brown is going to be a mother, but it is by no means the cause of (i.e., does not explain) her condition.

[73] C. G. Hempel and P. Oppenheim, *op. cit.*, p. 139.

have taken issue with the structural identity thesis, and have offered counter-examples to show significant non-pragmatic differences between explanation and prediction. The question remains a generally unsettled one with such writers as E. Nagel [74] and A. Grünbaum [75] continuing in what is essentially the Hempelian tradition and others like N. Hanson,[76] N. Rescher [77] and I. Scheffler [78] departing from it. But while non-pragmatic differences undoubtedly exist between *some* nomic predictions and explanation, such differences do not imply structural differences if by 'structural' we mean the logical characteristics of the inference in point, more particularly (1) the form of the premises, i.e., their consisting of a set of nomic and singular statements some of which may be statistical (2) the logical relation between these and the conclusion which, as we have seen, is either deductive or probabilistic. To fix our ideas consider once again our example, D_2, in which, by means of a covering law we infer the conclusion, E, that oxygen is present from the statement that live animals are present. Now, D_2 would be predictive if E were about finding oxygen at any future time and non-predictive otherwise. On the structural identity thesis one would have to say that if D_2 were non-predictive it would be an explanation. Such a conclusion, however, would have to be wrong. For whether or not D_2 is predictive, it will not be the slightest bit explanatory. The presence of animals is not *why* oxygen is present, as stated in the conclusion of D_2, it only provides a basis for the conclusion. What is more, the non-explanatory character of this inference does not result from any peculiarity of logical structure, in the relevant sense of 'structure', but rather, as we have seen from the simple fact that the nomic premise is not of the cause-effect kind. If it be thought that at least some of the essential differences between causal and non-causal laws are of an ultimately structural nature, we must reply that this may well be so but requires considerable elaboration on rather complex matters possibly involving the over-all structure of some piece of science rather than the structure of the particular inference or of the law itself. At any rate such

[74] E. Nagel, *op. cit.*, pp. 29-42.

[75] Grünbaum, A., "Temporally Asymmetric Principles, Parity between Explanation and Prediction and Mechanism versus Teleology", *Philosophy of Science*, ed. B. Baumrin, Interscience Publishers, 1966, pp. 57-97.

[76] Hanson, N. P., "On the Symmetry Between Explanation and Prediction", *The Philosophical Review*, Vol. 68 (1959), pp. 349-358.

[77] Rescher, N., "On Prediction and Explanation", *British Journal for the Philosophy of Science*, Vol. 8 (1958), pp. 281-290.

[78] Scheffler, I., *op. cit.*, pp. 43-46.

considerations do not figure at all in the structural identity issue and therefore need not keep us here.

The objections to the Hempelian identity thesis have thus been largely irrelevant. They have for the most part attempted to point up a structural difference between explanation and prediction as inferences in general. The Hempelian identity thesis, however, is reasonably understood as applying only to what we here call subsumptive inferences. Predictive inferences of the ampliative kind are, in a very obvious way, different in form from all subsumptive inferences, be these explanatory or not. It is seriously doubtful therefore that the objections to the Hempelian identity thesis have succeeded in establishing any claim against it. As to whether ampliative predictions are ultimately implicit forms of subsumptive ones despite their immediately apparent differences is a matter which will concern us in a later section. For the moment we turn to another consideration, namely that of what truth-status the premises of nomic inferences must have in order that one be entitled to call them explanations.

3. *Must the Premises of an Explanation be True?*

In our discussion of nomic inferences in general we found it inappropriate to require truth or even high degree of confirmation of the law-like premise or covering law, L. For, such epistemic requirements would literally rule out the vital scientific process of testing hypotheses through nomic prediction. We found it proper to require, at most, that L *not* be known to be false. Let us now consider the possibility of stronger epistemic requirements on L in the case of explanations.[79]

As we have already noted the notion that nothing has been explained unless the premises of the explanation are true goes back to Aristotle who says of "demonstrated knowledge" that "the premises must be true".[80] It would of course, follow from this that no inference can be known to be an explanation unless its premises are known to be true – an awkward consequence indeed when measured against actual *modern* science. For Aristotle this requirement on explanatory premises was made in the context of a rationalistic theory of knowledge according to which there are "primary" and "undemonstrable" principles that may be known with incorrigible certainty as "immediate" or "necessary" inductive truths.[81]

[79] Our general considerations regarding the epistemic status of the singular premise, C of a nomic inference apply without qualification to explanation. C which is a statement of initial conditions must be true, or highly confirmed in order that L not be misapplied.

[80] *Analytica Posteriora*, 72ᵇ, 25, *passim*.

[81] *Analytica Posteriora*, 71ᵇ, 2, 20.

Once known, these first principles can then serve as premises from which to establish *conclusively* the rest of science by deductive elaboration.

But today few if any scientists would accept either the doctrine of "inductive" certainty for any principle of nature or, for that matter, the claim that such principles are necessary truths. In actual practice, the "truth" of scientific hypotheses is very often not only unknown but undecidable for reasons that run deep and cast serious doubt on the applicability, in such cases, of the very notion of truth. This consideration makes the Hempelian truth-requirement obviously awkward. For as far as explanatory efficacy goes it puts a wide and central class of working scientific hypotheses in a state of limbo from which there may be no possibility of escape. Particularly true is this for theories, which are tied to observational evidence far less directly than are experimental laws. Moreover, it is a general point of logic that an indefinite number of alternative laws, theoretical or other, can be framed for the systemization of any given body of observational evidence. Indeed, the successful use of alternative and incompatible *theories* for the same body of fact is – methodologically speaking – a strikingly real possibility for which the history of science offers rich testimony. For that matter, even experimental laws which merely report observational regularities, statistical or universal, are based on only limited evidence. Ordinarily we sometimes *do* say that we know some experimental law, L, to be true. Reflection, however, easily cautions us in such a claim. And there are those who would insist that we can at best assign only probabilities to laws.[82]

A reasonable alternative seems to be that of weakening Hempel's requirement to read that the nomic premises of an explanation must be *highly confirmed*.[83] Those who favor the stronger requirement, however, reject this alternative because, it would seem to do violence to the ordinary notion of explanation by making it time dependent. That is, any account which was an explanation at one time might not be at another; for the extent to which any hypothesis is said to be confirmed is a function of available evidence and this, in general, varies with time. The reluctance to admit a time-relative notion of explanation is not clearly accounted for. Perhaps it is to be traced to the idea that explanations give the causes of things and causal regularities in nature, as expressed by laws, are not expected to change with time. But while the uniqueness, stability, and reality of causes may be a matter of easy agreement, it is not obvious that

[82] Reichenbach, H., *op. cit.*, pp. 434-442.
[83] E. Nagel, *op. cit.*, 43-44.

there is any advantage in refraining from calling any inference an explanation if, at any stage of *genuine* scientific practice it purports to "subsume" phenomena under causes, though its covering hypothesis may not be known to be true. To ease the Hempelian stricture, the notion of a proffered explanation has been introduced. But this seems not only to multiply distinctions unnecessarily but may also be misleading to those who read it as suggesting that hypotheses are, in general, only tentatively doubtful and may eventually become virtually certain. We see that where hypotheses are theories the sense and possibility of such certainty is seriously open to question as, indeed, would be any view that tended to idealize the search for scientific explanation as a search for timeless and incorrigible truths. In this general connection, it also seems appropriate to note that, as ordinarily understood, the term, 'explanation', even when applied in contexts pertaining to science, seems not to incorporate any very definite truth requirements for the premises concerned. Thus, ordinary usage provides no clear indication that the phlogiston approach to oxidation or the caloric accounts of heat, though no longer accepted, are any less appropriately regarded as explanations than are current interpretations of the same phenomena, even in contexts where the truth-status of the earlier accounts is held in question. Indeed, any explicit stipulation that no account may be called an explanation unless its nomic premises are true or highly confirmed, would seem to reduce ultimately to a matter of terminological adjustment – an adjustment for which there seems to be little or no methodological basis.

G. PATTERNS OF NOMIC INFERENCE

1. Nomic inferences can be differentiated in a variety of ways. And some writers, in order to supplement the traditional explanation – prediction account have found it necessary to talk also about "post-diction" (H. Reichenbach), "retrodiction" (G. Grünbaum), "positing" and "substantiation" (I. Scheffler). Our aim in what follows will be to review some of these distinctions and to suggest others for their possible methodological significance.

Reflection along the lines we have thus far followed strongly suggest that from the viewpoint of nomic inference in general, it is appropriate to contrast not explanation and prediction but explanation and non-explanation, prediction being but one of several pragmatic variations [84] or patterns

[84] Pragmatic variations are those due to the special circumstances (time, state of knowledge, etc.) of the scientist making the inference.

of inference each of which may be either of the explanatory or non-explanatory kind. Thus, whether a nomic inference is explanatory or not it may fall under one of several types. Among these are prediction as well as others which have been given such labels as *post-diction* and *retrodiction*, and still others for which no labels seem as yet to have been thought up. The term 'prediction' has itself been used with variable meanings, the relevant considerations being the time-order of occurrence of the initial conditions (c), of the event (e) to be inferred, and of the inference itself (s). The following patterns of nomic inference have at some time or other been labeled *predictions*. In the symbolization, the order of the letters taken from left to right represents the order of occurrence in time.

(1) The inferred event, e, is in the scientist's future and the initial conditions, c, are in the scientist's past, diagrammed :

c-s-e

As an example consider a satellite moving through known forces, whose future position, e, is computed on the basis of some known past position and momentum (c).

(2) Both e and c are in the scientist's future, with c earlier than e, diagrammed :

s-c-e

In this case the future position e of our satellite would be computed on the basis of the anticipated future position and momentum (c). The, as yet, unrealized initial conditions (c) can be said to be "known" only in the sense of a highly probable prediction. That is one based on conclusive evidence, as, for example, when one says that he knows that he shall not live to be one thousand years old, or that he "knows" from careful observation that some celestial body will have a certain location and motion at a certain time.

(3) Both e and c are in the scientist's future but, this time, c is later than e, diagrammed :

s-e-c

The example for pattern (2) is easily adapted to this case by simply requiring that to compute our satellite's momentum and position at some future time, t_1, we use information regarding its position and momentum at a future time, t_2, greater than t_1.

(4) Event, e, is in the scientist's past but comes later than c, diagrammed :

c-e-s

In this case we compute what the satellite's position and momentum *must have been* (e) on the basis of where it was at some still earlier time.

Pattern (1) represents what seems to be the most usual and natural conception of a predictive nomic inference.[85] Hempel's account, however, requires only that e be in the scientist's future.[86] This would then admit patterns (1), (2) and (3) as predictions. A. Grünbaum, on the other hand, construes prediction as any case for which c is earlier than e. In contrast to such prediction there would then be *retrodiction* for which e is earlier than c. Accordingly cases (1), (2) and (4) would be predictions despite the fact that in (4) e is in the scientist's past. Such a reconstruction of prediction is, as its author indicates, free of homocentric (i.e., pragmatic) considerations concerning the relative time at which the inference is made, viz., the time of s.[87] Nevertheless it appears somewhat unnatural to call case (4), i.e., c-e-s, a prediction. To begin with the term 'prediction' is ordinarily applied to statements describing what has not yet happened and may possibly never happen even given that c has already occurred. This suggests what seems to be a rather important, though non-structural, difference between ordinary, *ante-factum*, prediction and case (4). In ordinary nomic prediction one must wait and "see" if e eventually does in fact occur. At the time of such prediction therefore, the eventual occurrence or non-occurrence of e is no part the evidential base of the Law, L, underlying the inference in point. Indeed, e may not occur at all with dire consequences in case L is universal. For, in such a case, the eventual non-occurrence of e is sufficient to discredit decisively or, as we might say, *falsify* the underlying law, L. This test potential in any projective or futuristic prediction based on universal principles lies deep at the heart of all experimental method and any extension of the notion of prediction to include patterns of inference where the inferred event has already happened could tend to obscure this important function of all projective inference. Obviously case (4) differs markedly in this respect from ordinary prediction. If at the time of s it is known that e has in fact occurred and if, as one might expect, L has been so framed as to allow for this fact, then the inference in point can provide no basis for the predictive testing of L. It is a fundamental point of logic, that a universal statement can always be formulated to subsume any given body of information however extensive.[88] Such a statement, however,

[85] I. Scheffler, for example, takes it to be the only predictive pattern. I. Scheffler, *op. cit.*, p. 48.

[86] As noted by A. Grünbaum, *op. cit.*, pp. 57-97.

[87] A. Grünbaum, "Explanation and Prediction", *Philosophy of Science*, the Delaware Seminar, I, 1961-62, p. 72, New York, 1963.

[88] A simple but convincing mathematical analogy to this is the fact that through any array of given points on the Euclidean plane a smooth curve can always be drawn whose

would have little significance as a scientific law (or other nomic principle) if it did not lead to factually correct predictions. After each wrong prediction the law could of course be revised to accommodate the discrepancy. But it is a matter of common scientific practice to lend little credibility or even to regard as inadmissible any universal hypothesis requiring repeated revisions of this sort.[89] The predictive efficacy (in the usual *ante-factum* sense) of any universal nomic principle is therefore methodologically critical and strongly suggests that ordinary usage, which seems to be sensitive to the feature, be upheld.

Case (4) does bear some resemblance to prediction whenever it is not known whether or not e has occurred. While it is not logically possible to wait and "see" if, in fact, e does occur, it is nevertheless possible to search for independent evidence of e's having occurred. In such a case though the inference is made after the time of e's occurrence, it is not made with the knowledge of e's occurrence. It was, as we might say, inferred *post factum* but not *ex post facto*. Subsequent findings bearing on e's possible occurrence would then also have confirmational bearing on L in much the same way that a correct prediction would.[90] As an example we

---

algebraic representation, y = f(x), would be satisfied by each of the points, though f(x) might have to be a polynomial of very high degree but one whose coefficients can always be found.

[89] Lest this be an oversimplification it is only fair to note that scientific hypotheses are often a matter of stepwise corrections sometimes spanning long periods of time – perhaps decades and even centuries. The systematic evolution of the Bohr theory of the hydrogen atom and the supplementations of classical mechanics by special relativity are vivid examples. In such cases it would be misleading to speak of simply impugning the admissibility of the hypotheses in question. The pattern of divergence of predicted from actual values is, for cases of this kind, highly relevant in the evaluation. Roughly speaking for such cases the corrections required are refinements affecting, for example, the degree of articulation or detail as in the step-wise development of the Bohr theory, or delimiting certain portions of the range of valid application as in the growth of mechanics. The revisions may indeed be deep cutting in their fundamentality, but the over-all result is one of gradual and systematic convergence towards what seem to be the "right" answers rather than mere elaboration and "tacking" on of theoretical content to accommodate the latest exception.

[90] I. Scheffler (cf. *Anatomy of Inquiry*, p. 48, *passim*) applies the term 'posit' to this and to any other nomic inference in which e occurs before s but is not known to have occurred. But though this choice of term is somewhat suggestive of a similarity to prediction, it is somewhat unfortunate because of its other connotations in the area of ampliative (i.e., inductive) inference where it refers to a "good" guess type of probable supposition having a much more tentative character than, generally speaking, do inferences of the subsumptive (i.e., nomic) type. H. Reichenbach, for example, applies the term to the probability assigned an event by reporting the relative frequency with

may imagine calculating the date of an ancient eclipse and later finding reliable records showing that such an event did in fact occur at the calculated moment. Despite this last point, however, our other considerations caution against any over-all assimilation of case (4) to prediction. We find it recommendable therefore to follow Hempel in calling a nomic inference a prediction if and only if e comes later than S. Accordingly, we take (1), (2), and (3) but not (4) to be predictions.

2. Our list of variations of nomic inferences may be extended a bit more. There are obviously six possible permutations of our three elements 'c', 'e' and 's', of which we have already listed and discussed four with respect to whether each of them was a prediction or not. Listing again these four and adding the other two we have for our total array of possibilities.

(1) c-s-e
(2) s-c-e
(3) s-e-c
(4) c-e-s
(5) e-s-c
(6) e-c-s

Examples of (1)-(4) have already come under consideration. An example of (5) would be the calculating (i.e., inferring), for a given space-craft traveling through a known environment, of where it must have been at some past time on the basis of knowledge of what its dynamical state (i.e., momentum and position) will be at some future time. A similar example can be constructed for (6).

We have already characterized (1), (2), and (3) as predictions simply because in these patterns s comes earlier than e. For the constrasting case where e is earlier than s, viz., (4), (5) and (6), the term post-diction naturally suggests itself.[91] In general, we may pick out a distinct pattern according to the order in which any two of the three letters occur. Consider, now, the order of c and e. Following A. Grünbaum we can call any inference in which e is earlier than c retrodiction, but for the reasons given we resist calling the contrasting case (i.e., where c is earlier than e) prediction. In the absence of an established term, we can introduce a new label and call this case *prodiction*.

---

which events of its kind occur in some sequence of trials. The relative frequency is gotten as an "induction via simple enumeration", occurring at the level of primary or "elementary knowledge" and subject to progressive correction as knowledge advances. See H. Reichenbach, *The Theory of Probability*, p. 466, *passim*.

[91] Following H. Reichenbach, *Experience and Prediction*.

Our last characterization concerns the order of s and c, which give the two remaining cases. There seems to be little point in labeling these two cases. The more usual and more familiar case is (type 1) the one where the antecedent conditions, c, in an inference describe circumstances already realized at the time of the inference even if these are so current as to approach the date of the inference. But some inferences (type 2) are made on the basis of as yet unrealized or future circumstances, usually when these are expected to occur with overwhelming likelihood. Such inferences may conclude with events in the scientist's future either retrodictively or prodictively or they may even conclude with events in the scientist's past, in the latter case, always retrodictively. Thus a strategist in 1945 might have ventured a probabilistic prediction of Hitler's impending suicide partially on the basis of, but prior to, Germany's imminent defeat. On the same basis he might also have made some inferences regarding what Germany's manpower and economic resources must have been late in 1944, just before the last great German offensive in the Ardennes.

To summarize, we have classified nomic inferences in a variety of ways leading to the following pairs of contrasting types : explanatory and non-explanatory; predictive and post-dictive; prodictive and retrodictive; and a last pair corresponding to the time order of c and s which we call, "type 1" and "type 2". It is clear from the manner in which these distinctions were made that all four modes of characterization apply compatibly to every inference. Applying these to our list of patterns and remembering that any nomic inference may be explanatory or non-explanatory depending on the causal character of the premises, regardless of pattern, we have the following characterizations :

(1) c-s-e prodictive prediction of type 1
(2) s-c-e prodictive prediction of type 2
(3) s-e-c retrodictive prediction of type 2
(4) c-e-s prodictive postdiction of type 1
(5) e-s-c retrodictive postdiction of type 2
(6) e-c-s retrodictive postdiction of type 1

H. SUMMARY

Our concern in the past chapter has been with the systematic structure of science as a product of inquiry and, more particularly, with nomic inference as the basic unit of such structure. The standard practice is to regard considerations of this sort as studies in explanation and to grant prediction formal parity with explanation, differentiating it merely as a pragmatic variant of the latter. On our analysis, however, explanation is not to be equated to nomic inference, but comes out instead as only a species of it. Moreover – and contrary to approved opinion – we have concluded that explanation is not particularly central to science but, instead, shares its systemizing function with non-explanatory forms of nomic inference.

A nomic inference consists, essentially of : a law-like or *nomic* premise which may itself be a conjunction of law-like statements; a set of specializing (initial or boundary) conditions; and a conclusion. One may, at the very outset, distinguish two basic sorts of nomic inferences : deductive and probabilistic. In the first, the premises necessitate the conclusion; in the second, the premises merely render the conclusion probable. But whether probabilistic or deductive, nomic inferences may be characterized as either *explanatory* or *non-explanatory* depending on the character of their nomic premise. A nomic inference is an explanation only in case its nomic premise, L, is of the cause-effect sort. Otherwise it is non-explanatory, as for example, when L is an indicator law featuring symptoms rather than causes or a functional relation in which the operative parameters could not properly be regarded as either symptoms or causes.

But all nomic inferences, whatever their formal mode (deductive or probabilistic, explanatory or not) may be differentiated on the basis of a variety of temporal patterns of which *prediction* is only one. These patterns occur independently of the formal character (deductive, explanatory, etc.) of the inference because they are purely pragmatic distinctions, that is, they reflect only the temporal order of events in a manner which relates to the actual *time* of the inference. As a consequence, a nomic inference can be both explanatory and predictive, non-explanatory and non-predictive, and so on – results which are somewhat obscured by the standard analysis of scientific explanation.

Having examined nomic inferences, in general, and having distinguished several formal and pragmatic modes of these we turn, in our next chapter, to the following considerations : The first of these has to do with the fundamentality of nomic inferences; and we shall ask in this connection whether there are forms of scientific explanation which are *not* nomic inferences. It seems *also* appropriate at this point to consider teleological or functional representations of natural phenomena and to ask how these "fit in" with the over-all account. Our other concerns will be with getting closer views of two other major types of nomic inferences, namely, those with nomic conclusions and those of the probabilistic sort.

V

OTHER ASPECTS OF NOMIC INFERENCE

A. ARE THERE NON-NOMIC EXPLANATIONS?

1. Explanatory discourse often offers explanations that on the surface seem to be non-nomic in character. We are not referring here to those enthymemic inferences which we often abbreviate by mentioning only the initial conditions or "causes" leaving the covering laws as understood. What we have in mind instead is inferences – predictive, postdictive, or other, that seem to require no underlying laws or nomic premises at all. For example, asked why there was a certain occurrence of Q, we may answer with the following : Because (1) a P occurred and (2) in all, a definite number, n, of P's, call them (P_1, P_2 ... P_n) has been observed to occur; and without exception, each has been followed by a Q.

If each P relates closely in space and time to a corresponding Q, so as to suggest a causal connection between them, then an account such as this seems to take on the aura of a *bona fide* explanation. The premises (*explanans*) of such an argument may be expressed without explicit reference to any general laws or other nomic statements; they can be written as a finite set of singular statements. In this form the argument is non-nomic, having an ampliative rather than subsumptive character, and as such therefore does not satisfy the nomic (hence subsumptive) model for explanatory inferences.

But the outward form of this inference need not be regarded as irreducible and may be construed, perhaps more naturally, as *two* concatenated inferences. Take the *first* of these to be essentially an inference with an unstated implicit conclusion and write it as follows with the conclusion made explicit :

P_1 is followed by Q_1
P_2 is followed by Q_2
...
P_n is followed by Q_n

Therefore, all P's are followed by Q's

This is not a nomic inference at all. It is ampliative, therefore inductive in character, and by itself explains nothing. The second of our two component inferences is the following :

All P's are followed by Q's
A (new) P has occurred

Therefore a Q follows

Assuming the appropriate causal relationships between occurrences of P and Q, this inference is an explanatory nomic inference. It is concatenated to the first by using the latter's conclusion as a premise. In our particular illustration it happens to be deductive, but a quite similar construction can of course be given for a probabilistic version of our original example. This manner of construing such inferences has a certain naturalness perhaps because of its greater transparency over the non-nomic, single inference form. At any rate it seems more gratuitous or arbitrary to say that P_{n+1} is followed by a Q because P_1, ... P_n were each followed by Q's. More naturally we would say that a Q has occurred because a P has, and because of the nomic principle that all P's are followed by Q's. Asked how we know this latter principle we furnish the first of our two inferences.[1]

2. The force of explanations of the kind illustrated by the last example increases with increasing n. Nevertheless it is not uncommon, either in science or in everyday affairs, to offer an explanation whose explanans describes only one event, that is to say the special case for our last example in which n = 1. To illustrate this, consider the following : a man preparing for a masquerade party uses some dye to paint the ball of his nose. A moment later an angry rash appears precisely over the area taken up by

[1] I. Scheffler offers a comparable approach to the sort of inference under consideration. It seems to the present author, however, that his account is somewhat clouded by the attempt to distinguish between (as he puts it) "the causal explanation of events and the explanation of judgments or beliefs". But recourse to such troublesome epistemic or psychological notions as judgment, belief, and the justification of these seems quite needless here. What Scheffler describes as the justification of a judgment or belief on our account is simply an ampliative argument comprising the first of the two component inferences in our construction. See : I. Scheffler, *op. cit.*, p. 40.

the dye patch. Asked later why he has the rash on his nose, he offers the fact that he used a certain dye in precisely the same place where the rash is, though admitting at the same time that he has no prior knowledge about the way such dyes generally affect the skin.

At first blush, this example, more so than its predecessor, seems to diverge incorrigibly from the nomic model of explanation. The inference it illustrates is made in the state of elementary knowledge, that is, without either nomic principles or any patterned and extended sequence of events. Only one pair of events is involved, though a conspicuously close-knit one. Our two events are contiguous (in this case some might say coincident) in space and closely successive in time. And these are conditions distinctively associated with the notion that two events be causally related. The account seems to be offered as some sort of non-deductive inference but the precise logical relation between the *explanans*, which describes the earlier events, and the *explanandum*, which describes the later event, is, as it stands, far from clear. Nevertheless, we can assimilate the account to the nomic model similarly to the way in which we did it for the preceding example. Simply "tack on" the general law, L, that all (or nearly all) applications in certain amounts of such and such a dye to the skin produce rash. Our account thus becomes : (1) a deductive or probabilistic inference, depending on whether L is a universal or statistical law. Asked how we arrive at L, we give another inference, namely : (2) Dye of such and such a kind was applied in a certain amount to the forehead and a rash appeared shortly thereafter precisely where the dye was applied. Therefore, any application (or perhaps, a large percentage of applications of such and such a dye ...) will be similarly accompanied by a skin rash.

We have reconstructed our second example similarly to our first, that is, we have taken an account which while claiming to be explanatory, was, in given form, non-nomic in character, and we restructured it as two inferences (1) and (2), above. Inference (1) is clearly a nomic inference of the explanatory kind. Inference (2) is an ampliative inference whose conclusion is a nomic statement.

This reconstruction has much to recommend it. For it provides an account of the explanatory character of the original inference by construing one part of it as a nomic analogue based on causal premises, and also by tracing out a genuinely inferential structure, as a chain of two arguments, one ampliative, the other subsumptive.

The logically dubious link in the chain is (2) which concludes with a general law using for an inductive base a singular statement about merely one event-pair. Some indication of the inconclusiveness of (2) is that it can have as its conclusion either a universal or statistical law with about equally low cogency. Nevertheless (2) cannot be entirely dismissed for non-cogency. There seems in this case to be much good sense in asking : What are the chances that, immediately after applying a patch of dye to the skin, a rash will appear coinciding precisely with the patch, unless the dye and the rash were causally connected? The question, of course, strongly suggests that such chances would be slim indeed! [2] Our illustration therefore seems to represent a case of genuine explanatory inference, however weak may be its force.

At any rate our consideration of both the cases just illustrated lends much reason to resist regarding any of them as irreducible deviations from the nomic form. If genuinely anomalous modes of explanatory inference do arise in ordinary discourse, then these are not the ones. And, until unassimilable counter-examples answering to what we ordinarily understand by scientific explanation are found, the nomic model of explanation (requiring law-like premises) remains essentially unchallenged.

3. It seems appropriate at this point to consider a very common pattern of discourse which bears some resemblance to the nomic form of inference yet seems to depart from it in an important respect. Unlike the two cases just considered, this construction makes explicit use of nomic premises but does so in such a way as to diverge from the nomic model. Consider the following example : Preliminary study of the surface of a planet in a nearby solar system shows it to have water, a temperature ranging from about –70°F to 160°F, an atmosphere capable of filtering out ionizing radiation, and traces of polymerized amino acids. It is therefore taken as a reasonable (somewhat probable) expectation that there is life on the planet. Also involved in this judgment is other unexpressed information such as the following :

(1) Wherever there is life water is present.
(2) Life is found only for temperatures ranging from about –70°F to 160°F.
(3) Life can be sustained only in the absence of strong ionizing radiation.
(4) Wherever there is life there are amino acid polymers.
(5) The co-presence of the conditions mentioned in (1)-(4), above, do not suffice for the presence of life.

[2] This sort of question, with its familiar and forceful reference to certain obvious improbabilities, is common on all levels of discourse. Its precise explication is, however, no easy matter. Any attempt at a careful analysis of it would take us into some technical and somewhat controversial topics of inductive logic, such as antecedent probabilities and Bayes' theorem, which cannot be considered here.

This construction is a predictive one but can be viewed as nonetheless explanatory in character. What it offers constitutes not merely a set of *reasons* for expecting to find life but also a set of conditions that go into supporting, in a causal sense, the presence and even the origination of life. That is we are in a familiar sense told *why* life may be present.[3] For simplicity, however, we may set aside the distinction between causes and symptoms (hence between an explanatory and non-explanatory account) and schematize the construction as follows :

(1) If anything is an S, it is a P_1
(2) If anything is an S, it is a P_2
(A) ...
(n) If anything is an S, it is a P_n
Not all things which have all the properties, P_1, $P_2 \ldots P_n$, are S's.
(B) Individual, a, has the properties $P_1, P_2, \ldots P_n$
———————————————— ↓ Probably
(C) Individual, a, is an S

It is to be noted at once that although the account has a set of nomic premises (1)-(n), it is not what we have called a *nomic inference* in that the singular premise (B), instantiates *not* the antecedent but the *consequent* clauses of the nomic premises. In this form, therefore, it is neither a deductive nor a probabilistic *nomic* inference. Moreover, if offered in the guise of a deductive inference the argument would, as it stands, be a glaring fallacy, the well-known fallacy of affirming the consequent. As a probabilistic argument, however, it seems to have some cogency, that is, the premises seem to lend a measure of support to the conclusion, and this support seems to increase as n increases. The purported inference is a familiar discursive device. And though it has come to be used as a standard example

[3] Our special concern in this particular dicussion is with explanations that appear to be non-nomic in form. But as is easily seen the case under consideration may be generalized to include non-explanatory inferences or arguments in which the basis for a conjecture consists not of causes but of indications or symptoms as, for example, in the following : A particle of moon rock on preliminary examination is found to have a dark color, a hardness of roughly 6.8, a specific gravity of about 3.8, high brittleness with rough cleavage and a disposition to dissolve in hot hydrochloric acid. It is concluded that, very probably, the moon particle is the mineral olivine. Also involved in the consideration is additional knowledge, e.g., that within a range of approximation, olivine has the color, hardness, specific gravity, brittleness, cleavage, and solubility possessed by the unknown particle but that not all particles with these traits are olivine. The properties mentioned in this example do not. properly speaking, *cause* the particle to be olivine. They are merely *indications* that it might be.

of fallacious deduction, it is an equally familiar and seemingly natural pattern for establishing probable conclusions. Even the case for $n = 1$ is not an uncommon mode of argument. Finding, for example, that a certain solid is a conductor of electricity and knowing that all metals are conductors, seems to arouse some expectation that the solid is a metal, even though we know that not all solid conductors are metals, e.g., graphite.

But for all their apparent naturalness purported inferences of this kind find no established analogue under a universally accepted system of inductive or probability logic. It would be gratuitous therefore to simply accept such a construction uncritically as a standard form of legitimate probabilistic inference, albeit not of the nomic variety. Reflection shows that its cogency and apparent naturalness stem from the impression that the nomic premises tend to place significant restrictions on the possible ways in which counter-instances can occur, that is, instances in which the individual in question has the observed properties and does not have the conjectured property. Thus it would appear that if an object were a good conductor it could be a non-metal only if it was graphite. On the other hand if it was not a conductor it could be a non-metal by being any one of a wide range of substances. It seems therefore to be far more likely that a solid will be a metal if it is a conductor than if it is *not*. As a matter of fact, there is no likelihood whatever of its being a metal if it is not a conductor.

But, as it stands by itself, this form of inference guarantees little or nothing about the number of ways in which counter-instances could occur. We would not, for example, be inclined to put much stock on the following argument :

(1) All men accelerate when subjected to a net force.
(2) All men are fair conductors of electricity.
(3) All men have an over-all specific gravity of about *one*.
(4) All men contain the element potassium.
(5) A certain individual, a, is known to have all the properties attributed to men in (1)-(4).

Therefore (probably) individual, a, is a man.

Though, formally, this argument is indistinguishable from the others it has virtually no cogency whatever. Obviously the reason is that despite the four well-established nomic universals in the premises the range of possibilities for counter-instances, that is, the range of counter-possibilities, is vast. An object can be, so to speak, *almost* anything other than a man and yet have the characteristics possessed by a. Clearly the premises must include information that will serve as a basis for regarding the range of counter-possibilities highly restricted. For example, to the premise that all metals are conductors we might add that the only other solid having fairly good conductivity is graphite. But even when the premises (A) are expanded in some such manner, any probability judgment regarding the conclusion would involve highly problematic assumptions about how to appraise the likelihood of alternative "ways" of having a conjectured property. At any rate, granting the debatable claim that such an inference can be made cogent in this way, it need not be regarded as irreducible to the nomic analogue. Our argument schema could be reconstructed in a manner similar to that employed for our other two cases. Represent it by offering the following probabilistic nomic inference:

(A′) A large proportion of things having $P_1, P_2 \ldots P_n$ are S's.

(B) Individual, a, has properties $P_1, \ldots, P_n$

——————————————————— ↓ Probably

(C) Individual, a, is an S

Asked about the assertability of (A′) we present as a basis the premises (A), appropriately expanded to include statements to the effect that there are very few "ways" in which an individual can have properties P_1, P_2, ... and P_n without being an S. The nature of the logical relation between (A) and (A′) is far from assured. Nevertheless, whatever cogency the inference from (A) to (A′) has provides cogency or naturalness for the argument. If the occurrence of P_1, P_2, ... P_n, together, constitute causal support for the occurrence of S, then the argument in point has explanatory import. What is more, the suggested reduction may be viewed as constituting an assimilation of the argument to the nomic model.

We turn now to a mode of discourse occurring in science, though with diminishing prevalence in more recent times. It has in the past been regarded as a pattern of explanation differing irreducibly from the nomic model. But the question of its irreducibility is not quite a settled one. We refer here to what, traditionally, has been called the *teleological* (more recently, the *functional*) mode of explanation – a matter of concern in our next section.

B. FUNCTIONAL (TELEOLOGICAL) ACCOUNTS

The history of thought offers many instances in which both philosophers and scientists have responded to the question : "Why S?" by indicating some function or goal, F, which S serves to maintain or achieve. This sort of answer to the question "Why?" has come to be called a *functional* or *teleological* explanation, and a common manner of expressing it is to say that S occurs, in order to bring about or maintain F. Thus, asked : Why does a given organism, a, have lungs? – a functional answer might be : In order for it to breathe air. Obviously, this is a different sort of answer from one which attempts to give explanatory causes as, for example, In saying that a has lungs *because* of genetically determined chemical states which are regularly antecedent to various stages of development of lung structure in all individuals of a certain class. Breathing in air, however, is no such antecedent. As a matter of fact the lungs must be there before breathing can take place, rather than the other way around. Breathing, therefore, cannot be the *cause* of lungs in any present sense of causation; nor, therefore, can it serve to explain lungs if, again, we construe explanation in the sense we have maintained, namely in the sense in which explanations are about what we have called *causes*.[4] Well then, perhaps our teleological account is nothing other than a disguised nomic inference of the deductive but non-explanatory sort. Such an inference would be covered not by the nomic principle that lungs in organisms are caused by air-breathing but merely by the nomic principle that lungs in organisms are necessary for air-breathing. Though, strictly speaking, not an explanation, such a nomic inference might well have genuine systemizing power despite the loose way in which natural discourse offers it as an answer to "Why?" Unveiled, the supposed nomic inference would then be :

(I) Any organism that breathes air has lungs.
a is an organism and breathes air

Therefore a has lungs.[5]

We note at once that the nomic premise in (I) is false. Insects, for example, breathe air but have no lungs. This puts in serious jeopardy any claim that the statement :

[4] According to some historical usage the goal or function F promoted by S would be called a "final cause" of S. Our own use of the term, 'cause', excludes any reference to such prospective goals or functions.

[5] Symbolically

$(x)(Ox\ Bx \rightarrow Lx)$
$Oa\ Ba$

La

Where : Ox = x is an organism
Bx = x breathes air
Lx = x has lungs

(1) The function of lungs in organisms is breathing air. (Organisms have lungs in order to breathe air)

is equivalent to :

(2) Whatever breathes air has lungs. (Lungs are necessary for breathing air).[6]

For (1) as ordinarily understood is true while (2) is manifestly false. Indeed an old maxim comes to mind here about cat-skinning. Apparently, there are alternative ways of discharging some functions although it may be said of each of them that it enables the function to occur.

It might be objected that the fault lies in our example and that ordinarily one would say *not* that the function of lungs in organisms is air-breathing but that :

(3) The function of lungs *in vertebrates under natural conditions* is air-breathing.

Its presumed equivalent :

(4) Any vertebrate which breathes air, under natural conditions, has lungs, would then be true and the equivalence of (3) and (4) would not be ruled out by disparate truth-values. Such an objection, however, would be unavailing. Statement (1) though not restricted in scope to vertebrates seems to be a perfectly good example of a functional statement and, moreover, is true. It seems to be erroneous, therefore, to maintain the desired equivalence without some further qualification.

At this point one might be inclined to believe that the conditions for discharging a given function must be made sufficiently specific in order for a particular structure to be necessary for that function. Thus it may be pointed out that, as a matter of fact, under certain specifiable conditions, no plants can synthesize food unless they have chlorophyll. And again under specifiable conditions no human can survive unless it has a very strong heart. The equivalency thesis then would be required to deny that any statement about functions, which did not give the "appropriate" specifications, was genuinely functional or else it would have to limit its applicability to only those cases of functional statement which had been "appropriately" specific. Neither of these alternatives seems very attractive for the equivalency thesis, and much more would have to be said about them if they are not to remain trivial or inapplicable. But even granting full feasibility to this dubious program, the equivalency thesis yields unacceptable results as pointed out by I. Scheffler with a simple example along the following lines.[7]

[6] A claim for such an equivalence is made by E. Nagel, *op. cit.*, pp. 403-404.

[7] Scheffler, I., *op. cit.*, pp. 50-51.

Consider the statement :

(5) Breathing organisms (eventually) die.[8]

On the equivalency thesis this statement has full logical parity with :

(6) Organisms die in order to breathe.

Statement (5) is true; but (6) is intolerable as an equivalent.

A reasonable alternative seems to be to regard functional or teleological accounts as *not* being inferences at all. If so then neither would they be explanations. Accordingly, to say that organisms have lungs in order to breathe air is no more than to indicate that lungs play a role in the air-breathing of organisms. This view has much to recommend it. In particular, it avoids the implication that lungs are necessary for breathing, that is, that they are the only means for breathing in organisms, and it avoids even the slightest suggestion that a function like breathing in some obscure way "causes" an antecedent structure like lungs.

A final point that occasionally comes up in connection with teleological accounts: Philosophers have sometimes claimed that there is an element of teleology in even the most mechanical of phenomena. This is so simply because, in a sense, some events often determine others which are their temporal antecedents. Or, putting it roughly, later events seem to determine earlier ones. So that certain changing states are drawn towards later states which therefore would seem to have efficacy even before they have occurred. A simple example will illustrate this claim :

A stone is projected vertically upward from the roof of a building and is observed to have fallen 100 feet below its starting point after 5.0 seconds. On the basis of this and the law that, neglecting air-resistance, all bodies near the earth fall with a constant acceleration, we can determine that 1.0 second after being thrown, the stone must have risen 44 feet above its starting point.[9]

[8] Symbolized :

$(x)\ (Ox\ Bx \rightarrow Dx)$

Where : Ox = x is an organism

Bx = x breathes

Dx = x dies.

[9] The law of constant downward acceleration for falling bodies together with the definitions of all the kinematical quantities involved logically imply that the vertical displacement, H, from the initial point of projection be given by :

$$(1)\quad h = v_0 t + \tfrac{1}{2} g t^2$$

where v_0 is the initial vertical speed and g the *constant* acceleration of −32 feet per second per second.

Inserting in equation (1) the given information that $h = -100$ feet at $t = 5.0$ seconds we get that $v_0 = 60$ feet per second.

Undeniably information, E_2, about a certain event, e_2, made it possible, on the basis of a simple law, L, to calculate quantities, E_1, about an earlier event, e_1. But to say therefore that e_2 determined (caused, produced) its antecedent, e_1, would be grossly misleading and literally speaking wrong according to any familiar sense of the term 'cause'. By the same token, the inference from E_2 and L to E_1 does not *explain* E_1. What it does, instead, is to give systematic (in this case deductive) support or *substantiation* to E_1. Confusing *substantiation* with *explanation* in cases such as this may well create the aura of causal efficacy of future events on present ones, and consequently of an underlying teleological nexus in which the present occurs in order to serve future states which in turn shape it. Following such a viewpoint some might go on to say that the stone in our example rose 44 feet by the end of 1.0 second so that it could fall to 100 below its starting point after 5.0 seconds. Picturesque though this way of putting it may be, it is difficult to see what more it claims beyond simply saying that if the stone fell to –100 feet after 5 seconds then, given the constant acceleration of gravity, it had to have risen 44 feet after 1.0 second. At best, it seems, the teleological idiom may serve to direct our attention to the instant, $t_2 = 5.0$ seconds, at which the stone has fallen to –100 feet and to the fact that under the given circumstances rising to 44 feet at $t_1 = 1.0$ second was a necessary antecedent. Any further nuance singling out what happened at t_2 as a *causal basis* or *explanation* for what happened at t_1 is indefensible and, as has been already suggested, tantamount to a blurring of the notion of substantiating with that of explaining.[10]

Now, applying (1) again for t = 1.0 second we get :

$$h = (60\text{-}16) \text{ feet}$$
$$= 44 \text{ feet}$$

The non-negative value of h at t = 1 is interpreted to mean that the stone is above its starting point.

[10] Some principles of classical optics, mechanics and thermodynamics are sometimes expressed as general laws asserting that a certain quantity is an extreme value (minimum or maximum) during a physical process or even that some quantity tends always to increase or decrease throughout some process.

Examples are :

(1) *Fermat's Principle* which says that light is propagated between any two points, P_1 and P_2, in such a way that the *optical* path (length of the path multiplied by the index of refraction of the medium) is a minimum, as compared to all other geometrically possible, neighboring paths. Mathematically :

$$\int_{P_1}^{P_2} n\, ds = \text{minimum}$$

Where : n is the index of refraction of the medium (a quantity which may vary) and

C. DERIVATIONS : NOMIC INFERENCES WITH NOMIC CONCLUSIONS

1. Scientific knowledge is to some extent expressed by nomic inferences which show particular events to be instances of general laws. This, however, is only a part, not even a major part, of the total account which science gives us. We may give, for example, a simple or "low-level", nomic account of how it is that a particular object, O, is floating in liquid, L, by offering a nomic inference of roughly the following sort :

Any object will float in a liquid of higher density.
Object, O, is in a liquid, L, of higher density.
So, O is floating in L.

But a deeper account of the phenomenon could be given by telling *why* any object will float in a liquid of higher density – which could be done by deducing this law from more general hydrostatic principles involving depth, pressure and the resulting forces on the floating body.[11] The account

---

ds is an element of geometric length.

(2) *Hamilton's Principle of Least Action* which says that the actual motion of any system as compared to other geometrically possible ones is such that the *time* integral of the so-called "action function" is an extremum (in general, a minimum). The action function is the difference between the kinetic and potential energies of the system and is known as the Lagrangian of the system. According to a basic principle in the calculus of variations, this means that variations in this time integral should vanish for the actual path in point. And this is mathematically expressed as :

$$\delta \int_{t_1}^{t_2} (T\text{-}V)\, dt = o$$

where : T = the kinetic energy of the system
V = the potential energy of the system

(3) *The Second Law of Thermodynamics,* namely that all changes in an isolated system are such as to increase the entropy of the system and that, therefore, the system will change until the entropy has reached its maximum value. Mathematically the entropy is expressed as : $\int dQ/T$
where Q is the heat of the system and T its absolute temperature.

It is easily seen how such formulations could be given in a teleological mode with the physical processes depicted as goal oriented, that is, as drawn towards the maximization or minimization of certain quantities or towards certain states.

But once again, it seems reasonable to regard such formulations as matters of idiomatic preference rather than as expressing any special causal efficacy of "final" physical states on temporally antecedent processes. As a matter of fact, there is nothing particularly significant about the extremum mode of formulation. Since for any set of related quantities, it is in general possible to make a selection from among them or to define a function of some of them which will maintain an extremum value.

[11] An argument showing the underlying nomic form in full detail and stating all premises would be extremely tedious to develop. In the more quantitative sciences such

could then be deepened still more by going on to explaining hydrostatic pressure-depth relations in terms of the still broader principles of kinetic molecular theory. Each such extension of the account introduces new nomic inferences which, *unlike* the first, have, as their conclusions, laws rather than the description of a particular event. Nomic inferences which account for nomic regularities rather than particular events make up a major part of any modern and theoretically developed science. Such inferences are typically deductive in character and are usually referred to as *derivations*. At the most systematic levels of scientific expositions derivations may occur as theorems of *axiomatic* systems some of whose characteristics we shall return to in a later section.

We note, then, that nomic inferences with singular conclusions are relatively low level systemizations governed usually by nomic statements reporting natural regularities that are observed in a more or less direct manner. Whatever central interest basic science has in subsuming particular events under laws – be this by explanatory, non-explanatory, predictive, postdictive or any other kind of nomic inference – this interest stems largely from the concern for testing in order to establish or disestablish the laws. Apart from such testing, any active emphasis on nomic inferences with singular conclusion, especially predictions, is largely the preoccupation of the applied sciences, technology and common sense. Basic scientific exposition, on the other hand, is mainly concerned with the logico-mathematical elaboration of theory; and this means largely derivations or, as one might also say, the subsumptive systemization of laws *via* nomic inferences with nomic conclusions.

arguments are developed as mathematical computations rather than as explicitly deductive inferences, as for example in the following :

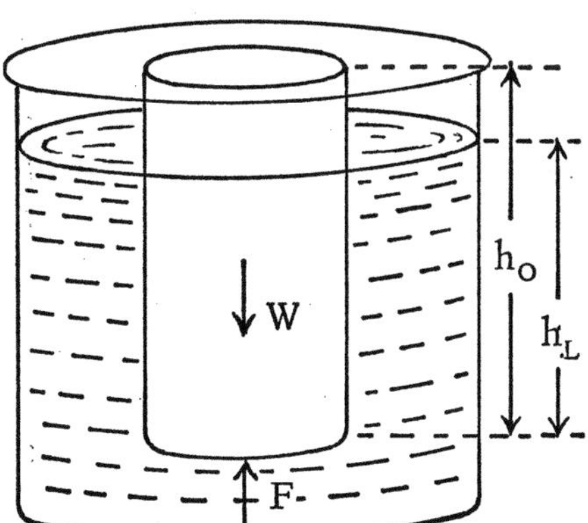

For simplicity assume a cylindrical homogeneous body, O, of density, D_O, and cross-sectional area, A, floating vertically in homogeneous liquid, L, of density, D_L. O will float when and only when the hydrostatic force, F, is equal to the body's weight, W. Thus, for O afloat we have : F = W. And if g is the acceleration of gravity then,

Density × volume × g = Pressure × Area × g. That is :

$$D_O A h_O g = D_L h_L A g$$

$$D_O h_O = D_L h_L \text{ or } \frac{D_O}{D_L} = \frac{h_L}{h_O}$$

It is clear that if $D_O \leqslant D_L$ then $h_L \leqslant h_O$

2. As in the case of nomic inferences with singular conclusions one might attempt to distinguish explanatory from non-explanatory nomic inferences with nomic conclusions. For example, asked *why* the pressure multiplied by the volume and divided by the absolute temperature for a fixed quantity of gas is a constant ratio, one could present the kinetic-molecular theory of gases and derive the result, with the help of appropriate assumptions. Or, asked *why* the displacement and acceleration for the bob of a simple pendulum are directly proportional to each other one might derive the relation dynamically by introducing the gravitational field and equations of motion. Few would deny these to be explanations, impulsive molecules and gravitational forces constituting the causal factors. In many cases, however, there seems to be no clear way of deciding whether a nomic inference with nomic conclusion is explanatory or not. Where inferences refer back to space-time systems (or models) governed by dynamical factors like masses, collisions, electric charges, forces, chemical or biological agency, etc. we readily speak of cause-effect sequences and explanation. But where the covering laws are mathematical relations between concurrent quantities, with no associated system of geometrically intuitable space-time "entities" and forces, one seriously demurs at applying terms like cause and explanation even though dynamical quantities like momentum, position, electric and magnetic field vectors, etc. may be associated with some of the variables. In quantum physics, for example, by means of complex nomic inferences the physically possible energy states of an atomic electron may be calculated. On one version of quantum theory this is done by setting up a highly abstract type of nomic premise known as the Schrödinger wave equation. Into the skeleton of the equation are inserted certain parameters corresponding to the classical energies of the system. This is an abstractly contrived procedure stemming largely from a structural analogy with formulations in classical mechanics. The desired results are obtained by solving the equation under certain appropriate restrictions. The wave solutions, in turn, imply the desired energy levels.[12] Similarly by starting with Maxwell's equations for the electromagnetic field as nomic premises and following the appropriate computational techniques, one can derive, among other things, the velocity constant of electromagnetic waves in *vacuo* (a nomic conclusion). It is difficult to see how, in both this and the case of quantum computations, one can single out causes and speak of the highly algorithmic computations as explanations. Indeed, as in the

[12] On another version of quantum theory – the so-called Heisenberg method – a mathematical matrix (a 2-dimensional array of quantities) is used for the computation. Into the matrix are inserted optical quantities (i.e., spectral intensities and frequencies) and then the computation proceeds by a set of stipulated rules.

case of nomic inferences with singular conclusions, there is a wide and important range of derivations from theory which are not readily classed as explanations in the cause-effect sense to which we have anchored this latter notion. The dubious centrality of explanation in mature science is particularly evident in the elaboration of much contemporary physical theory – be this in the area of classical, relativistic or quantum physics. Moreover, with the possible introduction of increasingly abstract and algorithmic, mathematical theory ("models") in the social sciences, the departure of scientific, subsumptive inference from the explanatory format may become far more explicitly underlined throughout theoretical science.

3. Virtually all systematically significant derivations seem to be of the deductive form. Possible exception of the following sort may come to mind :

Nearly all (kinds of) reptiles lay eggs.
Dinosaurs were reptiles.

———————————————— ↓ Probably

Dinosaurs laid eggs.

Another illustration of such exceptions may be taken from astronomy.

On the basis of certain details of location composition and luminosity, stars are generally classed under two groups, Population I and Population II. Consider, then, the following nomic inference :

For Population I stars of nearly *all* types, those having (about) the same mass have (about) the same luminosity.

Red giants having a certain composition, M, are all Population I stars.

———————————————— ↓ Probably

All red giants having composition, M, and having (about) the same mass also have about the same luminosity.[13]

But probabilistic inferences of this sort may be considered to be rather simple instantiations of some observational law and therefore rather negligible from the viewpoint of scientific systemization. We shall therefore regard them as more or less negligible exceptions to the deductive character of derivations.

Familiar examples of actual scientific derivations stating all premises and explicitly showing the full deductive structure would be extremely tedious to give. In the more highly developed sciences where derivations are what provide a pervasive webbing of systematic structure, subsumptive inferences

[13] The first premise of this inference is known as the *mass-luminosity* law.

The term 'luminosity' refers to the *actual* or *intrinsic* brightness of a star rather than to any *apparent* brightness as observed from earth. For a lucid introductory discussion see Baker, R. H., *Astronomy*, D. Van Nostrand : Princeton, 8th edition, 1964, pp. 415-416, 456-458.

are conveniently worked out as mathematical computations rather than as explicitly syllogistic or logistic deductions. An example would be a derivation of the laws governing vibrations of a mass on an elastic spring from Newton's second law of motion and a law about how the force due to such a spring varies with its displacement.[14] Another example would be a derivation of Kepler's law of equal areas from Newton's second law of motion and the law that the force of the sun on the planets is a central force.[15] Another example would be the derivation of various laws governing electromagnetic wave propagation from Maxwell's equations together with certain assumptions regarding the nature of the supporting medium (vacuo or other dielectric).[16] Explicit mathematical accounts of any such derivations would not be particularly profitable here. These can be found in standard physics texts.

4. Even the most general acquaintance with the structured results of scientific inquiry, quite apart from technical details, easily points up those aspects of derivations which tend to enhance their systemizing power. The following two features are notable :

(1) Several distinct universal laws appear in the premises. By virtue of this feature the regularity expressed by a law is shown to be the result of

[14] These two laws may be expressed as follows : Letting F = force, x = displacement, m = mass and k = a constant for the spring

(1) $F = -kx$

(2) $F = m\frac{d^2x}{dt^2}$

Together these yield the second order differential equation : $m\frac{d^2x}{dt^2} + kx = O$

This is known as the differential equation for undamped simple harmonic motion; and it yields as a solution, a description of the motion of the vibrating mass, m.

[15] A central force is one which is always directed towards a fixed point, O, and whose magnitude may be any arbitrary function of distance, $\vec{r}$ (a vector), from O. We then "combine" the law of central force on planets (LCF) with Newton's Second Law (NSL) as follows :

(1) $\vec{F} = m\frac{d^2\vec{r}}{dt^2}$ (NSL)

where m is the mass of the planet and the arrows indicate that the quantities are vectors.

Now take the vector product of both sides of (1). Since by LCF, $\vec{r}$ and $\vec{F}$ are in the same direction, $\vec{r} \times \vec{F} = O$ and (1) becomes $\vec{F} \times \frac{d^2\vec{r}}{dt^2} = O$

which may be integrated to yield $r\frac{d\vec{r}}{dt} = \text{constant}$

But the left hand side of this equation is twice the area swept out by the radius vector, $\vec{r}$, per unit time. Thus the planets sweep out equal areas in equal times.

[16] Maxwell's equations are four first order partial differential equations each of which may be interpreted as an abstractly represented universal law of the electromagnetic field. It would not be profitable here to inquire into the mathematical details of any derivation from these laws.

a set of separate factors interconnected by chains or networks of laws. The following is a simple example expressed in such qualitative terms as to be unlikely to occur in any *actual* scientific account. It may, however, serve to illustrate the interconnected character of premises in a derivation. Let us consider the following :

Assume that by 'metals' we mean substances having certain obvious characteristics like lustre, malleability, sonorous ring, etc.[17] Asked, then, why metals are good electrical conductors, one might answer with the following inference :

(a) All metals (M) have atomic structures characterized by a few outer electrons separated from the atomic nucleus by some inner ones (P). And they are either solids or liquids (R).

(b) All substances having the above atomic structures (P) also have free electrons (Q).

(c) All substances that are either solids or liquids (R) consist of closely packed atoms (S).

(d) All substances having free electrons (Q) consisting of closely packed atoms (R) are good electrical conductors (C).

(e) All metals (M) are good electrical conductors (C).

More schematically this inference could be represented as follows :

(a′) All M are (P and R)

(b′) All P are Q

(c′) All R are S

(d′) All (Q and S) are C

(e′) All M are C [18]

[17] That is, the definition takes no account of chemical, electrical and atomic properties, but is given in terms merely of what we might call common sense properties, largely familiar to the ordinary man.

[18] Symbolically this inference would go as follows :

1. $(x)(Mx \rightarrow Px\ Rx)$	
2. $(x)(Px \rightarrow Qx)$	
3. $(x)(Rx \rightarrow Sx)$	
4. $(x)(Qx\ Sx \rightarrow Ox)\ /\ \therefore\ (x)(Mx \rightarrow Cx)$	
5. $My \rightarrow Py\ Ry$	1, Universal Instantiation (U.I.)
6. $Py \rightarrow Qy$	2, U.I.
7. $Ry \rightarrow Sy$	3, U.I.
8. $Qy\ Sy \rightarrow Cy$	4, U.I.
9. $\sim My \vee Py\ Ry$	5, Implication
10. $(\sim My \vee Py)(\sim My \vee Ry)$	9, Distribution
11. $My \rightarrow Py$	10, Simplification and Implication
12. $My \rightarrow Ry$	10, Simplification and Implication

It seems clear that for the enhancement of systemization derivational inference must have more than one nomic premise. The logical incorporation of any particular law, L, into some body of knowledge by means of subsumptive inference becomes undoubtedly more explicit when the relations expressed by L are shown to be derivable from *several* independent factors related by the *several* interconnected premises of the inference.

(2) The importance of generality in the premises of a derivation is generally recognized by students of the subject. That the premises of a derivation must be more general than the conclusion is, of course, a truism, given that derivations are subsumptive inferences. For we have cited, as a rough characterization of subsumptive inference, the greater generality of premises over conclusion. More notable than this, though still rather obvious, is the fact that high generality in the premises of derivations greatly enhances the comprehensiveness of the resulting inferential systems. But though, in actual practice, scientific laws seem to be readily compared for their generality, a precise explication of what it is for one law to be more general than another is a matter of some complexity, and it need not detain us here. In this regard, however, note the following :

(a) In speaking of the generality of the premises, we have in mind the generality of the most general statement in the premises.

(b) The sense in which a law, L_1, is said to be more general than a law, L_2 is the sense in which one would say that laws about masses (e.g., the law of gravitation) are more general than laws about living things (e.g., the laws of heredity). In a rough manner of speaking one might want to say that L_1 is more general than L_2 if and only if L_1 is more inclusive than L_2. The greater inclusiveness of L_1 is due to its descriptive terms which include fewer specifications and hence apply to a less specialized subject-matter. L_1, as it were, "talks" about a wider class of things than does L_2. Thus, the law of gravitation which is about masses (in any form) is more inclusive, hence more general, than the law that every organism has at least one parent which is about *living* masses. The first law holds for all entities which the second law holds for, but not conversely. The second law simply does not apply to non-living entities and therefore would not ordinarily be said to "hold" for such entities.[19]

13. $My \rightarrow Qy$	6, 11, Hypothetical Syllogism
14. $My \rightarrow Sy$	7, 12, Hypothetical Syllogism
15. $(My \rightarrow Qy)\,(My \rightarrow Sy)$	13, 14, Conjuction
16. $My \rightarrow Qy\,Sy$	15, Implication, Distribution
17. $My \rightarrow Cy$	8, 16, Hypothetical Syllogism
18. $(x)\,(Mx \rightarrow Cx)$	17, Universal Generalization

[19] More precisely we should say that the biological law holds non-vacuously for organisms but vacuously for non-organisms.

It may be objected, however, that there are cases which this manner of construing generality does not seem to help decide. How, for example, does one compare, for generality, the laws :

'Humans are omnivorous' and 'War-time economies are inflationary'. There is an element of incomparability between the descriptive terms or *predicates* of the respective laws. In the first place the terms, 'human', and, 'economy', cannot be compared for relative specificity as, can for example, the terms, 'mass', and, 'living thing', the second of which connotes all that the first does, but more besides. As a matter of fact, the incomparability between 'human' and 'economy' is far worse than this. There seems to be no one set of individuals to each of whose members both terms are meaningfully applicable. If an entity is a human it would seem to be sheer nonsense to say it is also an economy. This difference between the two terms has even further consequences. It seems to cut away any possibility that there be a set or domain of individuals within which ranges of meaningful application for the two laws can be compared. Thus a precise explication of what it is for one law to be more general than another must, it seems be given in terms of : (1) conditions on the predicates occurring in each law, more particularly, on how the predicates of one law relate to those of the other and (2) requirements on some set, S, of entities all of which can be meaningfully characterized (though perhaps falsely) in terms of the descriptive content of both laws. The requirements in (2) must be designed to make possible a comparison of the "ranges" or extent of applicability (in S) of the descriptive language in each law, and the extent to which each law holds in D. We shall, however, not concern ourselves with such an explication, leaving the notion of comparative generality more or less "intuitively" understood on the basis of examples and our few reflections on the matter.[20]

The law that every organism has at least one parent may be formulated as follows :

$$(x) \{Ox \rightarrow (Ey)\, Pyx\}$$

where Ox = x is an organism

Pyx = y is the parent of x

This statement is a universal, material implication which, if instantiated with a false antecedent, will yield a truth by virtue of the definition of material implication. Thus, in this sense, the law in point "holds" though, in this trivial sense, for all entities which do not have the traits whose relations the law formulates.

[20] For a detailed explication along these lines see : E. Nagel, *op. cit.*, pp. 38-40.

In this connection the following point deserves some attention. A highly specific explication of what it might be for one law (L_1) to be more general than another (L_2) may, by itself, be a matter of genuine philosophical interest. The extent of explicitness needed for present purposes, however, is not at all clear. It must be remembered that

A set of premises may be so encompassing as to be capable of generating more than one or, as we might say, a system of derivations. The most comprehensive derivational systems are those stemming from theories. And the most systematic manner of elaborating such systems seems to be the *axiomatic* or *formal* mode of exposition. The structure of theories and of their formal elaboration will be taken up in a later section.[21]

D. PROBABILISTIC NOMIC INFERENCE

1. At the outset of our discussion of nomic inferences we distinguished two major kinds, *deductive* and *probabilistic*, both subsumptive in the sense that they "logically encompass" given statements "under" more general ones, what is described by the given statements falling under what is described by the more general ones as "special cases". Unlike their deductive counter-parts, however, probabilistic nomic inferences have conclusions which are not logically (i.e., deductively) guaranteed by the premises. That is to say, though the premises support the conclusion, it may nevertheless

---

all laws to be compared for generality in the context of our concern are logically related in such a manner that L_1 is a premise (or partial premise) for inferring L_2 either *deductively* or *probabilistically*. And it seems reasonable to suppose that the condition of inferrability of L_2 would rule out at least some of the incomparable cases. One might therefore want to show that a less specific sense of 'L_1 is more general than L_2' would suffice for implementing the requirement of generality in derivations, it being, of course, the case that L_2 is the conclusion of an inference whose premises contain L_1.

[21] In scientific treatises the term 'derivation' is often used for any deductive inference whatever as long as it has a nomic conclusion. Thus transformations in which premises and conclusion are logically equivalent to each other are sometimes referred to as derivations. For example, based on a simple transformation of Newton's second law of motion to an equivalent form, due to D'Alembert, the mathematical physicist Lagrange developed, for a wide class of physical systems, a highly convenient, but nonetheless equivalent form of Newton's law known as the Lagrangian equations. A notable text speaks of "the derivation" of these equations. See : H. Goldstein, *Classical Mechanics*, Massachusetts, 1959, p. 30. Our own use of the term 'derivation', however, will be restricted to subsumptive inferences for which it is required that the premises be more general than the conclusion. On an adequate explication of generality, it would seem that logically equivalent statements have equal generality. If so then equivalence transformations would be ruled out as non-subsumptive and therefore as non-derivations. At any rate, whatever the implications of the generality requirement, equivalence transformations in which the premises and conclusions logically imply each will not here be regarded as derivations. Scientific history offers many instances of valuable equivalence transformations yielding fresh theoretical and practical insights. Such mathematical innovations, however, by themselves, do not incorporate particular knowledge under broader knowledge and must therefore be distinguished from what we have called derivations, which do.

be the case that they are true and the conclusion false without contradiction. For such inferences, it may be said that the premises support the conclusion inductively. Examples P_1-P_4 in the chapter before last, illustrate probabilistic nomic inferences and also show that as in the case of their deductive counter-parts they can be either of the explanatory or non-explanatory sort depending on whether or not their covering laws are of the cause-effect type.

Probabilistic nomic inferences play a distinctive and important role in the systemization of knowledge both in the physical and the social sciences. Their scientific beginnings seem to go back to 18th century writings in statistical theory, stimulated in part by inquiries into betting odds based on relative frequencies, but concerned also with theory of errors in measurement and most notably with the statistics of micro-populations as postulated in statistical mechanics.[22] This new and important physical science sought to account for certain observed aspects of phenomena in terms of statistical laws governing large numbers of objects or *populations*. There is a definite sense in which the statistical models underlying these accounts are causal and therefore explain though this may not be precisely the same sense in which deductive inferences explain. Indeed it seems entirely consonant with both scientific and everyday discourse to allow for both *probabilistic* and *deterministic* senses of *explanation*. In contemporary physics, the statistics of micro-populations (i.e., swarms of molecules, atoms, electrons) provide major theoretical links between essentially mechanical models and thermal, solid state, or even electronic phenomena. So that – in a distinctive and important scientific sense – observed instances of temperature exchange, diffusion of gases, motion of colloidal particles, evaporation of liquids, flow of electric current, and very much more are (probabilistically) explained by the laws of statistical mechanics. In the behavioral or social sciences this sort of explanation and, more generally, of nomic inference has become, by far, the pervasive mode of systemization. Faced with a vastly complex subject-matter that resists fundamental description in terms of universal laws, the sciences of social phenomena have developed elaborate methods for both formulating and testing

[22] Among the earliest writers we note T. Simpson who as far back as 1775 tried to show the advantages arising by taking the arithmetic mean of a number of determinations in observational astronomy. Others who followed were D. Bernoulli, R. Adrian, A. Cournot, Gauss, A. DeMorgan, and A. Quetelet. Particularly notable among these is D. Bernoulli who contributed not only to theoretical statistics but also introduced the kinetic molecular theory of gases, later to be generalized as statistical mechanics.

For an excellent historical bibliography of early statistical writings, see : J. M. Keynes, *A Treatise on Probability*, London, 1929, pp. 431-458.

sophisticated statistical hypotheses. Indeed, it seems hardly an exaggeration to say that such hypotheses and the probabilistic inferences which they cover constitute the dominant aspect of whatever explanatory or other nomic structure is at present possible in these sciences.

2. Any attempt to give scientifically important and interesting examples of probabilistic inference taken either from the physical or the social sciences tends to bog down with unwieldy complexity and detail. We shall confine ourselves, therefore, to simple qualitative illustrations with merely *supposed* premises, which nonetheless reveal the structure of probabilistic nomic inferences without loss of generality. Thus, suppose it the case that a very large proportion of humans carrying an extra Y chromosome exhibit abnormally aggressive traits, and a geneticist is asked why Jones exhibits abnormally aggressive behavior. He could reply that it is because of having an extra Y chromosome. More explicitly structured, this is a probabilistic explanation of the following form :

Nearly all those carrying an extra Y chromosome are abnormally aggressive
Jones carries an extra Y chromosome
———————————————————— ↓ probably
Jones is abnormally aggressive

A non-explanatory sort of probabilistic nomic inference would be one such as the following :

A large proportion of those who have blue eyes and only blue-eyed siblings have a blue-eyed parent
Smith is such a person
———————————————————— ↓ probably
Smith has a blue-eyed parent

As illustrated by the examples, probabilistic inferences have statistical premises though not all arguments with statistical premises are probabilistic. The following, for example, is a *deductive* inference with a statistical premise :

The statistical probability that an American will die of a car accident during the 60th year of his life is 0.005
The statistical probability that an American has blue eyes is 0.1
Accident fatality is statistically independent of eye-color
————————————————————
The statistical probability of a blue-eyed American dying of a car accident is 0.0005.

Assuming that the term, 'statistical probability' is understood in one of the several allowable senses as, for example, the limit of a relative frequency of random cases, then it can be shown that the premises of the above argument deductively necessitate the conclusion.[23] Our concern, however, is not with statistical *deductive* inferences but with *probabilistic* ones whose premises give only *inductive* support to the conclusion. It may be noted, for our example, that not only are the premises statistical, but so is the conclusion. This should bring to mind that, as a point of logic, deduction alone cannot yield a non-statistical conclusion, say, about a particular event, from premises that are totally devoid of non-statistical content. Purely statistical statements are about classes of events such as populations or sequences, and they make no reference to any individual member. Hence deductive elaboration of such statements could never yield specific conclusions about particular members of the aggregate concerned. Our example illustrates the point. Its premises are purely statistical, i.e., lack non-statistical content. Hence, as might well be expected, the conclusion is also purely statistical. In probabilistic nomic inference, however, the conclusion may well be a singular, hence non-statistical one. Indeed this is the only sort of probabilistic nomic inference we shall be concerned with. Our considerations regarding the conclusion of a *deductive* inference with purely statistical premises do not, of course, apply to probabilistic nomic inferences, which, as we know, are *not* deductive in character.

The following schema suggests itself as a general representation of probabilistic nomic inferences :

(L) P(A,B) is high
(C) Aa
———————— ↓ probably
(E) Ba

In the above schema, the statistical premise, L, is conveniently represented

[23] This is merely a special case of a general theorem of the probability calculus about the joint probability of statistically independent events whose repective probabilities are known. The valid applicability of this theorem depends on the fact that the limit-frequency interpretation of the probability concept satisfies the axioms of the abstract probability calculus; that is, it is an admissible interpretation of the calculus. For a concise account of the abstract calculus see : A. N. Kolmogorov, *Foundations of the Theory of Probability*, New York, 1950, pp. 2-12.

Probability as a frequency limit for infinite sequences is an idealized theoretical notion which requires further qualifications for its application to actual (hence finite) sequences. But these considerations need not concern us here. For an account of such "practical" considerations see : H. Reichenbach, *op. cit.*, pp. 447-448.

For a concise account of one version of the limit-frequency conception of probability see : R. von Mises, *Probability Statistics and Truth*, Trans. J. Neyman, C. Sholl, and E. Rabinovitsch, New York, 1939.

as a statement of statistical *probability*. The advantage in this is that such probability makes explicit reference to the statistical frequency underlying L. This frequency is what provides the inductive basis for conclusion, E. Thus : –

P(A,B) = the statistical probability of a B given an A
Aa = a is an A
Ba = a is a B

As in the case of deductive nomic inference, L is a nomic or law-like statement but, in this case, always statistical in character. C is a statement of initial conditions, c; E a conclusion about a particular event, e. Deciding the explanatory or non-explanatory character of the inference rests on considerations quite parallel to those which apply to deductive inferences. The conditions, c, are or are not causes of e depending on whether L is or is not a nomic statement of the cause-effect type. In case conditions, c, are causes of e the inference is explanatory. If conditions, c, are indications or symptoms of e, the inference is clearly non-explanatory. Other more or less debatable possibilities may present themselves. L may be a statistical correlation with no aspects of asymmetry to suggest any cause-effect sequentiality between the properties (events, states) concerned. In such a case, there seems to be little point to speaking of any one as cause of the other, though both might clearly be effects of some common cause, construing the latter in the sense of an asymmetric relation. An example of this might be, say, a correlation between mathematical and musical ability, regarding which we might be willing to say not that one causes the other but that both are *effects* of some more "basic" disposition. Or, L may relate two quantities, A and B, which are co-temporal but are nevertheless not symmetric with respect to theoretic fundamentality, in which case we might want to say that one does cause the other. An example might be a certain kind of psychotic state and suicidal disposition.

It is also to be noted that the relation between the premises, T, (i.e., L and C) and the conclusion, E, is inductive as indicated by the vertical arrow to the right of the schema. The nature of such a relation, as we have already indicated, is still a matter of unsettled opinion. But granting that there is an important sense in which evidence supports hypotheses, it seems quite reasonable to grant also that such "support" admits of some quantitative characterization.[24] If so, we can then say that probabilistic nomic inferences are arguments which subsume particular facts under

[24] Not all students of inductive logic would agree that numerical metrics can be assigned for expressing the degree to which evidence supports or *confirms* an hypothesis.

statistical laws, by virtue of the (more or less) *strong* likelihood conferred by the premises T on the conclusion E. What the probabilistic mode of subsumption assumes is that the numerical value of the statistical probability asserted in L is a measure of the extent of inductive support given by T to E. This, however, is an assumption about how to construe the probability concepts involved. Or, putting it in other terms, it constitutes an assumed connection between statistical probabilities and inductive probabilities which must be explicitly "built into" the assumed interpretation of the probability concept. That this may be done with justification is a debatable question in the philosophy of probability, and one that is beyond the scope of our concern here. For our purposes, here, it seems sufficient to rest with the view that it is indeed possible and, moreover, seems quite *reasonable* to construe the notion of inductive support in such a way that if in our schema, P(A,B) = p, then T supports E to the extent, p.[25] This, then, would obviously guarantee that, if P(A,B) is very high, the inductive probability of E is also very high, or, as we might also say, e is practically certain to occur, given T. Our schema for probabilistic nomic inference may then be generalized to :

$$\begin{array}{l} P(A,B) = p \\ Aa \\ \overline{\qquad\qquad} \downarrow p \\ Ba \end{array}$$

But construed as systemizing arguments, probabilistic nomic inferences serve scientific discourse by inductively subsuming particular instances under statistical laws, as special cases of such laws. It is clear, then, that the higher the statistical probability asserted in L, the more logically subsumptive is the inference. The logical extreme of this is the case in which L is a universal (rather than statistical) law. In such a case the inference achieves maximum subsumption, or, as one might also say, maximum systemization. In this connection, the notion of *cogency* suggests itself. We may say that the closer to 1 p is, the more cogent is the probabilistic nomic inference. Where the inference is explanatory the feature of cogency is particularly interesting. In such a case the explanatory import of the inference increases with the probability expressed in its statistical premise, that is, as p approaches 1. For highly cogent inferences our schema may be written :

[25] R. Carnap expresses this "reasonability" by regarding the statistical probability as a fair betting quotient on the individual case. See : R. Carnap, *Logical Foundations of Probability*, University of Chicago Press : Chicago, 1950, p. 168.

$$\begin{array}{l} P(A,B) = p \approx 1 \\ Aa \\ \hline Ba \end{array} \quad \Downarrow p \sim 1$$

(Where '$p \approx 1$' is to be read : "p is very nearly equal to one")
In the case that $p = 1$, a probabilistic nomic inference is *maximally cogent*; where $p = \frac{1}{2}$ it is non-cogent; where $p = 0$, the inference is maximally cogent for the conclusion's *denial*, i.e., for $\sim E$.

3. Probabilistic nomic inferences present an aspect of indeterminacy which they seem to inherit from the statistical probabilities to which they are so intimately tied. Suppose, for example, the statistical probability of having a college degree for an American, about whom nothing more is known, to be 0.05. Write this :

$$P(A,C) = 0.05$$

In this formulation, A (being an American) is called the *reference* class; C (having a college degree) is called the *conjectural* class.[26] On such a probability, fair betting odds that J. Doe, a randomly selected American, has a college degree would be 1:19. But suppose now that J. Doe is a male and that for an American male the probability in point is 0.1. Fair odds that J. Doe has a college degree would then be 1:9. The new probability statement may be written :

$$P(A \cdot M, C) = 0.1$$

The reference class is now $A \cdot M$ (i.e., the class of American males). We may modify the reference class further by conjoining additional qualifications such as, being 25 years old (Y), being a suburbanite (S) and having a family income of over $25,000 per annum (I). Suppose these probabilities to be as follows :

(1) $P(A \cdot M \cdot Y, C) = 0.2$
(2) $P(A \cdot M \cdot Y \cdot S, C) = 0.6$
(3) $P(A \cdot M \cdot Y \cdot S \cdot I, C) = 0.95$

On the basis of these probabilities, the fair betting odds that Doe has a college degree are respectively :

(1) 1:4 (2) 3:2 (3) 19:1

assuming that he is found, successively, to belong to each of the reference

[26] Strictly speaking in the present context, A and C are attributes, but we may equivalently regard any of them as determining a class, i.e., the class of all individuals having the attribute.

classes. These illustrations make it obvious that in statistical probabilities there is a relativity to information, as contained in the reference class, which seriously affects betting quotients, or the extent to which one may reasonably bank on any prediction – such as, that Doe, in our example, has a college degree. Indeed, as the reference class was narrowed down with more and more qualifications the odds on Doe's education changed from 1:19 to 19:1, thus reversing themselves completely. Yet what we are predicting or betting on is one and the same event whose likelihood of occurrence, it would seem, should depend only on the objective circumstances concerning the event in question. Well, does this invalidate statistical probabilities as a basis for betting or predicting? The reasonable answer seems to be "no", provided that we rely on all available, relevant evidence in computing our probabilities. In general, some qualifications on our reference class will change the probability as in our illustrations. Other qualifications, however, may have no effect on the probability and they would therefore correspond to irrelevant evidence. Thus color blindness or the month of birth would (very likely) have no effect on the probability of having a college degree. In principle we might speak of a reference class that includes *all* relevant qualifications and of which the individual in question is a member. The resulting statistical probability might then be said to be the *statistical probability on maximal relevant evidence*. It is clear that such probabilities are the most reliable guides for prediction, or, as we might say, have *maximal predictive import*. Let us turn now to parallel considerations regarding probabilistic explanation. Consider the following inference :

> The probability that a Pakistani peasant, exposed to the variola virus in 1945, would have contracted smallpox is very high
> Individual, a, was a Pakistani peasant thus exposed
> ———————————————— ↓ probably
> a contracted smallpox

Schematically if :

> Kx = x is a Pakistani peasant
> Ex = x is exposed to variola virus
> Cx = x contracted smallpox

then the inference may be written :

> P(K · E, C) is very high
> (I) Ka · Ea
> ——— ↓ probably
> Ca

But, though nearly all, *not* all Pakistani peasants in 1945 were susceptible upon exposure. Some had been properly immunized and for these the probability of contracting smallpox upon exposure was close to zero. Suppose a to have been one of those who were immunized. Then the following argument applies :

$$
\begin{array}{ll}
 & Ix = \text{x has been immunized} \\
 & \sim Cx = \text{x did } \textit{not} \text{ contract smallpox} \\
\text{(II)} & P(K \cdot E \cdot I, \sim C) \text{ is nearly 1.} \\
 & Ka \cdot Ea \cdot Ia \\
 & \overline{\hphantom{Ka \cdot Ea \cdot Ia}} \downarrow \text{ probably} \\
 & \sim Ca
\end{array}
$$

Paradoxically, we have here rival probabilistic inferences, (I) and (II) – in this case, explanations. Both have the same general form, and both stem from supposedly well-established premises. Yet they have mutually contradictory conclusions. On one inference it is practically certain that a did contract smallpox. On the other inference it is practically certain that a did not contract smallpox. Obviously the conclusion in (II) is based on more information than the conclusion in (I). It seems most reasonable therefore to regard (II) as having the greater import for purposes of systemization – be this explanation, prediction, or other. Indeed, one would want a probabilistic nomic inference to stem from a statistical probability whose reference class is the narrowest, in terms of relevant information, to which the particular case being subsumed belongs.[27] This

[27] C. G. Hempel refers to this as "the requirement of maximal specificity". A more complete descriptive label might be maximal *relevant* specificity, since only relevant specifications on the reference class are involved in the requirement. Hempel shows this requirement to suffice for guaranteeing the uniqueness of any probabilistic inference with high associated probability. The demonstration goes as follows :

Consider the rival, i.e., non-unique probabilistic arguments about whether or not a has property, C.

$$
\begin{array}{lcl}
P(A,C) = p_1 & & P(B, \sim C) = p_2 \\
Aa & \text{and} & Ba \\
\overline{\hphantom{P(A,C) = p_1}}\ p_1 & & \overline{\hphantom{P(B,\sim C) = p_2}}\ p_2 \\
Ca & & \sim Ca
\end{array}
$$

where p_1 and p_2 are each close to 1.

Assume that both are maximally specific. Then,

$$P(A \cdot B,C) = P(A,C) = p_1$$

$$P(A \cdot B, \sim C) = P(B, \sim C) = p_2$$

Obviously, $P(A \cdot B,C) + P(A \cdot B, \sim C) = 1$

Hence $p_1 + p_2 = 1$

But this is a *reductio ad absurdum*, since both p_1 and p_2 are close to 1. See : C. G. Hempel, *Aspects of Scientific Explanation*, pp. 399-402.

is again the requirement of maximal specificity of the reference class.[28]

4. Let us consider, now, a familiar objection raised against probabilistic explanations. The objection may be easily generalized so as to apply to *all* probabilistic nomic inferences but it is most naturally expressed with reference to explanatory inferences. Thus if someone were to show surprise that Jones, who could ordinarily hold his liquor, became ill after just one drink, he might be told that Jones had not eaten anything since yesterday. More explicitly the argument would be :

A large proportion, say 95 %, of those who drink
heavily on an empty stomach become ill
Jones drank heavily on an empty stomach
———————————————————— ↓ very probably
Jones became ill

This would most certainly eliminate any surprise over Jones's illness. Nevertheless some might want to say that the explanation is incomplete because the question persists as to why Jones should be found among the general run of cases and not among the exceptions, i.e., among those who are not affected in the usual manner.[29]

But in what sense does this question, Q, persist? The obvious answer to Q is simply to repeat the probabilistic explanation. If the pattern for the occurrence of some trait B, given A, is statistical with a very high relative frequency of B's, then to ask why any particular case, a, falls under the general run of instances rather than under the exceptions is tantamount to asking why a is a B. But an answer to Q is simply the probabilistic explanation which has already been offered. Obviously the question "why?"

[28] Reflection suggests that implementing the condition of maximal specificity of the reference class can meet with a fundamental practical limitation. It may be the case that a statistical sequence cannot be expanded at will because of numerical limitations in the available subject-matter, as, for example, in much of social statistics which are largely field – rather than experimentally – based. In such cases, the more the reference class narrows, the fewer become the available members. This, of course, tends to weaken the reliability of the underlying statistical probability which presupposes long-run sequences of instances. Thus, though what we might call the systemizing import of a probabilistic nomic inference grows as the probability on which it is based stems from narrower classes, the *soundness* of the inference may drop off with the corresponding decrease in the length of the sequence of instances. Severe shortening of probability sequence introduces serious uncertainty regarding the truth of the reported statistical probability. How to balance relevant specification of reference class against loss of statistics is a fundamental consideration of statistical practice lying much beyond the scope of our concern here.

[29] The objection is raised in this form by I. Scheffler, *op. cit.*, pp. 31-35.

in Q has a sense which only a deductive or deterministic explanation will satisfy. For otherwise it would be a request for what has already been provided, i.e., a probabilistic explanation. But then, it remains entirely unclear as to how probabilistic explanations are being compared to deductive ones with regard to "completeness". It may of course be the case that deductive explanations, in excluding exceptions, are more convenient and psychologically more satisfying forms of systemization than are probabilistic ones which are logically tolerant of exceptions. But if one is to indicate anything at all about how probabilistic explanations are *structurally* less complete than deductive ones, then it is clear that one must do something more than simply go on reiterating Q.

The notion of more or less completeness in a probabilistic explanation may perhaps be understood in terms of what we have called the explanatory *cogency* and the explanatory *import* of a probabilistic explanation. The first of these increases as the statistical probability in the nomic premise increases; the second increases with the growth of relevant evidence which narrows down the reference class to which the individual case under consideration belongs. One might, it seems, imagine some manner of combining these two characteristics so as to yield a measure of "completeness" for a probabilistic explanation. Conceivably one might then wish to characterize deductive explanation as the limiting case of "maximum completeness". But then the demand for greater completeness would be either a demand for more evidence, i.e., for greater specification of the reference class or for a higher statistical probability (including even the possibility of unexceptionable universality). The first of these is methodological in character; it entails further inquiry until all relevant evidence has been exhausted. The second is a factual matter depending on nature's nomic offerings.

There is another possible sense in which Q may persist even after a *bona fide* statistical explanation has been provided. To ask why a random case has the usual rather than exceptional characteristic may be to ask why the underlying statistical pattern occurs in the first place. The author of Q essentially accepts the statistical account but then wants an explanation of the explanatory premises themselves. An analogous situation can be constructed for deductive explanation. Consider the case in which, after being offered a deductive explanation for the occurrence of a B, one were to persist in asking why the individual in point counted among the B's rather than among the non-B's, as a novel and falsifying exception to the underlying universal law, L. If the question is not to be interpreted as one which merely calls for repeating the original deductive explanation, it might be understood as a request for an explanation of L. In such a case Q can of course persist no matter how many times we answer it. One can always ask "why?" of any law, L, in the sense of asking that it be derived from more general laws, whether L is statistical or not. But when construed in this manner the alleged persistence of Q has no bearing whatever on whether probabilistic explanations are less complete than deterministic ones. In this last sense, there is, in principle, no upper limit to explanatory completeness. No set of laws, statistical or universal, can ever be "absolute firsts" in the nomic hierarchies that structure scientific inference.

One last brief reflection on a related objection to probabilistic explanation : In his critique of probabilistic explanation I. Scheffler claims that the statistical account seems even less satisfying than according to the previous objection :

"To be told for example that Jones had a heart attack because in his age group 75 % of men have heart attacks is, normally, to leave out of account just what we want to know, i.e., why *Jones* had a heart attack, how he was differentially constituted and how he differentially reacted. To say that there is a high incidence of cancer in Smith's age, occupational and geographic group is compatible with saying we do not understand why Smith was afflicted with cancer".[30]

Undeniably, we derive little if any understanding from inferences whose covering laws are simple correlations between age and the occurrence of some disease. Time by itself – for example, the age of an object, organism, etc. – is not ordinarily taken as a substantive cause of any effect (e.g., heart attack, cancer or other), though time is a common parameter in all the circumstances that contribute to any effect. Causes and effects occur *in* time, but we do not – except in the most elliptical locutions – count time itself as either cause or effect. True enough, we speak of the "ravages" of time and the "consequences" of old age. But these are only short-hand for the effects of a long string of causes. The above examples are, as their author points out, quite feeble as explanations. But the reason for this is not the one he urges, namely that they are probabilistic inferences. They fail to explain only because they do not stem from statistical laws of the cause-effect sort. At best the passage of time can be an index to the extent of some causal development. But to place on time-event correlations the burden of explanation is to confuse indicators with causes. Indeed even if the above examples were *deductive* inferences governed by universal rather than statistical laws, they would be virtually no less feeble as explanations.

[30] *Ibid.*, pp. 31-35.

How much more understanding, for example, would we have of Smith's having cancer if we knew that *all* men in Smith's age, occupational, and geographical group got cancer. Once again – though this time the law is universal – asserting it in connection with Smith's cancer in compatible with saying that we do not understand *why* Smith has cancer. The nomic inference based on this universal law could well serve for predicting and post-dicting but then the resulting predictions or post-dictions would *not* be *explanatory* ones.

E. SUMMARY

In this chapter we have been concerned with several specialized topics which can be grouped under two distinct sets of considerations. The first of these has to do with the inclusiveness of the notion of *nomic inference*, and the guiding question has been : Are there scientific explanations which do not conform and are not reducible to the nomic-inference model? The second set of considerations goes on to examine two sorts of nomic inference of special scientific importance : those having nomic conclusions and those which are *probabilistic*.

With regard to the question of non-nomic explanations, we reviewed several types which seemed to be non-nomic. But nearly all of these were found to be assimilable to nomic form in rather natural fashion. The exception was a type of allegedly "explanatory" account known as *teleological* or *functional explanations*. No satisfactory manner of reducing such accounts to nomic form was found. But this did not require us to regard functional accounts as explanations of an irreducibly non-nomic sort. For, as the analysis suggests, such accounts may be quite reasonably construed as *not* being inferences at all and therefore as non-explanatory modes of discourse. Accordingly, a functional account in which a structure is said to occur for the sake of some function, F, is no more than a statement to the effect that S plays an important (though, perhaps, not indispensable) role in F. This manner of reconstructing functional accounts has certain advantages. It allows for their informative character yet avoids having to say that S is necessary for F. (Indeed there are functional accounts for which S is not at all necessary for F.) But more than this, such a reconstruction avoids any suggestion that F in some special sense "causes" the antecedent structure, S.

For our next set of considerations we went on to examine inferences with nomic conclusions and probabilistic nomic inferences. The first of these have, by some writers, been called "explanation of laws". But, though the bulk of all logico-theoretical exposition in science (particularly in the more mathematical sciences) consists of derivations, not all – perhaps not even the most important – scientific derivations are explanations. Indeed, there is much in the formalisms of our most mature sciences to caution seriously against any claim that explanation is central to scientific systemization.

Several other points were noted in connection with derivations : In the first place, as systemizing inferences, their most important form is deductive rather than probabilistic. Secondly, for the enhancement of systemization, derivations should have more than one nomic premise. Finally, the premises of any derivation must be more general than the conclusion. In discussing this last requirement, however, the notion of generality was left understood in a more or less familiar sense with no attempt at precise explication.

Our next topic was probabilistic nomic inference whose importance has grown in the more recently developed branches of physical science and whose role in the social sciences has becomes primary. For such inferences, it makes sense to apply the distinction between explanatory and non-explanatory nomic inference, though, admittedly, the sense in which a probabilistic account explains is somewhat different from that in which a deductive one does. In our account of this important mode of inference we focussed on structural details. Briefly such inferences are characterized by having at least one statistical law in the premises and singular conclusions which are not logically necessitated but, instead, are made probable by their premises. But the subsumptive power of such inferences depends to some extent on the degree to which the premises make the conclusion probable. In this connection we found it convenient to speak of the *cogency* of a probabilistic nomic inference as a property which depends on the statistical probability expressed by the covering law. This cogency is a minimum when the pertinent probability is *one half* or "even" and a maximum for a probability of *one*. The relativity of probabilities to the reference class (i.e., evidence) was also noted. In this connection, we introduced the notion of *predictive import* for statistical probabilities. Such import reaches its maximum when statistics are taken under the ideal of *all relevant* evidence or, equivalently, under *maximal specificity* of the reference class. Correspondingly, for adequate *systemizational import*, probabilistic nomic inferences should stem from statistical laws in which the expressed probabilities have been determined for reference classes that have been specified or narrowed under a maximum of relevant evidence. This requirement, however, potentially conflicts with the soundness or *reliability* of a nomic inference due to loss in the quantity of statistical instances as the reference class is narrowed down. Statisticians, here, seem to face a methodological dilemma. And how to balance relevant specification against loss of statistics was left as a consideration outside our scope of concern. On the basis of the *cogency* and *systemizational import* of probabilistic nomic inferences one might wish to speak of the adequacy or "completeness" of a statistical explanation. The charge that any probabilistic explanation is intolerably incomplete may then be construed as a claim either that the explanation is insufficiently *cogent* to be subsumptively satisfying or that it has insufficient *import* to be systemizationally significant. In this general respect, deductive explanations might be regarded as limiting cases, i.e., cases of maximum *completeness.*

Having examined the structure of nomic inferences in the last two chapters it seems appropriate now to take up a consideration of nomic statements, since it is from these that all the systemizing inferences of subsumptive structure in science stem. Thus, the very next chapter (i.e., chapter VI) will deal with scientific *laws* both of the statistical and universal sort. The chapter after that one (i.e., chapter VII) will take up theories and certain related formulations such as models and analogies.

VI

NOMIC STATEMENTS (I) : SCIENTIFIC LAWS

A. INTRODUCTION : NECESSARY TRUTHS, LOGIC AND FACTUAL SCIENCE

1. Statements are the basic linguistic means for expressing knowledge. Yet, strictly speaking there are genuine statements which seem to tell us nothing about the world or, for that matter, about any subject-matter at all, including their own. Such statements necessarily assume a very limited role in systematic discourse. They cannot, for example, serve as relevant premises in any factual inference. Nor can they serve, as nomic statements, to express any law of nature. Nevertheless, their function in the systemization of knowledge is an important one and is closely tied to the crucial role of logical inference in all systematic discourse. To see how this is so, let us review some basic matters about statements in general.

It has become rather standard to divide all consistent, cognitive statements into two classes : those which are true solely on logical grounds and those whose truth or falsity cannot be decided by logic alone but require some reference to matters of fact.[1] Statements of the first sort are said to be *logically necessary*; those of the second sort are said to be *contingent*. Logically necessary statements seem, themselves, to be of two major varieties. Let us consider each of these briefly :

(1) For one variety we go to formal logic where we distinguish a class of statements which are true no matter how we understand or *interpret* their non-logical components.[2]

[1] We say here, "cognitive", to leave out of consideration a variety of statement-like expressions not primarily used to communicate knowledge as, for example, those expressing wishes or commands.

[2] Some statements have as "non-logical components", only other statements. They are discussed in that branch of formal logic known as the *sentential calculus*. Other statements have as their non-logical components descriptive terms, i.e., predicates and individual terms. Such statements are dealt with in the other main branch of formal logic, namely the *quantificational* or *predicate calculus*. Roughly speaking, an interpretation of

This means that the truth of such statements depends only on their logical form rather than on their descriptive content. Statements true by virtue of their form are variously referred to as *formally true*, *logically true*, *universally valid*, or, simply, as *valid*. Examples are such statements as :
(1) Either Socrates was guilty of corrupting the young or he was not.
(2) If all Cretans are liars, then it is not the case that there are Cretans who are not liars.[3]

The logical necessity of these examples is so obvious that they appear trivial. Formal truths, however, can have virtually any degree of complexity. The following statement, for example, is formally true, though its tautologous form is not quite obvious without some computation :

Homer is a poet; or, Socrates is wise, but Diana is not real; or, Socrates is not wise and Diana is real; or Socrates is not wise and Plato is a dialectition; or Homer is not a poet and Diana is real; or Plato is not a dialectition.[4]

There is much to recommend the view that the statements of pure mathematics are ultimately nothing more than logical truths, though many of them are neither obvious nor trivial.

(2) Ordinary language presents us with what seems to be the other variety of logically necessary truths. First distinguished by Hume and Leibnitz, and, later, by Kant, such statements appear to be true merely on the basis of relating the meanings of the predicates used. Moreover, they seem to be reducible to logically necessary form once we have analyzed or "unpacked" the established meanings of their predicates. Thus, the statement : 'All bachelors are unmarried' is shown to be logically necessary, by putting in

---

sentences belonging to the sentential calculus is an assignment of truth-values to the compoment (atomic) sentences. An interpretation of sentences belonging to the predicate calculus is any reassignment of meanings to the descriptive terms. A closer account of the notion of interpretation will be undertaken in the next chapter in connection with scientific theory.

[3] These may be expressed symbolically as :

(1) $p \vee \sim p$

(2) $(x)(Cx \rightarrow Lx) \rightarrow \sim (Ex)(Cx \cdot \sim Lx)$

Where 'p' = 'Socrates was guilty of corrupting the young'.

Cx = x is a Cretan

Lx = x is a liar.

The term, 'tautology', is sometimes used loosely for any formally true statement. But many authors refrain from applying the term to quantificational statements, e.g., (2), preferring to restrict its application to the formal truths of the sentential calculus, e.g., (1).

[4] This may be written symbolically as :

$$p \vee q \cdot \sim r \vee \sim q \cdot r \vee \sim q \cdot s \vee \sim p \cdot r \vee \sim s$$

And with appropriate analysis, it is readily shown to be true no matter what truth-values (true or false) are assigned to the sentential variables 'p', 'q', 'r' and 's'.

place of the term, 'bachelors', its *synonym*, 'unmarried men'. Our statement then becomes : 'All unmarried men are unmarried', whose formal truth is obvious. It appears, then, that statements of this sort are, in the final analysis, formally true and, therefore, logically necessary. And if so, their truth is entirely decidable by the logical analysis of meaning and in no way on the basis of contingent fact. Accordingly, Kant called statements of this kind *analytic* to distinguish them from others which he called *synthetic* and which are statements whose truth (or falsity) cannot be decided by logical analysis alone. An example of this latter kind would be : 'All bachelors are vegetarians' a statement whose truth status is surely not decidable without factual knowledge, that is, knowledge about the eating preferences of bachelors.

Statements whose truth depends only on meaning relations, and hence would seem to be properly characterized as analytic, may also arise outside of the context of ordinary usage. A statement may be rendered analytically true as a result of artificial stipulations which create or modify the meanings of terms to suit some purpose. "Definitional truths" of this sort are particularly prevalent in systematic discourse where meaning conventions are often very deliberately and sometimes very precisely set forth. Thus if in some domain of discourse, we agree that the term 'gold' – quite apart from its ordinary meaning – shall designate the element whose atomic number is 79, then, on such usage, the statement, 'the atomic number of gold is 79' is analytic, though when uttered in the context of ordinary discourse it might not be.

Not infrequently in scientific discourse a statement may be rendered analytic in an *essentially* stipulative manner though, actually, no *explicit* definition has been framed. In such a case the *analytizing* convention consists of holding some statement, S, true, come what may. That is, we decide that S shall be true whatever happens in the world. This may at first blush seem to be nothing short of self-deception. But reflection easily shows such a convention to be, after all, a definition. What we are requiring when we say that S is true independently of the facts can be nothing other than that the terms of S be so *adjusted* in meaning that they guarantee the truth of S on logical grounds alone. We have in such a case what might be termed an *implicit definition* of the terms constituting S. Thus if we decide that : (S) Two quarts of any liquid plus three quarts of any liquid shall equal five quarts of liquid, no matter what happens when we work with liquids, then we have essentially required that the meanings of some (possibly all) the terms in S be "stacked" so that they guarantee the truth of S on the basis of logic alone. In our example the stipulation is usually understood as one which affects the meanings of the terms 'two', 'three', 'five', and 'plus'. But precisely what these meanings shall be is left undetermined. Indeed, strictly speaking, the convention leaves it quite open as to whether the terms 'equal' and 'liquid' shall not also be affected by the convention. What is more, it is clear that there is more than one way (indeed an indefinite number of ways) of assigning specific meanings to the individual terms of S, compatible with the analyticity of S. It is because of this absence of semantical specificity that we term the convention in point an *implicit definition* of the terms of S.

But though the notion of analyticity seems clear enough it is not quite a matter of settled opinion. Indeed the analytic-synthetic distinction has been radically challenged in contemporary philosophical analysis.[5] In a set of complex and subtle reflections serious questions have been raised regarding the relation of *synonymy* which (for an important class of statements of ordinary language) must in general be presupposed, in order to be able to show that a statement is analytically true. Moreover, it has been noted in these considerations that in any system of discourse, statements vary in their degree of "factual immunity" depending upon how readily we would be willing to withdraw them in the face of "recalcitrant experiences".[6] We would, for example, unhesitatingly call, 'All bachelors are vegetarians' false given certain unfavorable evidence, but would want to call the laws of arithmetic true given any factual evidence whatever. There are other statements, e.g., some of the conservation principles of physical theory or the law that the velocity of light in vacuum is invariant, that we would not consider changing except, possibly, under the most compelling and extensive experimental evidence. Thus, it is urged, there is no sharp cleavage between analytic and synthetic statements. And, depending on their places and roles in some body of discourse, statements may be said to have varying amounts of responsiveness to factual evidence. Accordingly, it is misleading and untenable to speak of either the analyticity or the factual character of any statement apart from the entire system in which the statement is couched. In principle, as Quine puts it, "Any statement can be held true come what may if we make drastic enough changes elsewhere in the system".[7] The amount of sensitivity of a statement to experiential circumstances or, putting it differently, the extent to which

[5] W. O. Quine, *From a Logical Point of View*, pp. 20-46.

M. White in S. Hook (ed.), *John Dewey : Philosophy of Science and Freedom*. New York, 1950, pp. 316-330.

[6] Quine's term.

[7] *Ibid.*, p. 43.

we must change the rest of a language in order to hold a statement true in the face of adverse evidence, depends, not only on the nature of the evidence, but also, on the logical "place" which the statement in point occupies within the broader framework of discourse. If this is correct, then, the truth-character of a statement is not an eternally fixed matter reflecting the cold hard facts of being. It is rather a changeable affair answering not merely to the call of evidence but reflecting, as well, the prevailing linguistic conventions that underlie the system of discourse.

It would be futile to deny the force of this line of critique of the synthetic-analytic distinction. In reviewing the notion of what it is for two expressions of natural language to be synonymous it raises serious questions regarding the possibility of the distinction for ordinary contexts. The conventionalistic analysis which it offers is a strong challenge to those forms of empiricism that play heavily on the analytic-synthetic distinction without regard for its relativity to linguistic context. It is indeed an illuminating insight of such conventionalism that statements in any system of discourse enjoy varying degrees of immunity in the face of evidence, that the degree of such immunity depends on the relationship of statement to system and, most of all, that the nature of both relationship and system are ultimately matters of human decree. But, these serious objections to the dogmatic and uncritical application of what has been a widespread empiricist doctrine, seem to be no basis for denying outright that in any determinate and relatively stable body of systematic discourse, some statements can be shown to be analytic along the standard lines here indicated. Moreover, it seems also appropriate to regard these as a determinate species of logically necessary statements, ultimately reducible to the formal truths of logic.

The denial of logically necessary statements, be they of the formal or of the analytic sort, entails formal contradiction or inconsistency. For, the logical principles on which such statements rest are precisely those whose denial is logically absurd or self-contradictory. It is basically this characteristic of all logically necessary statements which guarantees that their truth, if decidable at all, is decidable by logical analysis alone, that is, independently of empirical evidence, or, as it is sometimes put, *a priori*. Logical analysis serves to reduce such statements to patently formal truths; and these, *ex hypothesis* cannot be denied without contradiction. This means that no state of affairs could possibly falsify a logically necessary statement. For the falsifying report would, itself, have to be self-contradictory. And how is one to understand any claim that such a report describes anything at all? Considerations of this sort have led empiricists to deny that logically necessary statements have any factual content whatever. From time to time some philosophers have maintained that logically necessary statements and the rules of logic on which they rest do say something about reality, the claim being that such statements express the most general laws of being. But, again, this evokes the question : What, in any ordinary or familiar sense, could one mean by a "law", however general, whose falsification entails *logical* absurdity? Empiricist thinking on this question tends to view the rules of logic as linguistic conventions having, therefore, no factual, descriptive force whatever.[8] To this it has in turn been objected that conventions are logically arbitrary and should therefore admit of alternatives. But – as the objection goes – one cannot speak, without absurdity, of any alternatives to the alleged "conventions" of logic. And does this not make it doubtful that they are conventions?

Unsettled though this issue may be, it is most unlikely that any practicing scientist would attempt to decide any matter of fact on the basis of logical criteria alone – e.g., internal consistency. Indeed no philosophical reconstruction of contemporary science could be complete without noting that the factual neutrality of logic is an empiricist assumption at the heart of contemporary science. For, to admit that logically necessary statements have factual content, would be to admit what seems profoundly incompatible with all scientific strategy, namely, that we can know something about the world *a priori*, i.e., independently of empirical evidence.

2. Let us now turn briefly to the second of our two main sorts of statements, namely those we have designated as *contingent*. These we recall are statements whose truth cannot be decided by logical considerations alone but require reference to matters of fact. Underlying this character of contingent statements is the fact that, unlike logically necessary statements their denial is not self-contradictory and hence cannot be ruled out on purely logical grounds. This being the case, the truth of contingent statements can be decided only on an extra-logical basis. A large part of philosophy has been the attempt to say what this basis is. Broadly speaking, *empiricism* maintains the necessity of observational evidence for establishing *all* contingent statements. On the other hand, the philosophic doctrine of *rationalism* allows that some statements which are not *logically* necessary may be certified *a priori*.[9] Only a casual acquaintance with intellectual

[8] For further discussion in an empiricist vein see : E. Nagel, "Logic Without Ontology" in *Naturalism and the Human Spirit*, ed. Y. H. Krikorian, New York, 1944.

[9] It seems appropriate to note here that in certain philosophic contexts the term, 'contingent' has a sense different from the one we give it here where we use it to refer to *consistent statements which are not logically necessary*. In some quarters of traditional philosophy it has been held that statements may be "necessary" in senses other than

history suffices to convince one that modern science has become more and more empiricistic with the passage of time. Contemporary science, buttressed by logico-mathematical techniques that are vastly richer and more powerful then those of the past, has concerned itself, more than ever, with discriminating between genuinely factual and purely logical questions. This concern is unmistakably more explicit and obvious than it ever was before the present century. And, underlying this concern is the ever more explicit commitment to the empiricist principle that factual claims must be grounded in experience. Indeed, few if any working scientists today would seriously maintain that any matter of fact can be decided independently of observational evidence as would be entailed by the view that statements, other than those which are logically necessary, can be known *a priori*.[10]

3. If logically necessary statements are factually empty then of what use are they in *factual* discourse as, for example, in science? Indeed, such statements play a crucial role in all discourse, a fact which should cause no surprise on recalling that logic pervades all systematic expression and that logical necessity is itself rooted in the formal principles of logic.

Broadly, the tasks of logic are basically two : (1) The first of these consists of spelling out the rules for constructing statements from given linguistic elements. (2) The second task of logic is that of providing rules for deriving a conclusion from premises, in such a way that if the premises

---

that of logical necessity. Accordingly, a contingent statement is one which is consistent but not necessary. On such usage statements may be considered necessary even when their denials are not logically absurd. What one might call "epistemic" and even "logical" absurdity have been invoked to claim the "necessity" of some statements. Alleged intuitions, rational insight, and unimaginability have been given as a basis for regarding some statements as necessary. But the gambit in these claims has been to maintain that there are statements – allegedly necessary ones – which though factually significant, can be known *a priori*. In this way philosophers have attempted to establish an alleged necessity and hence *a priori* status not only for logical principles but for metaphysical, moral and aesthetic ones as well. Indeed even some scientific principles have in the past been taken to be necessary and hence *a priori*. Examples are the principles of Euclidean geometry (construed as truths about the physical world), the doctrine of absolute space-time, and an assortment of symmetry principles. But despite the unmistakable rationalism in our scientific past it seems fair to say that no scientist today would be inclined to accept, consciously, any *factual* principle as an *a priori* truth, eternally immune to the interdicts of empirical evidence.

[10] The status of mathematical statements has been the subject of philosophic dispute even up to present times. In such polemics, however, two issues often seem to remain blurred : (1) Are the statements of mathematics logically necessary or not? (2) Do logically necessary statements and the principles of logic to which they are reducible have factual content?

are true then so is the conclusion. Such rules are said to be *truth-preservative* or *valid* rules of inference.

Now for a brief reflection on each of these :

(1) The first task of logic, namely setting forth the rules for formulating statements is accomplished in two steps. The first of these provides various rules for connecting statements in order to produce *composite* statements. These rules, at the same time, stipulate the manner of computing the truth or falsity of any composite statement on the basis of the truth or falsity of the component statements. These rules are essentially conventions, i.e., definitions. They prescribe the various ways in which statements may be connected (combined) truth-functionally, that is with resulting truth depending only on the truth of the component statements. Thus the connective "." (read : "and") is so defined as to require that the composite statement 'p · q' is true if and only if statement 'p' and statement 'q' are both true. The task of setting forth rules of this nature falls under that branch of logic known as the sentential calculus. Much of the utility of this branch of logic in both science and everyday life derives from the fact that the connectives for joining statements are defined in such a way as more or less approximates to the statement connectives of ordinary discourse.[11] The second set of rules for formulating statements establishes the manner of structuring statements from elemental, sub-statemental parts. This is essentially an account of intra-statemental structure. Thus from '(x)', 'F', and 'x' we may construct the universal statement. '(x) Fx' which we read : 'Everything is an F'. This phase of statement formulation is carried out in that branch of logic known as the predicate (or quantificational) calculus.

(2) The second task of logic is the central one from the standpoint of the role of logical necessity in both common and scientific discourse. It consists, we recall, of providing truth-preservative, i.e., valid, rules for deriving conclusions from premises. As a matter of fact, logic does more than this. In actual practice it provides not only rules that preserve truth but also some that preserve whatever truth-value a premise has – be it truth or falsity. Rules of the first kind are called rules of *implicative inference*. Those of the second kind are usually referred to as rules of *equivalence transformation*.[12]

[11] Serious doubts arise regarding the translatability of some connectives of ordinary discourse into those of the sentential calculus.

[12] A rule of equivalence transformation is essentially a two-way rule of implicative inference. For a given premise, P it allows us to derive validly a conclusion, Q, such that from Q we can also validly derive P.

One more point : The rules of logical inference or transformation make reference only to the logical form of statements, that is, to that part of their structure which is independent of their descriptive components. It is this which makes for their vast generality and applicability. If valid, therefore, any rule of logic is valid for an indefinite class of inferences having a certain form. But formal validity has its price. It can be secured only if the rules are, as it were, *factually sterile*. That is, they must be so framed that it is impossible for the inferences which they govern to result in conclusions that say more about the subject-matter concerned than do the initial premises. Specifically, implicative inferences can result only in a loss, or else, no change of descriptive content; while equivalence transformations can result in neither a gain nor a loss of descriptive content. To see this we need only reflect that otherwise from true premises we could (invalidly) derive a conclusion which could be false only because of the additional content it might have over the premise.[13] Similar considerations assure us that valid rules of equivalence transformation permit conclusions that say no more and no less than the premises. That is to say the conclusions of such transformations or *inferences* are *logically equivalent* to the premises. The use of logic in discourse is now apparent. Implicative inferences permit us to draw out the content of premises without loss of truth. We may lose content along the way, as we "unpack" or analyze the original statement. But at each inferential step we can be sure (assuming the validity of our rules and their proper use) that if the starting statements, i.e., the *premises* are true, then the concluding statement or *conclusion* is also true. The logical unpacking of premises may reveal parts or aspects of their content that might otherwise be too implicit for our detection and comprehension. Such derived results are said to be logical consequences of the premises in point. Similarly equivalence transformations when valid

[13] We shall not stop here to take up the question of precisely what it means to claim that a statement, 'S_2' says more about a given subject-matter than does a statement, 'S_1'. Such a discussion would take us into considerations of language structure lying much beyond our concern. For present purposes the following brief definition seems to be sufficient :

'S_2' says more about a given subject-matter than does 'S_1' if and only if 'S_2' is logically equivalent to '$S_1 \cdot D$' where 'D' is a statement and is logically independent of (i.e., neither logically equivalent to nor logically implied by) 'S_1'. This characterization suffices to show that if 'S_2' says more than 'S_1' about given subject-matter then it cannot be the product of a logically valid inference based on S_1. For if it were, the transformation would not be truth-preservative. To see this assume 'S_1' to be true. The truth-value of 'S_2' then is a function of the truth-value of 'D'. But 'D' can be either true or false depending on what statement it stands for. Hence 'S_2' can be false though 'S_1' be true.

and properly applied can lead to more convenient or more perspicuous forms of statements having precisely the same content as do the originals. The basic feature of such processes is, then, their factual sterility : that is, they permit the logical manipulation or *analysis* of statements without adulteration of content, and saying that such manipulations are truth preservative clearly entails that they have this profoundly essential feature of all valid operations on statements.

Mathematics is a convenient quantitative calculus for carrying out valid inferences, i.e., proofs, calculations, solutions, etc., both of the implicative and equivalence type. Its value in scientific discourse, as in the case of all other systems of logical argument, is rooted in its validity. And, again, this means that valid mathematical computation can be carried out without danger of adulterating the given factual content of premises. When, in science, we engage in calculation or other deductive inference (as for example in deductive nomic inference) we add nothing to our store of scientific knowledge, except perhaps in the psychological sense of revealing content that was formerly inscrutable and implicit, though, nonetheless, implicitly contained in our starting premises. Yet, this is an indispensable and even crucial phase of scientific inquiry. For, as we shall see, theoretical science can present us with factual premises so general and compact that it would seem psychologically (i.e., humanly) impossible to cognize their content in full, without the aid of logical or mathematical analysis. Logical analysis in the form of subsumptive deductive inference, for example, constitutes a major and underlying aspect of systemization without which no body of discourse could count as science. Indeed, no body of discourse could pass for science in any full sense of the term without the extensive use of logico-mathematical analysis for unveiling the implications of assumptions.

Let us return now to our initial question. How do logically necessary statements figure in systematic discourse?

One part of the answer is something like the following : To every logically valid inference there corresponds a logically necessary statement. The reason for this is, simply, a basic principle of formal logic according to which, any inference, I, having premises, P, and conclusion, Q, is valid if and only if there is a logically necessary statement having either the conditional form, 'if P then Q', or the biconditional form, "P if and only if Q", depending on whether I is an implicative inference or an equivalence transformation.[14] Logically necessary conditionals are the statemental

[14] This follows obviously from what it means to say that I is valid. For if I is valid

counterparts – one might even say, encapsulations – of valid arguments. Correspondingly, the factual emptiness of such statements parallels, in a very precise sense, the factual sterility of valid inference. For, a valid inference from premises, 'P', always concludes with conclusion, 'Q' whose content is already implicit in 'P'. Hence no new content has been generated. Similarly, the logically true conditional : 'If P then Q' is factually empty because, in effect, it represents an analysis of 'P', with 'Q' expressing nothing other than all or part of the content already implicit in 'P'. The correctness of this analysis, however, depends only upon how the contents of 'P' and of 'Q' relate formally (irrespective of their truth or falsity) and hence not at all on any matter of fact.

Logically necessary statements in the form of analytic statements which have been established by stipulation play a systematic role in what we shall later call the "essentialization" of factual statements. In effect, this amounts to a modification of meaning for the descriptive terms of a factual statement so that it is rendered analytic and hence logically necessary. Such statements are unable to serve as explanatory premises or as other factual premises of nomic inferences. Nevertheless they serve as premises for inferences which closely parallel subsumptive nomic inferences and which make possible what one might call an explicative mode of systemization. Explicative systemization in scientific discourse, because it rests on analytic (essentialized) principles is logically distinguishable from explanatory or other nomic systemization which rests on genuinely nomic – hence factual – principles. This line of considerations, however, will not detain us here but will concern us under the topic of *interpretation*, in the next chapter.

B. UNIVERSAL LAWS

Having addressed ourselves, in a preliminary way, to some basic distinctions regarding cognitive statements in general, we turn now to our more central concern; namely those statements which govern subsumptive inference and which are therefore of supreme importance in all systematic discourse, especially in science. These are statements of a more or less general sort which we have referred to as *nomic* or *law-like* statements.

---

then there is no interpretation of 'P' or of 'Q' which could render 'P' true and 'Q' false. But this is precisely what is both necessary and sufficient for the statement '$P \longrightarrow Q$' to be true on all possible interpretations, which is to say, for it to be *logically necessary*. A similar line of consideration easily shows that if I is an equivalence transformation, then it is valid if and only if '$P \longrightarrow Q$' is logically necessary.

The most cursory look at science easily shows that nomic inferences vary considerably with respect to systemizing power, predictive reach, explanatory depth and cogency, and even the kind of understanding that they seem to give. Some important differences as we have already noted depend on whether the nomic inference is probabilistic or deductive. But after this root division among nomic inferences has been appropriately noted, it is clear that the systematic import of any nomic inference can vary, and depends on the character of its nomic premise. Compare, for example, an inference in which we account for Smith's violence yesterday by citing the fact that he has a disposition to be violent whenever he is distressed, with an inference in which the account is given in terms of a law connecting such dispositions with genes. The first inference might go roughy like this:

(L) Smith has a disposition to be violent when distressed
Smith was distressed yesterday

Smith was violent yesterday

The second inference might be as follows :

(L′) Any human male carrying an extra Y chromosome has a disposition to be violent when distressed
Smith has an extra Y chromosome, is a human male and was distressed yesterday

---

Smith was violent yesterday

The second inference has vastly greater systemizing power and affords a deeper understanding of Smith's violence than does the first. And there is a corresponding difference between the nomic premises L and L′ heading each inference. L is about the dispositions of a particular person mentioned by name. L′ on the other hand is a far more general statement. Its subject term makes a much wider reference, namely all human males. Moreover some of its descriptive terms are basic theoretic ones. That is, they refer to such micro-theoretical entities as chromosomes, which are capable of accounting not only for aggressive behavior but for other traits as well.

As our example shows, even singular statements attributing dispositions to particular entities have some systemizing power. The reason for this is that every such statement, in an important sense, is a generalization encompassing an indefinite number of instances. As such it can be a premise in an inference for subsuming some particular state of affairs or occurrence as a special case. Thus Smith's violence yesterday when he became distressed was merely a special instance of the more general fact that *every* time

Smith is distressed he manifests violence. The statement expressing this fact has an aspect of universality in that it refers to *every* time Smith is distressed. Moreover it also has an aspect of numerical indefiniteness in that it sets no explicit limits to the *number* of such times. Logically, it implies neither the finiteness of Smith's life nor any limit on the number of times Smith may become distressed. Thus, singular statements of this sort have a nomic or lawlike aspect which somewhat justifies our loosely including the subsumptive inferences which they govern in the class of nomic inferences. But despite their aspects of universality and unrestrictedness, singular, dispositional statements afford only the most meager nomic inferences constituting, as it were, the lower limit of systematic or explanatory import and affording little if any depth of understanding. Nomic inferences take on distinctively scientific character when they are governed by premises that are loosely referred to as scientific laws. In our two examples the second inference is of this sort, its nomic premise, L′, being a scientific law.

Contemporary writers concerned with the systemizing power of general scientific statements have shown a marked concern for providing an explication of the term 'scientific law' that could serve to distinguish "genuine laws" from statements which, in certain respects, resemble them but which one would not want to call laws. Actual scientific usage applies the term 'scientific law' (more simply, 'law') rather loosely to statements of varying scope and generality, ranging from circumscribed descriptions connecting observable characteristics, to highly encompassing abstract theories. Indeed such usage seems to be only a rough guide on just how to decide whether a given statement is or is not a law. And no doubt, there are many statements on whose nomic status there would be divided opinion. Nevertheless some statements are clearly laws while others are clearly not. Moreover, having the form of a universal conditional, though necessary, is obviously not sufficient for a statement being a law. The statement, 'All the students in tonight's seminar are over six feet tall,' for example, has this form but would not pass muster as a law.[15] Clearly, this statement would not do as a premise for any satisfactory systemizing account or explanation of Jones' being over six feet tall, though he is a member of the class. Even statements of considerably greater scope such as : 'Every scholar in 15th century Europe knew Latin', though they might serve as premises of (low-level) nomic inferences would not make it as laws. It seems entirely

[15] This statement is expressed more transparently in the symbolic form :

$$(x)\ (Sx \rightarrow Ox)$$

which we read : For any x, if x is a student in the seminar (Sx), then x is over six feet tall (Ox). Written in this way, its structure as a universal conditional is apparent.

appropriate and philosophically interesting, therefore, to ask what gives some scientific statements the high systemizing power (and when they are cause-effect statements, the high explanatory power) which entitles them to be called laws.

At the very outset, it will be convenient to distinguish two sorts of statements that generally count as lawlike : *universal* and *statistical*. Though both types have important features in common, a separate consideration of each seems by far to be the more appropriate manner of account. Accordingly, we take up *universal* nomic statements first.

The following are examples of statements that have at some time or other been accorded the title of *scientific law* :

(1) The orbit of any planet around the sun is an ellipse with the sun at one of the foci.

(2) Lead melts at 327.4°C.

(3) In optical reflection the angle of incidence is equal to the angle of reflection.

(4) The volumes of gases involved in a chemical reaction are to each other as ratios of small whole numbers.

(5) Every atom contains at least one proton.

(6) Blue-eyed human parents always have blue-eyed childern.

(7) No two electrons in an atom can have the same four quantum numbers.

(8) Whenever the effective demand for consumers' and producers' goods falls short of a nation's capacity to produce, the tendency will be toward unemployment and economic depression.

These examples call attention to a number of features which from time to time have come under consideration for the purposes of characterizing scientific laws. We consider each of them as follows :

1. Conditional Universality

It is a matter of common agreement that the structure of any scientific law must be that of a universal conditional. A simple, traditional representation of this structure is the statement form :

(a) All S is P

Modern logic, however, makes the conditional aspect more explicit by expressing (a) as :

(b) If anything is an S, then it is a P.

where the conditional connective : 'if ... then' is expressed as the material implication of modern formal logic.[16] If we represent the indefinite terms

[16] Material implication, represented by the sentential connective '$\rightarrow$' connects two

'it' and 'thing' by 'x' we can express (b) more perspicuously as :

(c) For every x, if x is an S then x is a P.[17]

The expression, 'for any x', which, in modern logic is known as a universal quantifier or operator, is the formal ear-mark of all universal statements. The clauses introduced by the terms 'if' and 'then' are known as the *antecedent* and the *consequent* respectively.[18] We have already noted that having the structure of a universal conditional does not suffice for calling a statement a law.

2. Nomic Universality and Counterfactuals

It has been a deeply entrenched tradition to distinguish radically between accidental and essential or nomic universality. Thus, a universal statement simply reporting that two properties always go together as for example, being a crow and being black would not report any sort of "necessary"

---

statements 'p' and 'q' in such a way that the composite result : '$p \rightarrow q$' is false for p-true and q-false and true for all other possible truth-value assignments to 'p' and 'q'. This leads to the strange result that : 'all winged horses are well tempered' is true even though there are no winged horses, and to the even stranger result that for an actual law like : 'Lead melts at 327.40C', all occurrences not involving the melting of lead as, for example, the boiling of water would be confirming instances of the law. The paradoxical aspects associated with the formal representation of laws and due principally to the truth-characteristics of material implication will, however, not concern us here.

[17] More symbolically :

$(x)\,(Sx \rightarrow Px)$

[18] Our representation of universal nomic statements is a highly simplified one and though it does not detract from the generality of those considerations which concern us, it does not always suffice to render more complex laws. Thus while all laws require at least one universal for their formulation, some require more than one, as for example, in the following statement : 'Whenever copper is heated it expands'.

For every x and *for every y*, if x is copper and y is a time and x is heated at time y, then x expands at time y.

More formally,

$(x)\,(y)\,(Cx \cdot Ty \cdot Hxy \rightarrow Exy)$

It will be noted that,

Cx = x is copper; Ty = y is a time;

Hxy = x is heated at time y.

Exy = x expands at time y.

Some universal nomic statements require also another sort of quantifier. This is an expression of the form, *there is an x*, known as an existential quantifier. Thus consider (5) of our examples above, i.e., the law, 'Every atom contains at least one proton'. It may be paraphrased,

For every x, if x is an atom, then *there is a y* such that y is a proton and x contains y. More formally,

$$(x)\{Ax \rightarrow (Ey)(Py \cdot Cxy)\}$$

connection and would therefore, not be a law. The basis for such a distinction was seriously questioned by David Hume and has remained highly dubious since. Yet we find contemporary empiricists still reiterating what is essentially the traditional distinction without, it seems, adding explicitly to its intelligibility.[19] E. Nagel, for example, writes :

"The *prima facie* difference between lawlike and non-lawlike universal conditionals can be brought out in several ways. One effective way depends on first recalling in what manner modern formal logic construes statements that have the form of universal conditionals ... Such statements are interpreted in modern logic to assert merely this : any individual fulfilling the conditions described in the antecedent clause of the conditional also fulfills, *as a matter of contingent fact*, the conditions described in the consequent clause. For example, in this interpretation, the statement 'All crows are black' (which is usually transcribed to read : 'For any x, if x is a crow then x is black') merely says that any individual thing which happens to exist, whether in the past, present, or future and which satisfies the conditions for being a crow is in point of fact also black ... Universal conditionals construed in this way so that they assert only matter-of-fact connections, are sometimes said to formulate only a "constant conjunction" of traits and to express "accidental" or *de facto* universality".[20]

But, though Professor Nagel correlates what is expressed by non-lawlike conditionals with what is expressed by the conditional relation of formal logic, he does not quite tell us what a so-called "accidental" relation is except to quote such phrases as "matter-of-fact concomitance" for non-lawlike connections and "physical necessity" or "nomic universality" for lawlike connections.

In a later passage he writes :

Again, some universal nomic statements are not simple conditionals, at all. They may for example express a functional relation between two quantities each of which is sufficient for the occurrence of the other. For example, at constant temperature the pressure of a gas is inversely proportional to its volume. That is, if the volume of the gas is reduced the pressure is proportionately increased, and *vice versa*. Such a statement would be represented as a bi-conditional, i.e., as a conjunction of two conditionals in which the antecedent and consequent of one switches places with the other. The formal representation of some laws would be extremely complex requiring the use of multiple quantifiers of both kinds. Fortunately it is often possible to express scientific laws by means of mathematical equations which, in general, greatly abbreviate the expressions of formal logic with vast simplifications of both notation and manipulation.

[19] H. Reichenbach suggests a distinction between the probabilities of "causes" and those of "mere correlations" without spelling out the fundamental differences. See H. Reichenbach, *The Theory of Probability*, pp. 92, 167.

[20] E. Nagel, *The Structure of Science*, p. 50.

"The distinction between accidental and nomic universality can be brought out in another way. Suppose that a piece of copper, c, which has never been heated is called to our attention and is then destroyed so that it will never be heated. Suppose, further, that after the work of destruction is over, we are asked whether c would have expanded had it been heated, and that we reply in the affirmative. And suppose, finally, that we are pressed for a reason for this answer. What reason can be advanced? A reason that would generally be accepted as cogent is that the natural law, 'All copper when heated expands', warrants the contrary-to-fact conditional 'If c had been heated, it would have expanded'. Indeed most people are likely to go further and maintain that the nomological universal warrants the subjunctive conditional, 'for any x, if x were copper and were heated, then x would expand'.[21]

At this point, however, Professor Nagel does not tell us what one is to understand by the claim that a law "supports" or "warrants" some counterfactual. Nor is there any obvious answer. For the unsettled state of opinion on precisely what one says when he utters a counterfactual is a well-known chapter in contemporary philosophy.[22] An indication of what he has in mind might be found in a later account where he maintains that "... a counterfactual can be interpreted as an implicit metalinguistic statement (i.e., a statement about *other* statements and in particular about the logical relations of these other statements) asserting that the indicative form of its consequent clause follows logically from the indicative form of its antecedent clause when the latter is conjoined with some law and the requisite initial conditions for the law".[23] Accordingly, the counterfactual, 'if the piece of copper, c, *had been* heated then it *would have* expanded', says that for some law (L), the statement, 'L and a is copper and *is* heated', logically implies the statement, 'a expands'. L is the law, 'for any x, if x is copper and is heated, then x expands'. The logical implication in point does hold with no other (initial) conditions, in this case, being required.

But if, by this line of analysis, Nagel means to deny the status of law to such allegedly "accidental" universals as, 'all crows are black' then it decidedly fails. For, in the sense in which we understand him, this universal can serve to "support" the counterfactual, 'if a *were* a crow it *would* be

[21] *Ibid.*, p. 51.

[22] For a standard account of the puzzle of counterfactuals the reader is referred to N. Goodman, *Fact Fiction and Forecast*, Harvard, 1955, pp. 13-34. For a later attempt on the problem, see : J. L. Mackie, "Counterfactuals and Causal Laws" in *Analytical Philosophy*, ed. R. J. Butler, New York, 1962, pp. 66-80.

[23] E. Nagel, *op. cit.*, p. 72.

black'. The formal requirements involved in this case are no less satisfied than in the case of any other universal conditional. Indeed even singular dispositional statements can be used in this manner. Thus the counterfactual : 'If Smith *had been* distressed he *would have* been violent' could in quite the same manner be "supported" by the singular dispositional statement, 'Smith is always violent when distressed'. And surely no one would want to say that this latter statement is a law of nature.

Still more uneasiness arises with this approach to nomic universality. And, engaging as Professor Nagel's account of counterfactuals may be, it tends to neutralize any characterization of laws in terms of counterfactuals. For, any attempt at such characterization is clearly on its way to circularity the moment we attempt to explicate the notion of counterfactuals in terms of laws.

Thus Nagel's attempt to distinguish between nomic and accidental universality so as to exclude statements like : 'All crows are black', from the category of laws seems to fall short of its purpose. Indeed, Hume's account of the nomic relation in terms of nothing *objective* beyond the constant conjunction of traits was calculated to show the futility of any such attempt. The history of all subsequent philosophical analysis gives much reason to suppose that Hume's claim remains an essentially unchallenged thesis in the empiricist background of contemporary science.

But while statements like, 'all crows are black' and, for that matter, 'the ratio of charge to mass for the electron, in emu per gram, is 1.7598 ± 0.0004',[24] remain undifferentiated from universals having an allegedly more nomic flavor, it does remain possible to characterize laws in such a way as to exclude certain universal conditionals which are clearly not laws, as for example the statement, 'all the students in tonight's seminar are over six feet tall'. An obvious difference between this sort of statement and a law like 'the volumes of gases involved in chemical reactions are to each other as ratios of small whole numbers', is the scope of subject-matter involved. Let us turn, then, to consider this as a possible criterion for nomic universality.

3. *Unrestrictedness*

Compare once again, the universal conditionals :

(a) For any x, if x is a student in tonight's seminar then x is over six feet tall.

[24] See R. S. Shankland, *Atomic and Nuclear Physics*, Macmillan : New York, second edit., 1960, p. 33.

(b) For any x, if x is a set of gases reacting chemically then x is the set of gases whose reacting volumes are related to each other as ratios of small whole numbers.

In statement (a) something about height is predicated of (i.e., attributed to) the set of individuals in tonight's seminar. This set of individuals, which comprise the subject-matter or *scope of predication* of the statement, is restricted, by the subjet-terms of the statement (i.e., 'student in tonight's seminar'), to a finite set of spatio-temporally circumscribed entities. In statement (b), however, there is no such restriction. The statement predicates a certain set of relations for any set of gases reacting chemically regardless of when or where they exist and react. Statement (b) differs in this respect from even the statement : (c) 'All scholars of 15th-century Europe knew Latin', (mentioned as an earlier example of a non-law). This non-law though having a vastly wider scope than (a) is restricted in fundamentally the same manner, that is, to a particular time-place region.

In general we resist calling a universal conditional a law of nature if its scope of predication does not have space-time universality, that is if the entities to which it refers are restricted to a particular space-time segment of the world.[25]

[25] Our insistence on this sort of universality is particularly strong with respect to time (perhaps more than with respect to space) as evidenced by the closely related fact that we resist formulating any generalized conditional as a law of nature if the "causal regularity" which it describes is itself changing with time. Putting it in other terms, we tend to regard any general statement about natural regularities to be genuinely causal and lawlike only if its form is time-invariant. This is sometimes expressed by saying that the laws of nature are required not to be explicit functions of time, that is, cannot contain the time variable explicitly. This statement about our scientific predilections raises a number of still unsettled questions :

(1) Is the time invariance of our present scientific laws a matter of stipulation and convention or is it simply a matter of fact that nature usually favors us with the temporal simplicity and uniformity we desire?

(2) The time-invariance requirement tends to collapse, somewhat, when we consider – as an elementary point of mathematics – that for nearly all time-differentiable functions (there are some exceptions) time can be eliminated as an explicit variable by successive differentiation, as defined in the infinitesimal calculus. This means that the requirement of time-invariance must be specified with respect to a maximum order of time-derivatives allowed for expressing nomic regularities. In general present theoretical science makes no use of differential equations of order higher than second.

It should also be noted in this regard that differentiation with respect to any variable eliminates from a function all terms which are constant with respect to that variable. Differentiation therefore increases the generality and abstractness of a mathematical function and, as a result, its mathematical opaqueness as well. In a special but important sense, therefore, limitations on the order of nomic differential equations is one sort of

Thus, we would hesitate to accept as a law of nature a pattern of regularity which seemed to hold only for a particular time or only for a particular place. Imagine, for example, that it were true only for bodies falling freely over New York City that they fell distances proportional to the *cube* of the time on odd years and to its square on even years. This would mean that, for New York City, the law of falling bodies was changing with time and, moreover, changing in such a way that the acceleration, and hence the underlying force, was constant on even years and proportional to the time on odd years – a most singular and strange result from the viewpoint of causal uniformity. But, setting this complication aside, would we say that there was a special law of nature governing the behavior of falling bodies over New York City? Very likely not. Observing this odd circumstance peculiar only to New York City we would regard it as just that, an odd circumstance. Indeed by leaving it simply a singular fact about New York City we would invite the possibility of explaining it in terms of the universal law of falling bodies which requires constant acceleration and therefore proportionality to the square rather than to the cube of the time of fall. It might at this point be thought by some that whether we call the description of the New York City anomaly a law or a singular fact is merely a terminological matter involving no more than the application of a label. There is much merit to this objection. But if intended too literally it could be missing an important and subtle point about the use that laws are put to in scientific inquiry. Resistance to calling the anomalous phenomenon simply a "local law of nature" reflects our desire to explain the particular pattern in terms of a law that is not space-time restricted. We might for example tentatively assume that the New York City "neighborhood" is not a physically closed system. This would, then, imply that we should look for a nearby moving mass, or a changing density beneath the New York City surface or some other – say, electro-magnetic, source of perturbation which would explain the departure from normal regularities.[26] Failing to find the perturbing body we might well choose to stick to the universal law for falling bodies and regard the singularity in point as a special circumstance, representing not a departure from the law but rather a consequence of it, still awaiting explanation pending fuller knowledge of the appropriate initial conditions for applying the unmodified law.

But apart from their space-time restrictedness (a) and (c) still have what seems to be an unpalatable trait from the viewpoint of according them the

limitation on the generality of natural laws.

[26] This is the approach which has in fact been used for several planetary perturbations and has led to important astronomical discoveries.

status of laws. This has to do with the finiteness of their scope. Is this a disqualifying feature?

4. Indefinite Scope

The number of individuals comprising the subject-matter of (a) and (c) is definitely known to be finite. What is more, the finiteness is inferrable from the statements themselves. A knowledge of the meanings of terms like 'tonight's seminar' and '15th-century European scholar' is sufficient for inferring that there are definite numerical limits to the number of individuals comprising the subject-matter of these statements. Actually, we do admit to the status of laws some statements whose subject-matter is finite and even numerically quite limited. The laws of Kepler are a well-known example. There is even some reason to believe that the subject-matter of all laws may, as a matter of contingent fact be ultimately finite. But the finiteness of the number of orbiting bodies in the solar system and the possible finiteness of the universe are not inferrable from any of the laws in question. The descriptive terms in Kepler's laws or for that matter in any universal generally accorded nomic status do not set any numerical limits whatever to the subject-matter under consideration. Putting it another way, a law of nature of the form : "for any x, if x is S then x is P", provides no inkling at all of the numerical size of the set of entities it describes, even when we know precisely what the descriptive terms 'S' and 'P' mean. Laws of nature do not relate characteristics for some numerically distinct set of gases or of bodies or of light rays, etc. They describe any gas, body or light ray, whatever. The absence of numerical delimitations, either explicit or logically implied, for the set of individuals encompassed by a law may be expressed as the *indefiniteness of scope* of a law. Though actual usage of the term 'law' is vague with respect to several possible requirements for the title, such usage leaves little or no doubt that indefiniteness of scope must figure as a characteristic of nomic universality. This of course obviously rules out from the category of laws those cases for which the examined instances of the statement exhaust its entire scope of application. If all the instances that a universal statement is about have been exhaustively observed, then it would seem more appropriate to call it a factual report about the past rather than a law. For, one necessary function of any law is that of serving as a premise for predictive nomic inference. And predictive function is of course impossible with respect to any instance already known to have occurred. It would be nothing short of silly to speak of predicting that, if we measure Tom

Jones' height, we will find it to exceed six feet, on the basis of the universal conditional 'All eight members of last night's seminar (Tom Jones being one) are over six feet tall', and given that the evidence for this conditional is that all eight members were measured and found to be over six feet tall.

5. Causal Relevance; Non-randomness

Statement (a), in particular, has a feature which, it seems, would rule it out as a law even if it did not fail on other requirements. This is what we might call the randomness of what it reports or predicates about its subject-matter. We recall that (a) expressed as a universal conditional says :

For any x, if x is a student in tonight's seminar, then x is over six feet tall.

For simplicity, assume that we assert (a) in a context of discourse (sometimes said to be the *universe of discourse*) consisting only of American male adults. Then it is known as a matter of contingent fact, that being a student in any seminar has no bearing on an individual's height. Consequently any set of seminar students – and of course, therefore, the set consisting of tonight's seminar students represents a random or irrelevant selection of American males from the standpoint of their height. Since this is a point of some importance in the philosophical analysis of scientific discourse it bears somewhat more detailed consideration.

To determine the factors that have a bearing on whether or not an American adult male is over six feet tall we might, roughly speaking, proceed in some such way as this; Record as a sequence, Q, the heights of individuals (selected only on the principle that they are American adult males) as 'O' for over six feet tall and as '∼O' for not over six feet tall. Such a sequence might look like the following :

$$\sim O, O, O, \sim O, O, \sim O, \sim O, \sim O, O, O, O, O, \sim O, O, \ O, \sim O, \sim O, \ldots$$

Call the relative frequency of O after n instances, f^n. It is a point of logic that f^n will tend to fluctuate less and less, with each additional instance, as n becomes larger and larger. Moreover it is reasonable to expect that, as a matter of contingent fact, the long-run value of f^n will tend to stabilize to some determinate value within ascertainable, practical limits. In this sense, one can speak of a relative frequency in the long-run of O's among adult American males. Call this :

$$F(A,O)$$

And read it : The long-run relative frequency of O's among A's. The reference class (or attribute), A, is : being an adult American male. The conjectural class (or attribute), O is : being over six feet tall.

It 'follows from what we mean by 'relative frequency' that :

(1) $$F(A,O) = f_1$$

where :

$0 \leqslant f_1 \leqslant 1$ (i.e., f_1 may have values between O and 1, end-points included.)

Now let S be a possible attribute of the entities which have been observed in order to generate the sequence, Q, and consider another sequence, Q′, gotten from Q by selecting only those occurrences of A's which exhibit also the attribute S. Q′ is called a sub-sequence of Q. The subsequence, Q′, will now yield a relative frequency f_2 of O's among instances of a composite attribute, A · S, (read : A and S). Thus we have :

$$(FA \cdot S, O) = f_2$$

in case $f_1 \neq f_2$ then we say that S is *causally relevant* to the occurrence of O's among A's. If, on the other hand, $f_1 = f_2$, then we say that S is *causally irrelevant* to the occurrence of O's among A's. In such a case, we may also say that O's are randomly distributed among A's, relative to S. The conjoining of 'S' to 'A' in the expression 'F(A,O)' represents a narrowing down of the reference class A to the new reference class A · S.

In terms of these distinctions, that attending a seminar has no bearing on whether or not an adult American male is over six feet tall, may be expressed by saying that :

$$F(A \cdot S, O) = F(A,O)$$

where 'S' stands for 'attending a seminar'. Or we may also say that the occurrence of O's among A's is random relative to S. The irrelevance cf S to the occurrence of O's among A's or, correspondingly, the randomness of O's among A's relative to S is a matter of contingent fact. In particular our universal conditional (a) is stated in the context of advanced knowledge which tells us that the frequency of occurrence of height over six feet among adult American males is unaffected by selecting such individuals in accordance with whether they attend seminars, that is to say, the occurrence of such height is random with respect to seminar attendance. Accordingly, the membership of tonight's seminar or, for that matter, of any seminar attended by adult American males is a *random sampling* with respect to the occurrence of height over six feet (O). Whatever we report, then, about the occurrence of O's among the students of last night's seminar (S′) it cannot be a law of nature. For, though such a report is articulated in the universal conditional form,

For any x, if x is an S′ then it is an O

it does not in the context of wider knowledge report any causal connection between S′ and O. The uniform occurrence of O as reported by statement (a) must be taken to be a matter of pure chance or accident. It would clearly do intolerable violence to all familiar usage to ascribe nomic status to any statement expressing such chance concomitances.

6. Universal Conditionals with Vacuous Antecedents

Consider the universal conditional : (i) 'All ghosts know Latin' which is written as the material conditional : 'For any x, if x is a ghost then x knows Latin'. This statement is true for the awkward reason that there simply are no ghosts. This means that every x will falsify the antecedent of the conditional and hence (by the conventions of the sentential calculus) be true. Conditionals that are true for this reason may be said to be *vacuously* true.

Statements of this sort seem to be definitely objectionable as laws if only for the fact that, strictly speaking, they are not applicable. Such a law for example could yield neither predictive nor explanatory nomic inferences because without ghosts no initial conditions could be provided for inferring conclusions. Indeed this conditional, along with so many that could be dreamed up like it, is unqualified not only as a law but as any sort of scientific statement.

But not all conditionals with vacuous antecedents are scientifically useless. Some are *derivable*, with appropriate specializing conditions, as logical consequences of well established laws. An example would be : (ii) All men on the surface of any planet with the same density of the earth but three times the radius weigh three times as much as they do on earth. This statement seems to have a vacuous antecedent since, as far as we know, there is no man on such a planet. (There may not even be such a planet!) It is not, however, merely *vacuously* true. Its truth or, at least high confirmation, is assured from the law of universal gravitation on the basis of which it may be derived.[27] Nor is the statement entirely useless. Though there is not, and has never been, a man on any planet of the type described, it is not entirely assured that there will never be. That is, it is not conclusively known that (ii) is omnitemporally vacuous. Granting its

[27] Let : M_e = the mass of the earth
M_x = the mass of planet x
R_e = the radius of the earth
R_x = the radius of planet x

Then : (1) Weight of man on earth, $W_e \propto \frac{R_e^3}{R_e^2}$

(2) Weight of man on planet, x, $W_x \propto \frac{M_x}{R_x^2} \propto \frac{(3\ R_e)^3}{(3\ R_e)^2}$

Dividing (2) by (1) we have $W_x = 3W_e$.

other nomic features therefore, it seems reasonable not to deny (ii) the status of law because of its (presently) vacuous antecedent. Another point of difference between (i) and (ii) must be noted : The existence of ghosts is incompatible with the laws of nature (as known); that is, it seems to be physically impossible for there to be ghosts. Unlike (ii), therefore, the applicability of (i) is decisively blocked by what seems to be the nature of things. No such impossibility may be claimed for (ii).

Still another sort of conditional may be of concern in this connection. Consider the statement : (iii) 'Metals at absolute zero have no electrical resistance' or the statement : (iv) 'A body sliding on an endless frictionless plane slides on forever'.[28] Or : (v) For any sample of an ideal gas (i.e., one whose molecules are dimensionless and exert no attraction or repulsion on each other) : $PV/T =$ some constant. Or again, (vi) some economic laws formulated in terms of ideal markets, ideal managerial staffs and ideal flow of information. Strictly speaking, there are no instances of frictionless planes, ideal gases and ideal economic agents. In this strict sense, then, the corresponding conditionals have vacuous antecedents. Moreover universal conditionals of this sort are not always derivable from other laws, as for example, the gas law before the advent of kinetic-molecular theory. Such statements, however, are anything but dead lumber. On the other hand they may be said to constitute a central portion of all theoretical science. Idealization is a characteristic practice of theoretically mature science. And generalization in terms of limiting concepts lies at the heart of virtually all theory construction. Theories framed in terms of ideal processes, states, entities are vastly simpler and lend themselves much more readily to logico-mathematical treatment.[29] What is more, in an important practical sense, idealized universals such as (iii), (iv), (v), (vi), do not have vacuous antecedents in the sense in which universal conditionals about ghosts (such as i) do. Idealized conditionals are interpreted and applied in terms of realizable observations and measurements that can be made to converge, in the sense of practical approximation, to the idealized states as limits. Hence, though in a specious sense, vacuously

[28] Assuming there is no other net force on the body.

[29] The science of mechanics formulated in terms of point-masses, infinitesimal intervals and continuous variations is conveniently expressed in the language of the infinitesimal calculus and "well-behaved" mathematical functions. Mathematical calculi of finite differences and other alternatives to the infinitesimal analysis are generally speaking less simple and less powerful.

Similarly, one advantage to the limit-frequency interpretation of probability, an idealized concept, over some important alternatives is the immediate applicability of the highly developed theory of infinite sequences.

true, idealized universal conditionals are not only genuinely law-like in character, but they constitute a major portion of the nomic content of science.

7. *Empirical Import*

Reflection on actual scientific practice makes it clear that in order to be genuine scientific statements, universal conditionals must have empirical content. That is, they must assert something of an essentially matter-of-fact nature about the world, (i.e., time-space subject-matter). In satisfying this requirement such statements may be said to have *empirical import.* There are two ways in which statements may fail in this requirement : (1) They may be logically necessary; or (2) they may be empirically untestable.[30]

(1) Little need be said, by this time, about logically necessary statements. These were discussed in an earlier section and shown to occur throughout scientific discourse either as formal truths decidable, directly, on formal grounds or as analytic truths decidable on the basic of meaning relations though ultimately also on formal grounds. It was noted that whether they are formal or analytic, logically necessary statements are factually empty and dependent for their validation only on logical analysis. They are, as it is often put, decidable *a priori.* Thus, if nomic universals are to contribute to the factual content of what is concluded in any nomic inference, they may not be *logically necessary.* For to have factual import such statements must be susceptible either to falsification or to continued support by factual evidence.

In this connection it seems appropriate to note, as an elementary point of logic, that a logically necessary statement may be dropped from the premises of any deductive argument. Logical necessity, therefore, would deprive any universal conditional of a characteristic role of all nomic statements, namely that of serving as essential (uneliminable) premises of nomic inferences. This failure alone would rule out logical necessity as a feature of scientific laws.

(2) To require that a nomic statement be empirically testable is to require that it be possible to say in sense-experiential terms what sort of observa-

[30] The label, 'empirical import' seems preferable for the purpose of keeping the present considerations separate from the closely related but more general problem of *cognitive* significance or meaningfulness on which there has been a wide range disagreement and development of opinion over the past three decades.

tional results would count for or against it.[31] This requirement rules out statements of pure mathematics if these are conceived as non-empirical in character.[32] It also excludes statements generally belonging to traditional metaphysics, as for example, statements about monads (Leibnitz), a world spirit (Hegel), an *élan vital* (Bergson), noumena (Kant) or a demiurge (Plato). The question of whether statements of this sort have any meaning at all is debatable, its answer turning largely on one's philosophy of meaning and ontological preferences. But statements of this sort, as they are generally understood, do not have empirical significance. The reason for this is that the subject-matter they refer to is conceived in such a way as to make empirical detection impossible. It would, for example, be self-contradictory to speak of detecting "substances" or "things-in-themselves" in any *observational* sense of 'detecting', for these are, by definition, stripped of all sense-experiential attributes and hence of the possibility of being experienced. The same may be said of observing "causal links" as realities in their own right, quite apart from the states or events they allegedly connect. For by definition, all we ever observe *are* states or events. In general, scientists have little difficulty deciding about the testability or, as physicists usually put it, the *physical significance* of any nomic statement, though there will be problematic cases like statements about absolute motion, gravitons, and effects with orders of magnitude too small, in principle, to be detected.[33] But despite familiar difficulties in formulating precise criteria of admissibility along these lines, it is generally agreed that the descriptive terms in nomic statements must bear some sort of meaning connection to what one might call observational terms, that is, terms referring to what is more or less immediately and publicly detectable by direct experience (e.g., pointer readings, colors, tastes, etc.). This is essentially the empiricism at the heart of scientific method.

Reflection on what shall or shall not count as a scientific law brings to mind a list of possible minimal requirements which virtually no one would dispute. Our discussion under items 1-7, above, takes up a list (by no means exhaustive) of such requirements. Having satisfied these, however, a state-

[31] Those who with K. Popper deny the confirmability of laws would express this criterion only in terms of falsification. See : Popper, K., *op. cit.*, pp. 40-43.

[32] Some believe mathematics to be entirely reducible to logic. This would make all mathematical statements logically necessary, and therefore without empirical import. But even most of those who believe mathematics to be more than logic, would allow that empirical test, in the usual sense of this, is irrelevant.

[33] Such as orders of magnitude less than that of Planck's constant, as, for example, a length less than 10^{-27} centimeters.

ment may still meet with a wide range of opinion regarding its nomic status. Would a universal conditional containing a proper name, for example, qualify as a law? Some would say no. But, then, Kepler's "laws" which make reference to the sun would no longer be laws! Others would prefer that no statement is "genuinely" nomic unless it is deductively linked to other nomic statements in the systematic body of discourse. On such a view some of the generalizations of physics and chemistry would not have counted as laws before they were systematically tied to other generalizations. Examples would be the gas laws, the periodic law of the elements, the first law of thermodynamics and others. Similarly excluded would be many generalizations of biology, e.g., that every organism has a parent or that the life span of dogs is about ten years, or that the unit of biological structure is the cell or that the normal temperature of humans is 37°C, etc. By far most of the generalizations of psychology and some of the other behavioral sciences would have to meet with a similar fate. For in such sciences there is as yet relatively little systematic structure based on stabilized laws, statistical correlations and generalizations providing the main body of factual content.

And one can imagine far greater restrictions than these.[34] Indeed at such levels of option, it seems rather obvious that fixing the line between law and non-law is largely a terminological matter in which ordinary scientific usage provides only rough guidelines.

C. STATISTICAL LAWS

1. The Form of Statistical Laws

In the preceding section we took up laws of the universal form, the major concern being to develop a characterization for distinguishing universal laws from universals which would not ordinarily pass as laws. Several possible characteristics were listed which, together, could reasonably be viewed as

[34] Following trends in modern physical theory, it might be proposed that physical generalizations have some transformational invariance property or other, as for example, the property of being Lorentz-invariant. Roughly speaking a law, L, is said to be T-invariant if and only if there exists a set of reference frames, T, such that L retains the same mathematical form regardless of which reference frame of T it is expressed in terms of. Transformational invariance requirements of one sort or other can, indeed be profoundly linked to high elegance and great generality as well as logical simplicity of laws. But such requirements could be so restrictive as to deny all but a few generalizations the status of laws. Requiring Lorentz-invariance would, for example, mean that no generalizations in physics could be called laws except for Maxwell's theory of the

necessary for any statement to count as a universal law, though the sufficiency of this set of characteristics seemed somewhat less conclusive. Universal laws, however, are not the only kind of premises that "cover" subsumptive inferences. When subsumptive inferences are probabilistic, the covering nomic statements are statistical laws which, as a class, comprise the second of two major types of scientific generalization. We have already encountered them in discussing statistical nomic inference. We return to them now, after having discussed nomic universality, for some comparative considerations.

Unlike universal laws which assert that, without exception, all A's are B's, statistical laws assert that the probability of an A being a B is p. It seems appropriate to note, here, that the fundamental nature of probabilities and therefore of the precise manner of computing them is a matter of some contention. In actual practice, however, the probabilities occurring in laws of the kind under consideration seem to be uniformly interpreted as relative frequencies and, ideally, as limits of convergence of such frequencies for increasingly numerous instances.[35] It is for this reason that they are generally referred to as *statistical probabilities* and the corresponding laws as *statistical laws.* In general the probability, p, of statistical statements may take on values from *zero* to *one*, end-points included. Given the statistical character of p, however, a probability of *one* does not convert a statistical statement into a universal one. For, in principle, a sequence of relative frequencies can converge to the limit, *one*, though some A's are not B's. For similar reasons, a probability of zero does not convert a statistical statement into a universally negative one. That is, a sequence of relative frequencies may converge to the limit zero, though some A's are B's. Statistical statements are not, therefore, more general forms of universal ones, reducible to the latter for extreme values of p. They are radically different in the sense that while universal statements logically imply singular assertions to the effect that a particular A is also a B, statistical ones imply no such assertions. Consequently, though the premises of a statistical subsumptive inference may be true, they can only serve as an *inductive* base for

---

electromagnetic field and Einstein's theory of special and general relativity. (In Lorentz-invariance, T is the set of all reference frames which are moving uniformly with respect to each other.)

See : H. Goldstein, *Classical Mechanics*, Addison-Wesley Publishing Co., 1959, pp. 187-199.

[35] Strictly speaking, limits of convergence presuppose infinite sequences which, of course, are, in practice, unrealizable. The notion of a limit, however, may be regarded as an idealization capable of application in a finitized and approximative sense. Cf., earlier footnote under statistical nomic inference.

establishing the *probability* rather than the truth of the conclusions. And though there are those who would regard the *probabilistic systemization* thus achieved as something less than what is achieved by deductive systemization, it nevertheless comprises a vastly pervasive mode of structuring knowledge on virtually every level of cognitive discourse both scientific and prescientific. Thus, statistical laws play a crucial role in major portions of physical science, namely, statistical and quantum mechanics, while in the medical and behavioral sciences the probabilistic mode of nomic inference – both explanatory and non-explanatory – continues to be the major manner of systemization. On pre-scientific or common sense levels of discourse, explicit reference to probabilities is often not clearly made. Nevertheless whatever systemization occurs on such levels is often – perhaps for the most part – unmistakably probabilistic in character.

The form of a statistical law may be represented by the schema :

(1) The probability that an A is (also) a B is p.

where p is statistically understood, that is, understood in terms of the relative frequency of B's among A's. The statistical sense of p is, in practice, often made explicit by expressing (1) in any of the following, roughly similar, forms :

(1′) There is a tendency for a proportion, p, of A's to be B's.

(1″) The relative frequency in the long-run, of B's among A's is p.

(1‴) The limit of the relative frequency of B's among A's as the number of A's increases without limit is p.

The statistical interpretation of the probability term, 'p', is so pervasive that some authors take the primary form of statistical laws to be (1′)-(1‴) rather than (1).[36] Be this as it may, what is important to note is that what makes a law statistical is not the degree to which it has been confirmed, which is to say, its truth-status, but its form. It is erroneous to confuse probabilities assigned *to* a statement with probabilities mentioned *in* the statement. Thus, if a small proportion of all known physical objects contracted instead of expanding when heated, this would not make the statement, 'all physical objects expand when heated', a statistical law. Such a statement would still be universal in form but would be false. On the other hand, the statement, 'a very high proportion of objects, when heated, expand', *would* be a statistical law. Similarly, if, in some sense, one could establish the improbability of the law-like universal statement, 'All intelligent beings are earthlings', it would not thereby become a

[36] Nagel, E., *op. cit.*, p. 77.

statistical statement, but would, in its unchanged form, remain a universal one of highly questionable truth.[37]

2. *Nomic Statisticals*

The following are examples of statistical laws, expressed so as to illustrate some of the variants of the statistical form :

(1) The probability (relative frequency in the long-run) of tossing a head with a well-made coin is $\frac{1}{2}$.[38]

(2) The probability that in a sample of oxygen gas at –200°C a given molecule will have a velocity of from 200 meters per second to 201 meters per second is approximately 0.005.

(3) The probability that a molecule of air at an altitude of 100 kilometers will travel at least one meter before colliding with another molecule is approximately 0.36.[39]

(4) The proportion of cases recovering from pneumococcal pneumonia when properly treated with penicillin is 95 %.

But though all statistical laws have the form illustrated by our examples, does the possessing of this form by any statistical statement guarantee that it is a law? Here our considerations on the characteristics of nomic universals come to mind in connection with distinguishing nomic from what we might regard as "accidental" *statisticals*. Consider the following list of characteristics paralleling those which we discussed for nomic universality. Owing to the obvious parallelism only brief comments need be made here.

(1) *Statistical Conditionality* : The examples listed show that corresponding to the conditional form of nomic universals there is a somewhat parallel form for statistical statements. The following symbolization of (1) will serve to show this :

$$P(A,B) = p$$

This may be literally rend : 'The probability from A to B is p'. A and B are called the reference and conjectural traits (or classes), respectively, and

[37] It seems appropriate to note, here, that the assignment of probability to universal laws is a difficult and controversial topic. It is evident, of course, that no simple or direct statistical account of such probabilities is possible, since just one unfavorable instance falsifies any universal law.

[38] The property of being well-made is, of course, not to be understood here in terms of relative frequency but rather in terms of other physical measurements for establishing symmetry or balance.

[39] For laws similar to (2) and (3) and an elementary account of statistical mechanics see : D. Halliday and R. Rosnick, *Physics*, John Wiley and Sons : New York, 1966, pp. 571-615.

are roughly analogous to the attributes mentioned in the antecedent and consequent, respectively, of a universal conditional. The analogy with universal conditionals suggests the term *statistical conditionals* for characterizing statements of statistical probability.[40]

(2) *Lawlikeness* : Those attempting to differentiate allegedly "accidental" universals like 'all crows are black' from nomic universals might wish to carry the distinction to the case of statistical statements. If so, they would want to differentiate between allegedly accidental statisticals like, 'The probability of a swan being white, is greater than $\frac{1}{2}$', and nomic ones like, 'the probability of tossing a head with a well-made coin is $\frac{1}{2}$'. Any such attempt, however, must meet with the same difficulties that arose in connection with nomic universals. Indeed an analysis similar to Hume's for causal connections could be invoked to argue that *regular* statistical patterns correlating bird types with color are no less privileged than those correlating the structure of coins with the frequency of heads or, for that matter, the structure of atomic nuclei with their stabilities. Admittedly some such correlations can be deductively related to others – that is, explained, predicted or otherwise derived on the basis of others. But again, analogously to Hume's account of invariable "correlations" : What idea can one form of any nomic statistical connection except in terms of tendencies, patterns etc., that persist with more or less determinable stability? And even when we can explain these in terms of supposedly more fundamental ones, are we not eventually left with statistical regularities for which no explanation is at the time known? Some sub-nuclear statistics are today no more explainable than are the statistics of bird colors – although the former are undoubtedly the more fundamental in the *scala naturalis*. But of the two, and for the very same reason of fundamentality, it is the latter which would more likely be eventually explained and the former which may simply remain a *given* fact or – if one prefers – "accident" of nature.

(3) *Unrestrictedness* : Still, one has no difficulty in providing statistical statements having the form of our bonafide examples of statistical laws but which would surely not count as statistical laws.

[40] In even closer analogy with the conditional of nomic universals some writers have symbolized (a) as :

$$\underset{p}{(A \ni B)}$$

where the symbol '$\ni$' is referred to as a probability implication and superficially resembles the conditional, 'if ... then ...' of the sentential calculus – sometimes symbolized as '$\ni$'.
See : Reichenbach, *op. cit.*, p. 49.

For an illustration we need only modify our examples about the students in tonight's seminar. Consider a set of lots in a box, the set consisting of one lot for each student, on it being recorded whether or not the corresponding student is over six feet tall. Now, suppose that we draw lots from the box, noting the recorded height on each lot and replacing the lot as well as reshuffling the box after each drawing. If, as this process continues, the designation 'over six feet' comes up with a relative frequency which tends to stabilize to a definite ratio, say 1/3, then there is reason for the following statistical generalization :

(a) The probability of randomly selecting a student from last night's seminar, whose height is over six feet, is 1/3.

But though this statement has genuinely statistical form it is clearly no statistical law. And, one reason for this, as in the case of the universal conditional in the original, example is *restrictedness*. For no statement (universal or statistical) reporting a regularity confined to the set of last night's seminar students could, in any ordinary sense, be taken to be a law of nature. And, even if the statistics of the drawings were explained on the basis of the simply ascertained circumstance that, in fact, 1/3 of the students in last night's seminar were over six feet tall, the observed trend in the sequence of drawings would nevertheless count only as a "simple" (i.e. non-nomic) fact involving but one circumscribed neighborhood of the universe, namely our box of lots.

(4) *Indefinite Scope* : We have noted that for a universal conditional to count as a law it should not be possible to infer from the law itself that the individuals comprising the subject-matter are of a definitely limited number. A similar requirement seems appropriate where statistical laws are concerned, as in cases, for example, where the sequences refer to necessarily extinct populations such as 17th-century scholars. This requirement, of course, rules out from the category of laws any statement from which it is possible to infer that the population concerned has been *exhaustively* observed. As an example consider the following hypothetical case : Suppose that a *specifically* new mutation, M, recently originated, in captivity, in some unknown manner, was propagated until n individuals of the new species existed, and, later, became extinct. Suppose, also, that all n individuals had been examined and that m of them had been found to possess a certain trait, T. This, then, would be good reason to assert the following statement :

(b) All M's (that ever were, or will be) have been examined and m of them have been found to be T's.

Though (b) might in some sense or other count as a statistical statement, it would hardly pass as a statistical *law*. Statistical laws are statements of statistical probability and hence include the notion of indefinitely extendable sequences of trials. It may indeed be the case that all experimental sequences are ultimately destined to terminate and, beyond that point, to remain unextendable. But the open ended, general character required for calling a statement a law seems to rule out from the category any statement which, in itself, implies that the underlying sequences have exhausted the subject and hence cannot be further extended. If so, then the statement about the M's of our example would at most be not a law of nature, but a factual report about a dead species of the recent past.

5. *Causal Relevance; Non-randomness*

But, as in the case of universal laws (see, pp. 205-222), statistical statements, like those correlating height with attendance at tonight's seminar, would not count as laws regardless of all other disqualifications. Once again, the reason is (assuming our context of discourse to be only adult American males), that attending a seminar and, hence, attending last night's seminar, has no bearing on whether or not an adult American male will be over six feet tall. This fact, as we have seen (p. 217), may be ascertained by noting the relative frequency of heights over six feet (i.e., of attribute, O) for the subsequence constructed from the original one for adult American males, by selecting only those elements corresponding to adult American males who attend seminars (S). If the relative frequency of O's for the resulting subsequence, that is, the relative frequency of O's among A · S's does not differ significantly from the relative frequency of O's for the parent sequence we say that S has no bearing on O. But if attending any seminar (i.e., S) has no bearing on O, then neither has attending last nights' seminar (i.e., S′). Thus, though (a) is a genuinely statistical statement, it will not pass as a statistical law. The statistical probability of selecting an O among an S′ is indeed a determinate one but must be taken to be a matter of pure chance or accident. It seems appropriate therefore to characterize a statement like (a) not as a *nomic* statistical but rather as an *accidental* one.

6. *Statistical Statements with a Vacuous Reference Class*

An example of a statistical statement with a vacuous reference class would be :

(c) The probability that any ghost knows Latin is very high.

But in any attempt to draw a parallel between universal conditionals with vacuous antecedents and statistical "conditionals" with vacuous antecedents we run into a technical difficulty. Statements of the former sort are true though vacuously so.[41] Statements like (c), however, are of dubious significance if – as it seems – statistical probabilities must be understood in terms of sequences. For if the class of ghosts is empty then no sequence relating the two attributes concerned can be generated and the required probability becomes indeterminate.[42]

It is entirely possible, however, that probability statements with reference classes that, strictly speaking, are vacuous can be derived on theoretic grounds from premises containing some genuine statistical content. Statistical statements involving extrapolations to ideal attributes, as, for example, those involving ideal gases, absolute zero, ideal economic agents, etc., would be of this sort. Among others would be statements about objects assumed to be in the sort of places for which it is known that there are no such objects. A probability statement involving – say – the behavior of humans that, for example, have been in a space ship for twenty years, would be of this sort. Such statements may indeed be genuine statistical laws. And their scientific utility is entirely analogous to that of theoretically based, universal conditionals with vacuous antecedents. The reader may at this point wish to refer himself to our discussion of the latter, pp. 218-220.

(7) *Empirical Import* : Our discussion of empirical import for universal laws (pp. 220-222) applies entirely to the case of statistical laws. One point, however, invites some special consideration. Recall that logically necessary statements are factually empty in the sense that they are decidable on formal grounds only and are therefore compatible with any state of affairs whatever. Having no factual content, such statements cannot count as laws of nature – universal or statistical. Where probabilities are concerned, certain special statements, namely, the theorems of the probability calculus, also known as mathematical theorems of probability or, simply, as probability theorems, come under particular consideration. The great importance of these statements in probability computations often tends to obscure the fact that, in the final analysis, they are logically necessary statements and therefore have no factual import. In particular, when probabilities are understood as statistical metrics and, more specifically, as limits of relative frequencies, the theorems of the probability calculus become logically necessary statements which, though having statistical probabilities as their subject-matter,

[41] Assuming that we understand conditionals, in accordance with the stipulations of formal logic, then they are true if their antecedents are false.

[42] Arithmetically viewed, the relative frequency – hence the statistical probability – of an attribute occurring in a sequence having no elements is the indeterminate ratio, O/O.

nevertheless express no statistical fact whatever. That is to say, as in the case of any logical truth, they are devoid of factual content.

As an example of a logically necessary statistical statement consider the following :

If, *given the occurrence A*, the occurrence B and C are independent of each other, then the probability of both B and C is the product of the probability of B and the probability of C.

Symbolically :

(a) If $P(A \cdot B, C) = P(A, C)$, and if $P(A, B) = p_1$
and $P(A, C) = p_2$, then $P(A, B \cdot C) = p_1 p_2$

This formulation is a theorem of the probability calculus and may be shown to be analytic if interpreted statistically and in terms of limits of relative frequencies. An instantiation of it (also analytic) would be the following :

(b) If the probability of tossing a head with any coin is $\frac{1}{2}$, then the probability of two heads when two coins are tossed is $\frac{1}{4}$. This statement is indeed about statistical probabilities which as we understand them here are empirical relations. But, though in this sense it is statistical in character, it is, nonetheless, a logically necessary statement and therefore devoid of factual content – statistical or other. It cannot, therefore, pass as a statistical law; nor can we deduce any statistical fact whatever from it alone.

On closer inspection one readily sees that both (a) and its instantiation (b) do not have even the form exhibited by statistical laws, which, we recall, is "$P(A,B) = p$". They have, instead, the form : "If the probabilities of events A, B, C, ... are, respectively, p_1, p_2, p_3, ... then the probability of a composite event, Q, composed in some specified way of events, A, B, C, ... is p". This is the form of a universal conditional whose antecedent and consequent mention initial and derived probabilities, respectively. Initial probabilities are quantities which must be determined experimentally. They are matters of fact which must be provided as initial conditions for the application of any probability theorem. Derived probabilities, on the other hand, are quantities which may be deduced logically once we know the initial probabilities. They represent, therefore, no new factual content but, rather, a logical transformation of given derived probabilities. As a matter of fact, the probability calculus consisting of a set of axioms and derived theorems is a system of possible transformation of this sort, its theorems merely providing the means of calculating derived probabilities from given (initial) ones. These transformations reflect formal rather than empirical relations and therefore express nothing whatever about the world.

On these grounds alone, no probability theorem even when statistically interpreted can qualify as a statistical law of nature.

Failure to recognize the logically necessary – hence factually vacuous – character of probability theorems has been a source of some confusion in discourse about probabilities and, in some cases even, in practical applications of statistics. No purely logical truth can do as a law of nature, and the formal principles of mathematical probability are no exception. Any attempt therefore to implement such principles as laws whose instantiations are capable of yielding specific probability metrics describing actual events is bound to be groundless.[43]

[43] Particularly notorious in this respect have been confusions concerning at least two famous theorems of the probability calculus each of which have at times been read by some students of the subject in much the same way as they might read laws of nature having positive statistical content.

One of these is an historically important theorem due to Daniel Bernoulli and known as Bernoulli's Theorem. According to this theorem – if the probability of an attribute, B, in a sequence, S, of random instances, is p then as any set of n successive elements of S is expanded more and more, the probability p' that the relative frequency of B deviates from p by less than some fixed amount, d, approaches *one* as a limit. But though this theorem connects the abstract probability concept of the calculus with a relative frequency it is not a statistical law predicting any relative frequency at all from a merely abstract or uninterpreted probability p. The crucial point is that the theorem is merely a deductive (i.e., logically necessary) transformation of the *initial* or given probability, p, of a certain event to the *derived* probability, p′, of another event, namely, a certain relative frequency. The probability, p′, can have no more predictive content than the probability, p. And so, Bernoulli's theorem is no more than a logically necessary principle of probability logic. It is not a statistical law, and, by itself has no factual instantiations. Thus any attempt to extract statistical information from his principle, alone, is futile. (For a discussion, in some detail, of Bernoulli's Theorem see : H. Reichenbach, *op. cit.*, pp. 274-280.)

A second theorem of the probability calculus associated with similar confusions is the famous Theorem of Bayes which makes it possible to calculate the probabilities of so-called alternative causes B_1, B_2, ... B_+ of some attribute, C, given the occurrence of attribute. A, More specifically the theorem gives the following relation :

$$P(A \cdot C, B_i) = \frac{P(A, B_i) \times P(A \cdot B_i, C)}{\underset{n}{\Sigma}\left\{P(A, B_i) \times P(A \cdot B_i, C)\right\}}$$

But the probability on the left side of the above equation is only a transformation of the probabilities on the right side which must be available as initial probabilities. Once again, therefore, this theorem of the calculus like all the others has by itself no factual content. It is merely a means for deductively transforming given statistical information. Such a formulation therefore is not a statistical law or other nomic statement about nature. (A somewhat detailed discussion of Bayes' Theorem may be found in : H. Reichenbach, *op. cit.* pp. 92-95.)

D. SUMMARY

Our concern in this chapter has been with scientific laws. And the discussion has been developed in line with the empiricist view that no logically necessary statement (or its negation), that is, no statement which is entirely decidable on the basis of purely logical procedures can have any factual content. So that, a scientific law in order to have factual import must be a species of contingent (i.e., non-necessary) statement. In the context of these introductory considerations, it seemed natural at this point to pause to examine the role of logically necessary statements and more generally of logic itself in factual inquiry. For, once granted that logical truthts are devoid of all factuality, the question of precisely how they come to figure so crucially in factual inquiry seems to take on a special urgency. Moreover, the manner of answering this question is a key to understanding the role of mathematics, as an analytic discipline in empirical science.

Turning more directly to the subject of scientific laws, we noted the wide variety, as to scope and generality, of scientific statements which have systemizing power and which can serve as premises of nomic inferences. These range from singular dispositional statements about Smith's bad temper to the most encompassing abstract theories. And though there are some such statements which would clearly not pass muster as laws and others which would, just precisely where to draw the dividing line between non-laws and laws is to some extent a matter of verbal option. Still, students of the subject have proposed a number of features on the basis of which to characterize a given statement as a law. But before reviewing these, we distinguished *universal* from *statistical* laws with a view to considering each of these types one at a time. We then considered a number of bases for regarding any statement a universal law. It was easily seen at the very outset that having the form of a universal conditional was necessary but not sufficient to mark any statement as a law. A standard approach which attempts to characterize laws in terms of conterfactuals, was then examined and shown to founder in circularity. But other criteria such as indefinite scope of subject-matter, unrestrictedness in space-time, causal relevance or non-randomness, non-vacuousness, empirical import, deductive linkage with other lawlike statements could reasonably be viewed as necessary characteristics of any universal statement which is to count as a law, though the collective sufficiency of the set of characteristics considered seemed somewhat less conclusive.

We next turned to the second major type of scientific laws, namely, statistical laws – first encountered in our discussion of probabilistic nomic inference. Their distinguishing form as statistical statements were considered and this was found to be analogous to the universal conditionality of universal laws and also a *necessary* but *non-sufficient* trait for nomic status. The analysis of nomic characteristics for statisticals was then developed in a manner entirely analogous to that pursued in our account of universals, and again, similar conclusions were reached in connection with such criteria as scope, unrestrictedness, empirical import, etc.

Having examined the general structure of nomic (i.e., lawlike) statements, universal and statistical, we go on, in the next chapter, to consider a highly important class of *systems* of nomic statements, namely, *theories* – our considerations under this subject including such constructions as *models* and *analogies*. It was suggested very early in our discussion that theory is the feature which, perhaps, most decisively distinguishes scientific from everyday empirical knowledge. Let us turn, now, to this crucial ingredient of science.

VII

NOMIC STATEMENTS (II) : THEORIES, MODELS, ANALOGY

A. THEORY AND OBSERVATIONAL LAWS

The term, 'scientific theory' brings to mind a swarm of unsettled topics like hypotheses, laws, models, analogies, abstract entities, and more. But throughout the diversification of topics and views one can make out a significant convergence of opinion regarding the basic features of that part of scientific discourse which goes under the name of *theory*. Theory serves science in primarily two ways. In the first place, it provides very general accounts of various chunks and layers of the world, though we must be careful to note, here, that not all theories offer space-time descriptions that we can structurally visualize. In the second place, theories serve to systemize various matters of fact about the world. In this latter capacity, they provide nomic premises for subsumptive inferences whose conclusions are the matters of fact to be systemized. Very broadly speaking, then, we might say of theories that they are systems of statements about the world, some of which are of the nomic sort, though distinct in kind from observational laws with which they may be instructively contrasted.

In ordinary discourse, the term, 'theory', has the connotation of *unconfirmed explanatory hypothesis* as, for example, when we speak of a detective's theory as to who committed an unsolved crime.[1] This sense of the term 'theory' is often carried uncritically into discourse about *science*. Some of us no doubt recall our junior high school lessons in scientific method in which we were told that a scientific theory is an assumption (presumably law-like) whose truth is not yet established but which, when sufficiently confirmed, will become a law – meaning by this, a general fact of nature. But there is much to recommend against this employment of

[1] The notion of reflective thinking in search of understanding but not yet at the point of factual conviction is also suggested by the Greek term, *theoria*, which means contemplation or speculation.

the term, 'theory', in contexts like the present one. For it tends to distract from a scientifically important notion of theory which has little explicitly to do with how well-confirmed an hypothesis is. This is the sense in which an assumption is said to be *theoretical* not because of its degree of confirmation but, rather, because of the kind of vocabulary it has. Admittedly, the law-like assumptions that scientists usually call "theories" do seem, generally, to have lower confirmational status than other sorts of laws, as suggested by the old school distinction. But, if this is so, it is not because they have not yet won the status of established fact. This they can never do. And any expectation that repeated experimental successes will ever transmute theory into fact must stem from a failure to distinguish between two fundamentally different though intimately related components of scientific discourse. If theories are confirmationally "disadvantaged" – as it were – it is because of something much deeper than anything reflected in ordinary usage or in the old school distinction between "law" and "theory". The terminology or, as one might also put it, the conceptual content of what generally counts as scientific theory differs in very important respects from that of statements reporting observed regularities in nature. The appropriate distinction here seems to be not between law and theory but rather between two kinds of laws : *theoretic* laws (which constitute theories) and *observational* laws. The latter's vocabulary makes direct reference to what we are inclined to call *observables*, that is, things which are sometimes said to be directly perceivable, with or without the help of instruments. Quantities like temperatures, pressures, lengths and weights as determined by scale readings, by patches of color, by sounds, etc., are of this sort. And general statements about such quantities, comprise at any given time, the standing body of scientific fact. Examples of such laws would be Snell's law of refraction which reports that for a ray of light passing the interface between two media the ratio of the sines of the angles of incidence and refraction is for the two media independent of these angles;[2] or the law that whenever nitrogen and hydrogen react to form ammonia the volumes involved are in the ratio of 1:3:2, respectively, or the principle that, if in a free market the supply of any given commodity is fixed, an increase in demand will result in an increase in price. Theoretical

[2] Remembering that the concept of *ray* is an idealization, one could, with good reason, maintain that Snell's law has theoretical content and might therefore be classed under theory, though, of course, it makes reference to much that is directly experimental in character. This example, along with many others that could be given, is debatable and serves to caution against attempting any "hard cut" between theory and experimental law.

vocabulary, on the other hand, though it occurs as descriptive terminology as, for example, 'electron', 'molecular velocity', 'radiation field', 'suppressed wish', 'quantum mechanical state', etc., does not make direct reference to so-called observables. Indeed, for some formulations, it seems highly dubious that one can appropriately speak of theoretical terms as making any reference at all to existential subject-matter; though ultimately they must be coordinated, however indirectly, with experimental ideas. This sort of vocabulary occurs typically in a wide class of theories sometimes referred to as *mathematical models* (these will come up again in later discussion) and examples are abundant in mathematical physics.[3] But even where theoretical vocabulary does seem to make direct reference to physical "realities" as in the case of terms like 'electron', 'molecular velocity' and 'radiation field', the subject-matter of reference is not directly observable in the sense that tables and chairs are. Thus, though theories, like all empirical hypotheses, must be supported by factual evidence, they cannot be established in quite the same sense that statements about weight and volume ratios in chemical reactions or pressure, volume, and temperature relations for gases or color intensities emitted by glowing gases can. Theoretical entities are, in an important sense, only indirectly observable. We observe planets and pendulums themselves but only the *tracks* (or other effects) of beta particles, and only the conduct which we take to be indicative of motives. One might wish to maintain that even in the case of ordinary objects like tables and chairs we observe only effects, that is, colors, textures, odors, etc., and that hence there is no basic difference between theoretic and any other entities. To this, however, may be replied that, typically, in the case of theoretic subject-matter, what we *perceive* is, in basic respects, quite distinct from what we suppose we are *detecting*. Though we speak of an atom's diameter we never, in fact, qualitatively perceive a particle with a distinct diameter whose length, in turn, we can measure in the manner in which we ordinarily measure lengths, i.e., by visual comparison with standard lengths. Instead we do cross-sectional scattering experiments in which we count scintillations or pulses. We speak of molecular velocities but never "see" any individual molecules in motion. Instead we determine diffusion rates. Cloud chamber tracks are not moving alpha particles and spectral lines are not orbit-jumping electrons, though we may suppose some such tracks and spectral lines to indicate the presence of alpha particles or the occurrence of electron jumps. Theoretic subject-matter is

[3] E.g., the state-variables of quantum theory, terms alluding to displacement "currents" in electro-magnetic theory, to "centrifugal" forces in classical dynamics, etc.

detected, not perceived. We perceive or "see" falling bodies and planets, but only *detect* gravitational fields. And the latter mode of ascertainment involves a variety of systematically postulated connections which relate whatever is actually perceived to the entity which the perception is supposed to be a sign of. If it be insisted that not only tables and chairs but also atoms and electrons are perceived then the appropriate reply seems to be that if *such* entities are in any sense perceived then *they are not perceived as conceived.* We assume the existence of minute space-time particles or of electro-magnetic waves but do not – in fact, cannot – by their very nature, *perceive* them *qualitatively* as particles or as waves. Thus, to say, without further qualification, that one "perceives" such entities, could be misleading if not altogether at variance with what we ordinarily mean when we say that we *perceive* something. Undeniably, defensible claims can be made to the effect that laws about observables, i.e., observational laws, also involve theoretic assumptions. Thus in the observational law that for a given amount of gas, at fixed temperature, the volume is inversely proportional to the pressure of the gas, there are theoretical ingredients in the notions of pressure and temperature. The concept of pressure includes more than the notion of mere gauge-readings. It involves the notion of force which, in general, is theoretic in character.[4] Similarly statements about temperature and temperature determinations involve a variety of theoretical notions and assumptions as, for example, that of heat exchange and thermal equilibrium between gas and thermometer, or certain assumptions about the heat capacity, conductivity and dilational properties of the thermometer. It would seem futile to resist the claim that there are no clear-cut criteria for separating theoretic from observational statements. But it would be hardly short of quixotic to deny the distinction altogether. Indeed, as we have already noted in an earlier chapter, theoretic assumptions are what take us from everyday knowledge to science. The former does not go much beyond formulating observational laws. Precisely formulated though these may be, they are narrowly circumscribed in scope and largely reportative of what is actually "seen" to happen. Theory on the other hand is what gives science its characteristic ability to subsume wide ranges of phenomena under relatively few and very general assumptions about *underlying* subject-matter. It is theory which brings accumulated factual

[4] We do in certain cases claim to "perceive" forces directly, in terms of the effort exerted to resist them. Serious objection, however, may be raised to identifying any measure of such effort with force as a dynamical quantity. In general, we do not perceive forces but *detect* them by means of the deformations or the motions which we suppose them to produce.

know-how to the levels of what we call science. The characteristics of theory which give it its systemizing power are a central concern of the philosophy of science and will be the object of our considerations in the sections which follow.

B. THE FORMAL STRUCTURING OF THEORIES : AXIOMATIZATION

Theories, like all nomic formulations, function as general premises of nomic inferences in the subsumptive systemization of scientific knowledge. More specifically, they are *sets* of general statements and, typically, consist of one or more existential assumptions for introducing appropriate theoretical subject-matter, together with *laws* that this subject-matter must satisfy. Such subject-matter must be introduced by postulation. For as we have noted it does not arise in ordinary experience. It may be postulated as genuinely existential subject-matter lacking the requisite sense-properties for direct observability, as in the case of micro-entities, psychological states, physical fiields, etc. Or it may consist of theoretically convenient artificial "constructs" and idealized or limiting cases such as frictionless planes, perfect vacuums, ideal gases, mass points, etc. which, if they correspond to any physical subject-matter at all, do so only in some explicitly approximative sense.

In actual scientific practice, the logical structures of such systems are only partially explicit even where mathematical elaboration provides considerable formal structure to large tracts of theoretic exposition, as in the more developed sciences. Consequently, the reconstruction and further systemization of theory, even long after it has been in use, is a going concern of both logicians and practicing scientists. In this respect, the attempt to achieve full structural explicitness results in what is known as *axiomatization*. And this consists of regarding a theory, T, about a domain of subject-matter, D, as having essentially two components or aspects, namely, an *abstract* or *formal calculus* and an interpretation in D of such a calculus. These are sometimes called the theory's *syntactical* and *semantical* aspects, respectively.[4] Briefly characterized, a formal calculus gives the theory's structural or "internal" characteristics. Its interpretation relates it to the domain, D. This, however, is not to say that in actual scientific

[5] This is terminology developed in the logic of artificial languages where the distinction between matters of formal structure or *syntax* and matters of referential meaning or *semantics* have figured centrally. The application of this terminology here appropriately suggests the close parallelism between a formalized theory and an artificial language.

practice, working theories are explicitly formulated as abstract calculi and their interpretations. The analysis merely suggests that, in principle, any theory may be explicitly reconstructed or, even formulated at the time of its creation, as an abstract calculus together with its subsequent interpretation. In framing an abstract calculus for the theory, T, one is customarily said to be *formalizing* T. An interpretation is sometimes called a model for T – though we shall refrain from using this last label because of its variegated senses to be taken up in a later section. The general method of expositing theory in terms of these two basic components, namely a calculus and some interpretation, we shall refer to as *axiomatization.*[6]

But now, for a closer look at these two aspects of theory : the first, namely the calculus, is an abstract deductive system which delineates the theory's logical contours as follows :

(1) It contains a set of *undefined* or *primitive* terms, also called *elements*, each having its own explicit form. An element, for example, may have the form of a one-place predicate like the term 'electron' which requires only a single individual for its application as in 'x is an electron'; or it may have the form of an n-place predicate like the terms 'congruent' and 'between' which require more than one individual for their application, as in saying : 'x is *congruent* with y', or x is *between* y and z. Also, there is stipulated the manner in which these elements may be combined with each other to form allowable expressions or formulas of the system. In axiomatizations of the empirical sciences, logic and major portions of standard mathematics are generally assumed without explicit formal acknowledgment. Thus, for example, such mathematical notions as those of number, of numerical equality, of certain operations on numbers, etc., along with the concepts and principles of logic, are used freely, though these are not explicitly introduced as primitives or axioms of the calculus. Primitive terms may be designated by either letter-symbols or words like 'mass-point', 'line', etc. The latter suggesting some intended future *application.*[7]

[6] In addition to its use in the exposition of scientific theiory, axiomatization is held to be the ideal mode of exposition in mathematics where the method has seen its major development and applications.

[7] Recall in this general connection that in order for a theory to introduce theoretical subject-matter, it must include some statements of the existential form. The stipulating of primitive vocabulary for a theory generally achieves the formal phase of this objective. A typical formulation introducing the term, 'P', for the notion of (dimensionless) particle in an intended axiomatization of particle mechanics might, for example, be this :

Let p be any member of a finite, non-empty set, P.

The potential existential import of this set-theoretic form of statement is evident, and it hinges on the stipulation "non-empty". Such formulations, however, are purely abstract

(2) Some formulas of the calculus have the logical structure of statements. Several of these – preferably a mere few – in some cases because they bring to mind simple, intuitively plausible truths about an intended domain of application – are designated as the axioms of the calculus. The set of axioms then serves as a starting-point, or premises, for deriving logical consequences or *theorems* of the calculus.

A system, thus constituted, comprises the essential logical framework of the theory-to-be. The axioms and the derived theorems, one might say, are the logical skeletons of the theory's basic assumptions and of its deductive consequences. Similarly, the primitive terms which these contain are the formal silhouettes of the basic concepts of the theory, thus providing, in structural outline, the theory's basic vocabulary, a vocabulary whose formal delineaments are established by the stipulations of (1) and (2), above. Accordingly, it is sometimes said that the elements of a theory are "implicitly defined" by the axioms and the rules of an abstract calculus.[8] This, however, is somewhat misleading. For though "The stipulations of an abstract calculus fix the formal features or syntax of any theoretic vocabulary, they leave the subject-matter reference of all such vocabulary quite open until determined by a meaning assignment or *interpretation*. And the number of possible interpretations is indefinite. Thus, if the rules and axioms of any formal calculus are to be considered definitions of any sort, it should be kept in mind that such "definitions" do not invest any term with *referential* meaning. They establish the logical forms of terms and the formal relations which such terms may have to each other. But they do not confer on terms the status of what would, properly speaking, be called genuine *predicates* or terms that designate attributes of things.

---

and actually make no reference to existential subject-matter. Their existential content is eventually "filled out" when the abstract calculus is finally interpreted. It is then that such formulations become full-fledged existential statements. Thus, unlike informally constructed theories in which the existential assumptions are made explicitly and in one fell swoop, axiomatized theories introduce their existential content in a stepwise fashion which requires both abstract formulation and later interpretation. Contrast our formal illustration, for example, with an assumption like the following, possibly taken from the kinetic theory of gases :

There are sets of identical particles of which a gas consists and which do not exert forces on each other except during collision.

[8] Hilbert, for example, when stating his geometric axioms of congruency, says : "The following axioms define the notion of congruency or displacement". See : D. Hilbert, *The Foundations of Geometry*, Trans. E. J. Townsend, Open Court : La Salle, Illinois, 1938, p. 12.

This being the case, an abstract calculus does not, *per se*, make a theory about any subject-matter. For with undefined vocabulary its axioms and theorems are not statements at all, but only statement-forms saying nothing about anything and therefore capable of neither confirmation nor falsification.[8] Because of its formal character, such an abstract system may be regarded as essentially belonging to the formal disciplines of logic and mathematics. But unless appropriately interpreted it cannot serve as empirical science.[9]

Traditional examples of formal calculi are : Peano's formalization of arithmetic, using the primitive terms 'zero', 'successor', and 'number'; and the various axiomatizations of Euclidean geometry, using primitive terms that correspond to such notions as point, between-ness, congruence, equidistance, line and plane, the choice of primitives among these depending on the particular axiomatization. More recently, parts of physics have been reformulated axiomatically. Axiomatizations, for example, are available for quantum mechanics [10] and for rigid body mechanics.[11] An abstract axiomatization of particle mechanics may be framed in terms of expressions like 'particle', 'force', 'position', 'mass' and 'time' along with vocabulary belonging to set theory and the arithmetic of real numbers, e.g., 'set', 'function', 'finite', 'real number interval', 'differentiable in a certain set', etc. It would, then, contain a set of axioms prescribing certain mathematical restrictions on some of the parameters governing a system of particles, S (e.g., differentiability, real number values, etc.) and providing the mathematical forms of the basic laws of particle mechanics (i.e., Newton's second and third laws). A further elaboration of the calculus would introduce new terms like 'momentum', 'kinetic and potential energy', etc., all defined in terms of the primitives. When this has been done one can then proceed to derive a wide variety of abstract formulas or theorems. If sufficiently elaborate assumptions are made concerning the term 'force', then still more theorems are derivable having a formal resemblance to such

[8] A given set of axioms may happen to be *inconsistent* and shown to be so. This kind of formal "invalidation", however, is no part of what we mean here by 'falsification'.

[9] It seems appropriate to note at this point that in axiomatizing a given theory it is, in general, possible to select primitive terms and to designate axioms in a variety of ways. For example, several axiomatizations of Euclidean geometry have been given in modern times, each starting with a different set of primitives and axioms, then using these to formulate all other geometric notions and to derive the other principles of geometry.

[10] Lindsay, R. B. and Margenau, H., *Foundations of Physics*, John Wiley & Sons : New York, 1936, pp. 405-420.

[11] Adams, Ernest, *Axiomatic Foundations of Rigid Body Mechanics*, Dissertation, Stanford, 1955.

familiar mechanical laws as the conservation of momentum, the conservation of mechanical energy, and even Kepler's laws. It is of course understood that despite the suggestive terminology (mass, time, position, etc.) such a system would, as it stands, be only a set of algorithms for abstract calculation – ultimately only a pure formalism.[12] In order to make it empirical science, one would have to interpret its vocabulary in observational terms. Examples of abstract calculi for scientific theories are not readily found in standard scientific texts. This is in part due to the fact that, typically, scientists do not, at the outset, construct their theories as abstract calculi to be subsequently interpreted for application to existential subject-matter. In modern times, not until very recent science has there been some theorizing using mathematical techniques resembling the abstract axiomatic method. Axiomatizations of mechanics and other scientific theories have, almost always, been only reconstructions, offered by logicians rather than by working scientists. And these have, for the most part, come after a long history of successful application of non-formalized versions.[13]

The detailed elaboration of an abstract calculus for any scientific theory would take us far beyond our present concern. But a simple example will

[12] An axiomatization of particle mechanics may be found in : Suppes, P., *Introduction to Logic*, Van Nostrand Reinhold : New York, 1957, p. 294.

In essence, it consists of the following set of primitive terms :

p	(may be read as "particle, p")
q	(may be read as "particle, q")
m(p)	(may be read as "the mass of p")
r(p)	(may be read as "the position of p")
t	(may be read as "time, t")
f(p,q,t)	(may be read as "the force, f, on p due to q at time, t")
g(p,t)	(may be read as "the net force, g, other than f, on p at time, t")

Then assuming the vocabulary of arithmetic and set theory, there are also the following axioms :

(1) The non-empty set, S, is finite.

(2) The set, N, is an interval of real numbers.

(3) $(p \in S) \rightarrow \{r(p) \text{ is doubly differentiable on } N\}$

(4) $(p \in S) \rightarrow \{m(p) \text{ is a positive real number}\}$

(5) $(p \in S) \cdot (q \in S) \cdot (t \in N) \rightarrow \{f(p,q,t) = f(q,p,t)\}$

(6) $(p \in S) \cdot (q \in S) \cdot (t \in N) \rightarrow \{r(p,t) \times f(p,q,t) = r(q,t) \times f(q,p,t)\}$

(7) $(p \in S) \cdot (t \in N) \rightarrow m(p) \{D^2 r(p,t) = \sum_{q \in S} f(p,q,t) + g(p,t)\}$

[13] Axiomatization is regarded by many mathematicians as the essential aspect of mathematical method. Moreover, in the newer branches of mathematics, it seems to have become the pervasive mode of exposition. Indeed it is largely to mathematicians that we owe most of its development. Nevertheless, much of standard mathematics, like science, has been axiomatized long after it was developed in only quasi-formal fashion.

help set the framework for further discussion. Consider, therefore, the following miniature calculus in which we assume the language of formal logic.

(1) As our elements we introduced the primitive predicate, R, and the individual variables 'x', 'y', 'z', ... (The variables might have been simply assumed without comment as part of our legacy from formal logic.) 'R' is a two-place attribute expression having the explicit form, Rxy, which we read : *x has relation, R, to y.* (It is of course understood that the expression, 'Rxx' will also be allowed. It represents the case in which x has the relation, R, to itself.)

(2) We adopt the following axioms :

- (i) For any x, any y and any z, if Rxy and Ryz, then Rxz. (This expresses the so-called transitivity of R.)
- (ii) For any x and any y, if not-Rxy then Ryx.

With these we can prove a variety of theorems involving the relation, R, as, for example, that :

- (a) For any x, Rxx (This expresses what is called the total reflexivity of R.)
- (b) For any x, any y, and any z, if Rxy and Ryx and not-Rxz, then not-Ryz.[14]

2. *Interpretation*

As has already been noted, an abstract calculus, K, whose axioms are P cannot serve as a particular scientific theory until a non-empty domain of individuals, D, – one might wish to say, here, an *ontology* – has been *coordinated* with K. This is done by explicitly providing specific rules or conventions for assigning appropriate meanings to the undefined terms of K. Such rules are variously called coordinating definitions, semantical or meaning rules, operational definitions, or rules of correspondence. Roughly speaking, such rules tell how to replace the undefined terms in K by descriptive terms meaningful in D, so as to convert the axioms, P, into a

[14] More symbolically our axioms and two theorems may be succinctly represented as follows :

Axioms :

(i) $(x)(y)(z)\ (Rxy \cdot Ryz \rightarrow Rxz)$

(ii) $(x)(y)\ (\sim Rxy \rightarrow Ryx)$

Theorems :

(a) $(x)\ Rxx$

(b) $(x)(y)(z)\ (Rxy \cdot Ryx \cdot \sim Rxz \rightarrow \sim Ryz)$

set of statements, P′, about the individuals in D.[15] The set of all assigned terms constitutes the second of our two root components of a theory and is called an *interpretation* of K. Alternatively, one may prefer to call the set of resulting statements, P′, or even the domain, D, the interpretation or – one may speak of interpreting P instead of interpreting K. All of this, however, is a matter of inessential option. Interpretations of abstract systems may be grouped under two general categories. The subject-matter may be either of the purely logical sort such as numbers, abstract sets, the relations of logic, etc., or it may be of the empirical sort such as mass-points, measurable lengths, electric charges, etc. We call the first, *logical interpretations*, the second, *material interpretations*. (The term, *empirical interpretation* is also quite appropriate here and seems to be somewhat more in line with prevalent usage.) Our own miniature calculus is subject to a variety of interpretations both of the logical and material sort. For a logical interpretation, choose the domain of interpretation, D, as the set of natural numbers, i.e., the set of non-negative integers, and let the relation, Rxy, be the numerical relation, $x \geqslant y$. Axioms (i) and (ii) then become assumptions about the relation designated by '$\geqslant$' which is meaningful in arithmetic. Moreover, it should be noted, the interpretation we have given of our mini-calculus satisfies its axioms; that is, it converts them into statements involving the arithmetical relation, $\geqslant$, which are *true* of the assigned domain, i.e., of the natural numbers.

To exemplify a material interpretation of our calculus, take D to be the domain of material (i.e., space-time) states of affairs, events, or objects, and let Rxy be the relation : *x is regarded by humans as equally or more desirable than y*. We express this more conveniently as : *x is weakly preferred to y*. Axioms (i) and (ii) then become assumptions about human preferential behavior. And they may or may not not square with the way humans actually make preferential choices.

At this point it seems appropriate to illustrate a device which is important in the axiomatic construction of theories and to which we shall return for some further consideration in a later section. This is what is known as

[15] A more explicit manner of putting it would be something like the following : The correspondence rules tell how to assign to each predicate letter of K an n-place relation in the domain of interpretation, D, and to each operation letter of K an n-place operation in D, and to each individual constant an individual of D, the individual variables of K ranging over the set D. For more precise and generalized notions of interpretation along these lines, see E. Mendelson, *Introduction to Mathematical Logic*, Van Nostrand Reinhold : New York, 1964, p. 49. A less formalized discussion will be found in P. Suppes, *op. cit.*, pp. 64-79.

formal definition or simply *definition*. Thus, from the viewpoint of describing human preference we might want to introduce two more or less familiar notions, namely, *preferential neutrality*, or indifference, and *strong preference*, not as primitive or elemental concepts but as derivative ones. Accordingly, we stipulate the following definitions :

(1) Let 'Nxy' be shorthand for 'Rxy and Ryz'

(2) Let 'Pxy' be shorthand for 'not-Ryx'

Some of the theorems of our interpreted calculus may now be expressed in terms of the newly defined relations Nxy and Pxy. A few of these are :

(a') For any x, Nxx

(b') For any x, any y and any z, if Nxy and Pzx, then Pzy.

(c') For any x, and y, and any z, if Nxy and Nyz, then Nxz

(d') For any x, any y, and any z, if Nxy then neither Pxy nor Pyx.[16]

Our interpreted system may now be said to be about the notions of human preference and indifference to alternatives. But though our definitions have added vocabulary to our theory of behavior, they have added no new content. For each of the defined expressions, or the *definiendum*, is entirely replaceable by the corresponding old expression, or the *definiens*. Thus the expression 'Nxy' may be entirely replaced by its defining expression 'Rxy and Ryx'; and similarly for 'Pxy'.

3. Epistemic Status of Interpretations

At this point we might want to ask the following question about interpretations of abstract systems : Are we interested only in those which render the axioms true? The answer seems to depend largely on our purposes. If by interpreting a calculus, K, we intend to construct a theory about some domain, D, then it would seem quite natural to consider only those interpretations which convert the axioms of K into statements which are true in D. That is, what we would want is a body of true statements about the subject-matter, D. Now, when the interpretation is of the logical sort the resulting statements would be truths of logic or of mathematics. And few

[16] More symbolically expressed these theorems would be stated as follows :

(a') $(x)\,Nxx$

(b') $(x)(y)(z)\,(Nxy \cdot Pzx \rightarrow Pzy)$

(c') $(x)(y)(z)\,(Nxy \cdot Nyz \rightarrow Nxz)$

(d') $(x)(y)(z)\,(Nxy \rightarrow \sim Pxy \cdot \sim Pyx)$

It may be noted that (a') and (b') are equivalent to theorems (a) and (b) but seem to be somewhat simpler and more plausible or transparent on our intuitive notions of indifference and strong preference.

would nowadays deny that these are *necessary truths* – also called, *logical truths*. In general, the validations of such truths, when they can be carried out, are conclusive ones, which have led to much established logic and mathematics.[17] Hence it seems quite natural in the constructing of mathematical and logical theory to restrict one's concern to *true* logical interpretations. But, then, the following must also be kept in mind : One does not always interpret an abstract calculus for the immediate purpose of constructing a particular theory. An interpretation whose domain is D may serve other purposes. It may for example be used to show that the axiom set $A = \{A_1, A_2 \ldots A_n\}$, of K is consistent. In such a case, as in the case of theory construction, the set of converted axioms, $A' = \{A'_1, A'_2, \ldots A'_n\}$, consists of statements that are *all* true in D. But an interpretation may also be used to show that some one axiom, say, A_j, is independent of the others. This is done by providing what is essentially a *false* interpretation of K, more specifically, one for which A'_j is false and all the others, i.e., $A'_1, A'_2 \ldots A'_{j-1}, A'_{j+1}, \ldots A'_n$, true.[18] In the case of an empirical interpretation – even where the immediate aim is to construct an empirical theory – another sort of difficulty arises in restricting one's concern to "true" interpretations. For in such a case, the statements of A′ have matter-of-fact content and may – at least in certain cases – be more properly characterized as *probable*, in some appropriate sense, rather than simply as "true". A deductive system with such premises would certainly not be scientifically otiose and could well be of vital interest for scientific purposes. Of particular interest is the case where the domain, D, coordinated with K by an empirical interpretation, consists of theoretical subject-matter, e.g., electrons, atoms, genes, etc. The calculus, K, then becomes an empirical theory whose theoretic assumptions are the statements of A′. But even granting that there is a clear sense in which a theoretic statement can be said to be simply true, there is serious question regarding how one can ever decide that such a statement was true. Indeed given the role of theories in nomic inferences – for example, the use of theory in prediction – any insistence that a set of theoretical assumptions must

17 This is not intended to suggest that there are no statements in logic or in mathematics which we lack the knowledge or ingenuity to decide on or which are quite undecidable, in principle.

18 It is a basic theorem on deduction that a statement, 'q', is *not* logically independent of a statement, 'p'; that is, 'q' is a logical consequence of 'p' if and only if the implication : 'if p then q' is *logically* true. But this means that 'if p then q' must come out true on all possible interpretations, which in turn means that there may be no interpretation that makes p true but q false.

be true in order to be scientifically operative seems profoundly contrary to basic scientific practice. Indeed, such a requirement would, it seems, disqualify virtually all known theory. It seems appropriate, therefore, when assigning a *material* interpretation to any calculus, for the purpose of constructing an empirical theory (and this is primarily our concern, here) to allow only interpretations which are not *known* to be false. Any requirement stronger than this would be an idealization so divergent from scientific practice as to hold little more than merely academic import. For, in scientific practice, theoretic assumptions can play crucial roles before any sort of epistemic status is significantly established for them.

4. Essentialization

Logical interpretations, by themselves, cannot yield scientific theories. The deductive systems which they give rise to treat of purely logical or mathematical subject-matter only and remain, therefore, a part of logic or mathematics throughout. In order to result in scientific theory an interpretation must make reference to existential (time-place) subject-matter and hence must be of the material sort. The interpreted axioms then become a set of statements about some time-space domain, which is to say, about some portion of the world. For example, a body of statements about optical phenomena may be obtained from abstract Euclidean geometry only if we coordinate physical subject-matter with its primitive terms. Thus we may coordinate the notions, *very small light-source* and *light-path in a homogeneous medium*, to the terms 'point' and 'line', respectively. Some of the axioms will then become statements in the empirical science of geometric optics, that is, statements about the domain of optical phenomena.[19]

We have noted that, for the purpose of theory construction, it would be desirable for an interpretation of calculus K to satisfy the axioms, A, of K, that is, to convert them into statements, A′, which hold in the domain of interpretation, D. In the case of material interpretations, however, we found that this requirement had to be replaced by the considerably weaker one that *A must not be known to be false*, in order to allow for the role of theory in empirical science. This, of course, does not rule out our being primarily interested in theories which promise, however slimly, to satisfy the axioms in the sense of succeeding in future applications. The underlying

[19] To complete the interpretation we would also have to assign physical significance to other primitives of the abstract geometry, such as the predicates, 'plane', 'congruent', 'between', etc., depending on the particular axiomatization, of which there are several.

assumption in all these considerations has been that the set, A′, resulting from the material interpretation of K consists of *contingent* statements, meaning by this that the truth-status of these statements rests on empirical fact rather than on logic alone. But though this assumption would seem quite natural for any statement whose subject-matter is empirical, it meets with difficulty in the face of the following consideration. There seem to be two ways in which a material interpretation can satisfy a set of axioms, A : A material interpretation may satisfy the axioms by converting them into statements which because of their logical form are not *necessarily* true and hence whose truth must be decided on the basis of contingent fact. Or, it may satisfy the axioms in another way, that is, by converting them into statements whose truth, though not decidable on purely *formal* grounds, seems nonetheless to be necessary by virtue of the meanings of the constituent terms.[20] For terminological convenience, label the first kind of interpretation *contingent*, the second *analytic*. Statements which seem to be true merely by virtue of the meanings of their constituent terms are common in ordinary language, a familiar illustration often given for them being : 'All bachelors are unmarried'. In modern philosophy statements of this sort have been traditionally labelled "analytic" because their truth seems to be entirely decidable by the logical analysis of meaning, alone, and hence without recourse to contingent fact. The claim that such statements are logically necessary is usually buttressed by maintaining that their denials are self-contradictory. This is tantamount to saying that their falsity is logically impossible, or, equivalently, that their truth is necessary. It is customary to contrast such statements with so-called *synthetic* statements whose truth cannot be established by logical analysis of any kind and hence requires a factual base. The necessary character which analytic statements seem to have has accorded them the status of logical truths – a status which is essentially the same as that which formal logic attributes to statements that are true not by virtue of the meanings of the predicates used but, instead, *regardless* of such meanings and, hence, true by virtue of logical form only.[21] Contemporary philosophical analysis, however, has seriously challenged the tenability of the synthetic-analytic distinction by bringing into question the notion of synonymy on which it rests.[22] Involve-

[20] It is important to note that this is a definite possibility despite the fact that such terms refer to empirical subject-matter.

[21] More precisely, a logically true statement would be one which is true for all possible interpretations in a non-empty domain of individuals. Such statements are also alternatively labelled "valid" or "universally valid".

[22] To show that any statement is analytic it is in general necessary to interchange

ment with this thorny problem in the philosophy of language is, however, much beyond our concern here. Fortunately, the *formalization* of theory, i.e., the construction of abstract calculi can be carried out without reference to statements whose truth depends only on the meanings of predicates used. Moreover, formal logic and mathematics – and these are ordinarily presupposed in axiomatizations – can also be reconstructed without such reference.

The question of analyticity, however, is unavoidable in connection with interpretations. One can easily imagine assigning an empirical meaning which will satisfy the axioms in such a manner as would render them analytic in whatever sense 'All bachelors are unmarried' is analytic. For despite all demonstrated difficulties at the heart of the traditional analytic-synthetic distinction, there are linguistic contexts in which it is possible to identify statements which are decidable as true solely on the basis of the meanings of the predicates used. In appropriately fixed linguistic contexts, therefore, (i.e., contexts with fixed meanings of terms) such statements are true *come-what-may*. They are therefore factually empty for, any statement compatible with all possible states of affairs, can describe *no* state of affairs.

This leads to the rather awkward result that a material interpretation of an abstract calculus can satisfy the axioms yet yield a body of statements or theory that, strictly speaking, says nothing about its subject-matter. Our considerations indicate that such will be the case when the material interpretation is of the *analytic* sort. To illustrate we return to our geometric example in which we interpreted the terms 'point' and 'line' as : *very small light-source* and *light-path in a homogeneous medium*, so as to convert some of the geometric axioms into contingent statements in optics. This time, however, interpret the term, 'point', as *point-source of light* and the term, 'line', as *ray of light*. This meaning assignment, though seeming to be approximately synonymous with that of our last geometric example, can be understood in an essentially different sense. We may construe the predicate 'point-source of light' in such a way as to include all the properties imposed on points by the physically interpreted axioms of Euclidean geometry. Similarly we construe the term 'light-ray' so as to include all the properties of a straight line path as prescribed by the

expressions which though symbolically distinct, must, nevertheless, be claimed to be synonymous, that is, of the same meaning.

For the well-known critique of the synthetic-analytic distinction, see : W. O. Quine, "Two Dogmas of Empiricism", *From a Logical Point of View*, Harper & Row, 1961, pp. 20-46.

physically interpreted axioms of Euclidean geometry.[23] Taken in these senses, such meanings convert the axioms of abstract geometry into statements which are true merely by virtue of the meanings of their component terms. The resulting interpretation, therefore (unlike the previous one in which the property of satisfying the interpreted geometric axioms did not figure in the assigned meanings) is an *analytic* material one. Our material interpretation converts the axioms of abstract geometry into statements with genuinely empirical subject-matter. Moreover, the interpretation satisfies the axioms, i.e., renders them analytically true. But interestingly enough they say nothing about their subject-matter. Devoid of factual content they cannot serve as relevant premises for any explanation, prediction or other nomic inference with factual conclusion. For nothing but factually empty statements follow from factually empty statements. Hence such statements may be validly dropped without affecting factual conclusions.

As a second example of an analytic material interpretation consider once again our mini-calculus. Leaving the domain of interpertation unchanged, this time let Rxy be the expanded notion : *x is rationally weak-preferred to y*. The term 'Rxy' then becomes a predicate of *rational* preference rather than merely of preference, where by this we mean that included in its meaning is the characteristic of satisfying the two axioms of our calculus. Obviously when interpreted in terms of this expanded notion, the axioms are rendered analytic. It is, of course, clear that the defined concepts of indifference and strong-preference, on our revised interpretation, become notions of *rational* indifference and *rational* strong-preference, understood similarly to our conception of rational weak-preference.

Admittedly there is an air of triviality in these considerations for, obviously, any contingent, material interpretation may be rendered analytic by simply expanding the meaning assignments to include the characteristic

[23] Thus, for example, the term, 'light-ray' can be understood so as to include within its meaning the Euclidean property that – if, in a plane, a light-ray, a, intersects two other light-rays, b and c, so that on one side of a the two interior angles are, together, less than one straight angle, then rays b and c will (physically) meet on that side of a. This corresponds to the well-known axiom of parallels given in Euclid's *Elements*. Similarly the term, 'light-ray' would also be understood to include the property that – if A and B are two distinct points on a straight line and if A', is a point on the same or another straight line then we can always find one and only one point B' on the same line with A', so that the segment, A'B', is precisely as long as AB. To satisfy the axioms of geometry analytically, the concept of *light-ray* would have to be rich enough to include attributes corresponding to all the other axioms about straight lines in the particular geometry concerned.

of satisfying the axioms. If, in fact, scientific concepts were expanded capriciously merely to satisfy some body of statements the results would, indeed, have no scientific significance. In actual practice, however, scientific concepts are expanded, not arbitrarily, but, as we shall note later, only in accordance with accumulating knowledge of the domain of subject-matter concerned. It might be remarked, here, that it is not always obvious that an interpretation will satisfy a set of axioms analytically; and considerable analysis may be required to establish the fact.

But, now, does all this mean that, for scientific purposes, all empirical interpretations of the analytic sort are useless? An affirmative answer here, in the face of actual scientific practice, would amount to an absurdity. For it would be a sweeping dismissal of impressive portions of established scientific theory.[24]

Obviously enough, whether an interpretation of an abstract system, K, is contingent or analytic depends entirely on the conceptual content of the meanings assigned to the undefined terms. Hence where vocabulary can be freely created, it is possible to add enough conceptual content to the meaning assignment to render the axioms of K analytic and hence to convert a contingent interpretation into an analytic one. The redefinition of terms has thus modified the logical status of statements and has done so in what would seem to be a radical manner; for a contingent statement says something about its subject-matter whereas an analytic one does not. Yet, this sort of conversion is not an unfamiliar one in the development of language and occurs not only in science but on virtually all other levels of discourse as well. It is a result of the growth and enrichment of ideas occurring sometimes naturally, as gradual change in ordinary usage, and sometimes artificially, as careful stipulation in the framing of scientific concepts. For example, the ordinary notion of *a light-path in a uniform medium*, if coordinated with the term 'line' when interpreting geometry, will yield sentences, T, which are contingent truths of geometric optics. But if we define the expression, 'light-path' in such a manner as to include

[24] An historically important example of an empirical theory which may be viewed as an *analytic empirical* interpretation of some abstract calculus is the limit-frequency explication of probability. Briefly, it consists of coordinating, to the undefined term, 'probability', of the abstract probability calculus, the limit of a relative frequency. Such an interpretation renders the probability axioms analytic. It may by the way be noted that the concept of *limit*, because it rests on that of *infinite* sequences is only an idealization. But, like many other such idealizations in science, there seems to be no unsurmountable difficulty in regarding limits of infinite sequences as genuinely empirical in character in the sense of being instantiatable by appropriate experimental approximations. Point sources, frictionless planes, and straight lines are other such idealizations.

within its connotation the notion of a wave-front each point of which is the source of a new wave, we have enriched it so as to imply rectilinearity of propagation in a uniform medium.[25] Thus on this expanded notion of *light-path*, the sentences, T, are converted from contingent to analytic truths. The conversion from contingency to analyticity as a result of the growth of concepts is readily illustrated with examples from other levels and areas of scientific discourse. Thus until not very long ago, the term 'gold' could be said to refer to a rather inactive chemical element which, in its native state, was a yellow metal having a certain high density, malleability and ductility, along with marked insolubility in virtually all acids.[26] In the early part of this century the young British experimentalist, Moseley, working with x-rays showed how to assign a positive integer (the atomic number) unambiguously to each chemical element. And for gold this number was found to be 79. Accordingly, the sentence : 'The atomic number of gold is 79', originally expressed a contingent (i.e., non-analytic) truth. Due to its fundamentality and uniqueness, however, the notion of atomic number has by now been absorbed as a defining characteristic for that of element. More, precisely, in calling an element "gold", a contemporary physicist or chemist would, no doubt, consider himself logically committed to saying that its atomic number was 79. Indeed, if a scientific investigator were to find some very malleable and ductile, yellow metal with the other physical characteristics of gold (insolubility, specific gravity of 19.3, etc.) but having an atomic number other than 79, he would – despite all his perplexity – surely find himself logically compelled to deny that it was gold. If this is so then the word-string, 'The atomic number of gold is 79', though seemingly unchanged, no longer says what it said before. Unlike before, what it now expresses is a logically necessary truth rather than a factual contingency. Again, before anything was known about the structure and physiologic importance of the human nervous system (call it 'N'), usage of the term, 'human', could not have taken account of neurological features. Or, as we might put it, N was not a defining characteristic of the term 'human'. It seems safe to conjecture therefore that a certain muscular, and skeletal structure sufficed – at least as far as merely physical characteristics were concerned – for calling an animal 'human'. At present, however, it is highly dubious that one would apply

[25] This so-called *Huygens construction* assumes theoretically full connotations for the term wave so as to include what is known about waves such as the facts of interference and of wave-velocity in a uniform medium.

[26] Aqua regia (a proportioned combination of hydrochloric and nitric acids) being the notable exception.

the term to any type of organism that did not possess something like N, even though it had the muscles and the skeleton (cranium included) of the species we call "human". Hence, the statement: 'All humans have attribute N', previously contingent, would, on our expanded notion of what it is to be human, be analytic. And similar accounts could easily be suggested involving such terms as : 'fish', 'acid', 'metal', 'educated', 'democratic', and an indefinite number of others from both technical and everyday contexts.

But does this amount to saying that the growth of concepts entails the general collapse of factual knowledge into empty tautology? How do statements (or systems of statements) rendered analytic by the enrichment of concepts continue to be applicable to matters of fact? It would, indeed, seem to be a strange feature of language development if the price of semantical growth for its descriptive terms was the factual content of some of its statements.

For convenience and lack of a better term let us apply the term, *essentialized*, to any statement which, having once expressed a factual contingency has been rendered analytic by "assimilating" or, as we might put it, semantically incorporating known characteristics of its subject-matter under the meanings of the predicates used. At first blush, this sort of phenomenon might appear to reflect a paradoxical aspect of systematic knowledge. The systematic growth of concepts would seem to result in the vanishing of factual significance as matter-of-fact statements suffer transformation into empty truism. Yet, our sciences continue to flourish in both depth and scope while scientific notions seem to grow ever more in refinement and complexity. How, then, do essentialized statements continue to be applicable to matters of fact? This question takes on particular urgency in science, where the meanings of terms are often explicitly – sometimes elaborately – stipulated with resulting conversion of genuine matters of fact into mere truths of definition. It would be a strange state of affairs, indeed, if scientific concepts were made to grow more and more only so that scientific statements could say less and less. Is it, perhaps, the case that though essentialized sentences are *a priori* or "necessary" truths they are nevertheless, in some sense, ultimately factual, hence "synthetic", in character? We have here shades of an old and sticky doctrine with which it would not be profitable to engage ourselves. What seems sufficient to note is that one need not grant the factuality of any logically necessary statement in order to maintain that essentialized statements – far from being otiose verbiage – do, as analytic statements, have a definite role in scientific discourse. For, though a descriptive term 'C' may have become so inclusive as to render some statements about C's analytic, that *there are such things as C's* is by no means analytic. What is more, *it can never be*! [27] Thus, expanded concepts are, as it were, the semantical repositories of increased knowledge of things. Undeniably, the resulting essentialized statements become mere truths-by-definition despite the genuine factuality of their contingent antecedents. Nevertheless, essentialized statements constitute an important form of cognitive language. Their role on both scientific and other levels of informative discourse is an explicative one, serving to unpack the conceptual content of premises or, as one might also put it, to help make explicit what is already implicit in statements of existential fact. Thus, if the significance of the term 'C' is at any time expanded to include the attribute, D, as a defining characteristic, then, on the expanded conception of C, anything which is a C is necessarily also a D; that is, the statement, (1) 'All C's are D's' is analytic and therefore factually empty. What is *not* analytic, however, and hence *does* have factual content is another statement, namely : '(2) There are such things as C's'. Moreover – and this is the central point – once (2) has been established, by appropriate experiential methods, then the more conceptual content we have packed into C, the more (2) asserts. And, though (1) alone as an analytic statement, has no factual consequences, the conjunction of premises (1) and (2) does have. The logical consequences of (1) and (2) are, indeed, the same as those of (2) alone, since (1) is analytic and therefore may be dropped as a premise. But (1) together with its consequences (also analytic) constitute an unpacking or analysis of the contents of (2). They make explicit what it is to be a C. Factual import arises by virtue of (2) which tells us that there are factual instances corresponding to the meanings which (1) and its consequences make explicit.

Our considerations regarding what we have called *essentialization* make it plausible to regard it as a distinct and pervasive aspect of the systemization of knowledge, even if only because it entails the growth of concepts.[28] No doubt there are those who find analytic interpretations less fortuitous and therefore intellectually more satisfying than contingent ones – perhaps, because of the logical necessity generally attributed to analytic statements.

[27] For this existential statement to be analytic, the term, C, would have to be construed as though its designation included *existence*, the latter hence being regarded as a genuine attribute of things. Kant's classical analysis to show that the term, 'existence', in its ordinary sense, is *not* a predicate, suffices to establish the untenability of any such conception of C. See : I. Kant, *Critique*, pp. 331-337.

[28] The term *conceptual systemization* suggests itself here to differentiate from *nomic systemization* with which we have been concerned at some length, in previous sections.

Preference for "essentialized" systems is sometimes reflected in saying of certain truths (e.g., all metals are conductors) that they stem from the "essential" character of the subject-matter concerned (in this example, the characteristic presence of free elections without which some would not apply the term, 'metal'). But in such locutions the "essential character" of something corresponds to nothing other than the *defining characteristics* of the subject-term predicates applied to that thing. These are ultimately a matter either of established usage or of artificial stipulation. In either case, the essentialization of any statement is ultimately a matter of semantical choices which are logically quite arbitrary whatever be the underlying preferences or other pragmatic considerations. The term *explicative theory* strongly suggests itself for what results from *analytical* empirical interpretations of abstract calculi. Following this nomenclature, the theories resulting from *contingent* empirical interpretations of abstract calculi might be called *nomic* theories, the latter functioning as the factually significant premises that govern nomic inference thereby figuring directly in the *nomic systemization* of scientific knowledge.

5. *Definitions*

A measure of adequacy of any axiomatic system is the possibility of constructing, in terms of its elements, a variety of more or less complex expressions corresponding to notions that are central or otherwise interesting in some intended domain of application. When such a construction has been made a simple term is introduced to serve as convenient shorthand in place of the more complex one. Any such simple term is said to be *defined* by the complex term to which it is equated. An example of this occurs in our miniature calculus, where we defined the familiar concepts of indifference and strong preference in terms of the primitive notion of weak preference. From a purely logical viewpoint such definitions ultimately add nothing other than some optional expressions to the basic logical machinery of the system. For it is always possible to drop any expression which has been introduced by a definition without loss of formal content provided the expression is replaced by the formula which defines it. Nevertheless, definitions of this sort may be very important and even crucial steps in the systematic development of any science. In the first place, they may serve to explicate, i.e., analyze precisely, in terms of given theoretical notions, some term already in general use but understood only vaguely. Such clarification can help enormously in the further development or even, application, of a theory. An example is the everyday notion of a body's linear *momentum* defined precisely (though not in strictly axiomatic context) in rigid body mechanics, [29] or the notion of *fish* defined in terms of more elemental concepts of zoology.[30] Precisely defined, the notion of momentum serves as a starting point for more general and powerful reformulations of the principles of classical mechanics, such as the Hamiltonian form of the classical equations of motion in which this defined concept figures explicitly. A definition of the sort here in point can usher in a new expression, based on elemental ones, which promises to be theoretically interesting. Or it may single out an expression that is already recurrent in theoretical calculation, even when it does not correspond to any familiar notion. In so doing the resulting definition may set the stage for a "new" key notion in the further theoretical and experimental development of some science. An example is the concept of potential or "positional" energy. Defined, with the advent of particle mechanics as a relatively new notion, it ultimately provides a conceptual basis for one of the great conservation principles of classical mechanics, namely, that of the conservation of mechanical energy.[31]

The history of science offers many examples of concept formation and clarification in terms of given basic notions. In actual scientific practice these have generally not occurred in strictly formal contexts. They illustrate in all essential respects, however, the process of definition in terms of primitives, with which we are specifically concerned here and which could be regarded as an idealized version of the more usual mode of such definition. Indeed, there is no denying that the strategic use of this device is a profoundly basic "art" for concept building in science. Without doubt it is one of the most crucial and creative phases of scientific activity. In the axiomatic mode of exposition it may occur either as part of the abstract calculus or in the full theory after interpretation.

C. MODELS

Thus far in this section, we have distinguished two fundamental aspects of scientific theory : an abstract calculus which provides the logical or formal skeleton of a theory and an interpretation which provides the subject-matter content for converting the statement forms of the calculus

[29] Often expressed by the vectorial symbol, p, it is defined by the vector equation p = mv where m and v are the body's mass and translational velocity, respectively.

[30] The zoological concept of fish departs considerably from the pre-scientific sense which roughly equates *fish* to *aquatic animal.* Nevertheless the two concepts do have enough in common, it seems, for regarding one as an explication of the other.

[31] It is found useful for a particle situated at position, r, in a conservative force field. And it is defined as the negative of the work which would have to be done by the field

into genuine statements. Closely associated with scientific theory and, in certain contexts, with the aspect of interpretation, are what scientists and technologists label as *models*. Any systematic characterization of models is hampered, in part by the inherent complexity of the subject, but also – and to a large extent – by the fact that the term, 'model', as it occurs in scientific discourse, is used rather loosely to convey a variety of notions.

Despite this, however, there are some basic remarks that one can make in general where one or another sort of scientific model is concerned. To begin with, in speaking of models, we say that something, M, is a model *of* (in some cases *for*) something else. That is, the term, 'model', is relational and therefore requires references to more than one entity in it applications. The specific nature of this relation varies. It may be *representational* as in the case of scaled down aircraft models substituting for actual ones in flight research studies, *descriptive* as in the case of the Bohr model which may be said to describe the atom, *explanatory* as in the case of the colliding particle model which accounts for some of the behavior of gases, or other as in the case of a Heisenberg matrix which systemizes certain facts about spectral lines.

In any systematic discussion of models it seems helpful to keep the following distinct items in mind : In the first place, there is the domain or subject-matter, S, which a model, M, is supposed to help study, explain, or otherwise systemize. S may be observational subject-matter as in the case of optical phenomena *explained* by a wave model or it may be theoretical subject matter as in the case of the atom *described* by the Bohr model. In some cases (not all) M is said to be *a model of S*. Thus we speak of the colliding particle model of a gas or of the ether-wave model of light. But though the Bohr Atom is supposed to explain the occurrence of certain types of spectra, one does not ordinarily say that it is a model of such spectra. A possible reason for this is that in the case of the Bohr model one may distinguish a proximate subject-matter (the atom) and a more remote domain of systemization (spectral frequencies). But, for the present, this consideration need not detain us. Secondly, there is the structural content, S (e.g., orbiting electrons, atomic shells, ether-waves, mathematical expressions, etc.) in terms of which a model, M, is itself framed. This remark would have to be somewhat modified for what we shall refer to as *concrete models* e.g., models of ships, airplanes, ètc., which, unlike other

of force on the particle if the latter were to move from some standard position, r_0, to position, r, where the notion of work is, itself, to be defined in terms of fundamental theoretical concepts. For a standard mathematical discussion see Joos, G., *op. cit.*, pp. 81-82.

scientific models, consist of *concrete*, three-dimensional objects put together with *actual* pieces of wood, string, containers, etc. Models of this sort, however, figure little or not at all in scientific theory and our consideration of them will be only a passing one.

Finally where there is a deductive system, K, as, for example, an axiomatized calculus, the term 'model' may be applied to designate a certain sort of interpretation, M, of K. In such a case M is said to be *a model for* K. An example would be the standing wave (atomic) model as an interpretation of (i.e., model *for*) an axiomatized, abstract version of quantum mechanics, say the Schrödinger wave-equation version.[32]

The items which in scientific contexts are called models can be grouped under *three* broad headings :

(1) those which consist of concrete, three-dimensional objects as, for example, engineering models of ships, bridges, or other structures, miniatures of geologic formations, classroom models of planetary systems, etc.

(2) Those which consist of a set of assumptions for *explaining* or otherwise systemizing some body of subject-matter, S, and

(3) those which consist of readily imagined ("intuited"), but admittedly fictitious constructions that, in limited respects, *resemble* some subject-matter, S. Such models, by virtue of the strong analogy they bear to S, may help to visualize relations in S or even to suggest truths about S. For convenience we label the first *concrete* models, the second *nomic* models, and the third *analogical* models and stop for a closer consideration of each of them.

1. *Concrete Models*

Concrete models are relatively unimportant from the viewpoint of explanatory or other systemizing theory and therefore require only brief consideration here. Their greatest utility lies in the more experimental phases of scientific inquiry and in technology. As the label suggests, a concrete model is a three-dimensional object which represents another object, S, called the *prototype*. In general, M surrogates for S because it is more manageable by virtue of having a convenient size (sometimes larger, sometimes smaller), or because it has a more perceptible or otherwise more palpable form. It can do this because it so resembles S in relevant respects that by studying M one can draw some conclusions about S. The

[32] For a simple quasi-formal version of such an axiomatization see C. W. Sherwin, *Quantum Mechanics*, Holt, Rinehart and Winston : New York, 1959, pp. 12-28.

resemblance may be geometrical similarity as in models of bridges and other structures. Or it may be of a less obvious sort amounting, in some cases, merely to certain aspects of analogy or correspondence between M and S, as in water-wave models (ripple tanks) of optical phenomena or in electric circuit models of vibrating mechanical systems. Scientists and engineers using concrete models do not always have them actually present. They may work with diagrams or pictures of them or may even just have them in mind when thinking about a prototype as when thinking about electrical circuit elements (e.g., electromotive force, electric current, electrical resistance, etc.) in terms of their hydrodynamic analogues (e.g., hydrostatic pressure, rate of flow, resistance to flow, etc.). Concrete models may, to some extent, involve theoretical presuppositions regarding the phenomena they represent. Ripple-tank models, for example, which represent optical effects by means of water waves, are intimately tied to the wave nature of light. Such representations, however, are either applications of known theories or suggestive of new ones. They do not constitute any part or aspect of a theory's structure; nor do they enter directly in the systemization of scientific fact. For functions of this kind we must turn to our second sort of models, that is, to models that are not themselves concrete objects. These, like scientific laws, figure directly in the systemization (prediction, explanation, etc.) of scientific knowledge, and, as might well be expected, are intimately connected with scientific theory. For such reasons, the label "nomic model" seems appropriate. And it is with these, primarily, that the remainder of our discussion will be concerned.

2. *Nomic Models*

We ask now what are nomic models? Are they theories, components of theories, supplements to theories, or other? The answer seems to be *yes* to all of these. Thus, depending on the occasion, the term 'model' seems to do duty in one or another of the following senses :

(1) *The Theoretic Model* : To begin with, scientists often use the term to refer to a set of theoretical assumptions about some subject-matter. In such instances the term, 'model', designates just about the same sort of thing that also goes under the name of *theory*. Let us refer to nomic models in this sense as *theoretic models*. Examples are the corpuscular and wave *models* of light, the colliding particle *model* of gases, the helix *model* of DNA, the pure competition *model* of price making, the Heisenberg matrix *model* of the atom, and so on. In these and the many other examples that could be listed from both the natural and the behavioral sciences, the term,

'theory', would certainly work in place of 'model'. Nor would it do here to suppose that the term, 'model', is appropriately used only for those representations that offer space-time schemas which can be geometrically visualized. The last of our examples, namely, the Heisenberg matrix model of the atom would not satisfy this criterion. Matrices are mathematical, not space-time, entities. Similarly one might speak of the Schrödinger model for quantum mechanics, referring to the mathematical formulation known as the Schrödinger wave equation. In the social sciences and in the science of genetics mathematical models and a species of these known as probability models have become a matter of common reference.

Some differences of nuance between the terms 'model' and 'theory' have been the basis for claiming that a model, in the sense of a set of theoretical assumptions, is a distinctively different sort of entity from a theory. Thus one author urges the following differences between models and theories.[33]

(1) Models are more limited in range, more simplified for expediency, and less accurate than are theories, perhaps even known to be false by their authors. Unlike theories, they are not intended to describe things as they *actually* are. Admittedly theories are also something like this. That is, as he puts it, "There is some leeway here though not of the magnitude allowed in a theoretical model".

(2) A model of object, X, postulates an inner structure of X for explaining various properties exhibited by X. A theory need not postulate inner structure but may simply give relations which hold between these properties or between them and the properties of some other object.

(3) In models much that is proposed is typically derived from some more fundamental theory whereas theories sometimes make central assumptions that are not borrowed or derived from any other theory.

But how tenable is this proposed contrast between model and theory? With regard to point (1) it should be clear that whether or not anyone *knows* that a particular model is, in fact, false implies nothing about its structure. That anyone should *know* a model to be factually false does, indeed, logically imply its falsity. But, again, this would tell us nothing about its structure. Thus even were it allowed that models, unlike theories, could be known to be false, this would have to count as a relatively uninteresting, pragmatic, rather than structural, feature of some models, that is, the resulting distinction would reflect a state of knowledge of individuals rather than any logically inherent character of models. Nevertheless, pragmatic distinctions, however uninteresting they may be to the philosopher

[33] Achinstein, P., *Concepts of Science*. Johns Hopkins Press : Baltimore, 1968, p. 217.

of science, cannot, in general, be ruled out as a basis for distinguishing some things from others. It might seem possible therefore to speak differentially of a model, M, as a set of assumptions which, unlike those we call a theory, can be known to be false. Such a characterization, however, must fail if M is to *explain* anything at all. Even the weakest epistemic requirements on explanatory premises could not allow that they be *known* to be false. (Cf. Chapter IV, section F, 3.) For, in what sense could one claim to be explaining any event or property by making assumptions that he knows to be false. Explanations are causal accounts in terms of what one believes *are* – or, at least, *may turn out to be* – causes. To know that a set of assumptions is false is to know that what is being assumed can certainly not be a cause of anything. In (2) above it is maintained that models postulate in order to explain. Hence the view under consideration is, in this respect, inconsistent. There are formulations which permit the desired epistemic "leeway"; but they might better be classed under the special heading of "analogical models" which though having no explanatory function can nevertheless enhance our conception of a certain subject-matter and may even make possible some predictions about it.

In connection with these last considerations on the nature of models it seems appropriate to note the marked tendency in contemporary scientific contexts to apply the term, 'model', and in some cases even other terminology for what otherwise might be generally referred to as scientific theory.[34] And there seems to be good reason for conjecturing that such terminological preferences are more than adventitious trends. They may well reflect a growing *conventionalism* (as it is called) in the climate of opinion regarding the nature of scientific theory. It is a commonplace of logic that more than one set of assumptions can cover any given body of evidence. What is more, it has been urged, with some cogency, that no hypothesis need be abandoned in the face of new counter-evidence. For, it is argued, any set of assumptions can be "saved" by making suitable linguistic adjustments elsewhere in the systematic context. To some, this might mean that accepting any set of theoretic assumptions is more a matter of

[34] T. S. Kuhn, for example, introduces the term, 'paradigm' for fundamental scientific achievements that are innovative and open-ended enough to attract a broad and sustained following. What sort of more specific or more precise notion Kuhn has in mind when applying this term, is not quite clear. His own characterization of a "paradigm" and the abundant examples he offers make it difficult to imagine what, other than theoretic "achievement", he can be referring to when applying the term. Moreover, except, possibly, for innovations of a strictly methodological sort, scientific innovations have been almost entirely matters of theory creation. See T. S. Kuhn, *op. cit.*, pp. 10-23.

tentative expediency than a commitment about the "actual" nature of things. The term, 'model', therefore strongly suggests itself for application to any set of theoretical assumptions. For it is with models, *in one sense of this term*, that we represent subject-matter without literally describing it. But whether the conventionalist thesis is right or wrong, terminological preferences inspired by it would surely apply to *all* systematic assumptions rather than to any special class of them. Such terminological preferences imply no differential nuances for setting models apart from theories. And they can in no way suggest that some systems now referred to as "models" are structurally any different from others which have been more traditionally called "theories", although they may serve to emphasize or call attention to certain aspects of all theoretic assumptions.

According to (2) models always postulate an inner structure whereas theories sometimes give the relations holding between the properties of an object or between these properties and those of another object. Admittedly theories need not postulate inner structures of things (i.e., theoretic subject-matter) in order to systemize scientific fact. They may take the form of mathematical relations with correspondence rules for coordinating the symbolic results of calculation with observable events. But, then, such mathematical formulations are also readily referred to as mathematical models which we take to be one of two basic types of nomic model to be discussed shortly. Thus, the distinction between theory and model, in one or another sense of this latter term, seems again to collapse to one of verbal preference. What is more, the suggestion that theories, unlike models, may simply formulate relations between "exhibited" properties is highly questionable. Any set of statements expressing such relations would properly be called neither a theory nor a model but a set of *observational laws*. Except for mathematical theories (or models), which largely dispense with appeal to visualizable inner structures, theories postulate "unobservable" (i.e., theoretic) subject-matter in order to account for observational phenomena.

We turn now to (3) above, which offers as a means for separating theories from models, the claim that whereas models are typically derived from more fundamental theory, theories sometimes make central assumptions which are neither derived nor borrowed from any existing theory. Thus, the free electron model of conductors presupposes electro-magnetics and the kinetic model of gases rests on mechanics. Electro-magnetics and mechanics, however, rest on no other physical principles and hence are theories rather than models. But here again there seems to be no compelling systematic basis for urging the proposed differentiation between theory and model. At best the argument offers no more than a basis for distinguishing various levels of generality for theories. What it is claimed should be differentiated from theory and labeled model is much more naturally regarded as less general theory. The kinetic model of gases, for example, is essentially the application of Newtonian theory to a postulated system of invisibly minute particles, perfectly elastic upon impact, vanishingly small compared to their mutual separations, and subject only to contact forces. The model goes beyond Newtonian theory only by adding certain statistical assumptions about the space-velocity distributions of the theoretic particles and by coordinating the absolute temperature of a gas with the average kinetic energy of its postulated constituent particles.[35] Indeed, there is ample reason to regard both Newtonian dynamics and the kinetic model as paradigms of physical theory; albeit they differ vastly in degree of generality. Both postulate theoretic subject-matter (i.e., idealized mechanical states, forces, invisible particles, perfectly elastic impacts, unobservable statistical distributions, etc.). And, in this case, neither seems to make any less definite an existential claim about the structure of physical reality. It would seem entirely appropriate therefore (and perhaps less misleading than the proposed differentiation) to bring both formulations under the rubric of theory. The term, 'model', would then serve merely to reflect terminological preferences for the purpose of communicating a certain emphasis. In this connection it would surely not be beside the point to note that the expression, 'the kinetic *theory* of gases', is quite apparently no less prevalent than its equi-significant alternative 'the kinetic *model* of a gas'.

(2) *The Interpretational Model* : Students of deductive systems and of artificial languages have developed a concept of model that differentiates it from the notion of theory. This concept makes a model a distinct part or aspect of theory and provides us with a second variety of nomic model, which we label the *interpretational* or *semantical* concept of model. For it comes very close to what, in connection with the axiomatic reconstruction of scientific theory we have called an *interpretation.* On this notion, a formulation, M, would not only be said to be a model *of* a given phenomena (e.g., the wave model of light) but also a model *for* a certain theory (e.g., the ether-wave model *for* Maxwell's electromagnetic field equations, or the graviton-particle model *for* gravitational theory. There is, however, no precise agreement on what sort of thing such a model is supposed to be.

[35] The absolute temperature is assumed to be directly proportional to the average kinetic energy of the particles.

One author, for example, says that whatever set of concepts satisfies axioms of an abstract system is a model *for* that system.[36] Another simply equates the notion of a model to that of true interpretation.[37] In still another account, when a theory is axiomatized in terms of a certain predicate, a model for the theory is simply an entity satisfying the predicate.[38] In some cases, it is specified that for an interpretation to yield a model, the meaning or subject-matter coordinated with a calculus must be part of a system "outside" of or independent of the calculus. Another way of putting this is to say that whatever serves as a model must have an antecedent structure of its own. E. Nagel, for example, allows that 'model' may also designate a *system* of elements which comprise the subject-matter that the interpretation coordinates with a deductive system, K. On Carnap's account, an interpretation of (or model *for*) a calculus, K, is a semantical *system* M (i.e., a well-systemized language having meaningful terms and genuine sentences) such that if the vocabulary of M is appropriately used to replace the abstract terms of K, then all the sentences of K become sentences of M.[39] Here a model, as an interpretation, is clearly required to be a system of some sort or other. This requirement of antecedent structure takes a model to be a *species* of interpretation rather than simply any interpretation. Accordingly, to construct a model we would coordinate with a calculus not what is merely a set of appropriate predicates or entities, but, instead, a set of elements which belong to an explicitly systematic context of their own. That is, the assigned set has a distinct order or structure which can be specified quite apart from the calculus to which it is assigned.

(3) *Substantive and Mathematical Models* : Nomic models, whether they be of the theoretic or the interpretational sort are conveniently grouped under two headings. The first of these which we label *substantive models* are framed in terms of a subject-matter which, though postulated as having space-time attributes is nevertheless conceived in such a manner as not to be directly observable in any ordinary sense of the term, 'observable'. Examples would be the colliding micro-particles of a gas or the ether-waves which transmit optical, radio, and thermal signals. In many cases the attributes of such postulated subject-matter are idealized beyond what is known to be physically possible as, for example, in speaking of perfectly

[36] A. Tarski, *Introduction to Logic*, New York, Oxford University Press, 1965, p. 123.

[37] E. Mendelson, *Mathematical Logic*, *op. cit.*, p. 51. Also E. Nagel, *op. cit.*, p. 96 and P. Suppes, *op. cit.*, p. 21.

[38] *Ibid.*, p. 253.

[39] Carnap, R., *Introduction to Semantics*, Harvard University Press : Cambridge, Mass., 1948, p. 204.

elastic collisions or of *purely* competitive economic behavior. But such idealized objects are nonetheless taken to be close approximations to objects which are assumed to have concrete space-time existence, the idealization being ultimately for the sake of theoretical simplicity and mathematical manageability.

In nomic models of the second variety no underlying theoretical subject-matter is postulated at all. The model instead is merely a mathematical formulation, essentially, an algorithm or symbolic device for calculating observational results. We follow the common practice of referring to models of this type as *mathematical models* (theories). An example would be the Heisenberg matrix model listed among our earlier examples. It is framed in terms of a two-dimensional *array* of numerical variables, i.e., a *matrix*, which may be computationally manipulated and ultimately coordinated with observational subject-matter in specified ways.[40] Other examples of a mathematical model or theory are Maxwell's equations for the electromagnetic field and Schrödinger's wave equation for quantum states both of which most usually occur in mathematical form and without any particular concrete representations.[41] Mathematical models, in postulating no underlying existential subject-matter, cannot provide explanations of the phenomena to which they apply. Nomic models, when they do explain, do so by providing causal accounts of observed regularities in terms of a supposed underlying structure of things. Mathematical models, however, in supposing no such structure, can serve only for making formal calculations capable of yielding predictions, postdictions, and other pro-

[40] W. Heisenberg, *Zeitschrift für Physik* 33 (1925), p. 829.

[41] It is true that both Maxwell and Schrödinger were deeply motivated by analogies with concrete phenomena in constructing their mathematical representations. But the final formulations which resulted were essentially independent of any reference to these analogies. Maxwell's famous "displacement current" was indeed the notion which led him to one of the crucial terms in his equations namely : $\frac{1}{c}\frac{\delta E}{\delta t}$ (where 'c' is a constant and the vector, E, is the electric field strength). What is more, the idea of a displacement current stemmed from the notion of an all pervasive ether which was both mechanical and electro-sensitive or polarizable, as it is sometimes put. But the theory, as he himself formulated it, made no clear and systematic reference to any particular substantive model. Indeed, it is highly doubtful that any physicist would now accept the concept of a pervasive mechanical and polarization ether, though Maxwell's equations still serve as basic mathematical representations of electro-magnetic phenomena.

Similarly, Schrödinger's equation which forms the basis of one version of contemporary quantum theory did spring – as we have already noted – from rather arbitrary analogies with vibrating strings. But, except for historical considerations, such analogies do not figure in any systematic presentation of the theory.

jective (but non-explanatory) inferences on the basis of given observational data.

As has already been noted, the distinction between a mathematical and a substantive model applies whether the model is a theory (i.e., a theoretic model) or an interpretation of a theory (i.e., an interpretational model). In the latter case, we have either a mathematical or a substantive model depending on whether the interpretation is formulated primarily as a mathematical construction or, instead, makes explicit reference to a system of space-time, theoretical subject-matter such as electrons, electro-magnetic waves, ideal economic agents, and so on.

It is important to note at this point that when a model has been provided as an interpretation for an abstract calculus, the latter does not, *ipso facto*, become a full-fledged scientific theory. Generally, the model must first be given observational significance. This means it too must be interpreted by the assignment of observational content to its theoretic vocabulary. Moreover, this must be done whether the model is of the mathematical or of the substantive sort. That mathematical models require operational (observational, experimental, etc.) interpretations is quite obvious since such models are, at the outset, only logical devices having the potential to serve as computational algorisms or formulas only when the appropriate parameters in them are given observational significance. But substantive models must also be assigned empirical significance. The hypothetical subject-matter in terms of which they are framed is assumed to have space-time attributes and can therefore be readily visualized. But it is conceived in such a manner as to be unobservable except through the mediation of correspondence rules which link certain elements of the model to observational effects. Putting the matter somewhat differently, though a substantive model constitutes an interpretation for some abstract calculus, thus providing it with semantical reference, this reference is, in general, not sufficient to give the calculus observational (empirical) significance. The reason for this is the nature of the model itself. The system of entities to which the model refers though, conceived in terms of space-time relations, is, by its very nature, physically inaccessible to the senses. Thus, for example, we do not and cannot literally "see" a configuration of electrons or an electro-magnetic wave. More particularly, though electrons may be assigned time-space coordinates, velocities, orbits, etc. we never *directly* perceive them in these relations and the possibility of doing so is precluded by the very laws of nature themselves. Indeed, despite the fact that electrons are, in certain senses, said to have extension, they cannot be said to have color. Their supposed minuteness precludes their reflecting visible light in the manner required for visibility. So minute is the electron that, according to principles of microphysics, one determines its motion only at the cost of definiteness in its location and hence even of anything which one might call its "shape". Thus, if such postulated entities are physical objects at all, then they are indeed very different from ordinary sense objects. This difference is a deep one and is reflected in the need to postulate rules that make it possible to detect them. Without such rules the term 'electron' would have no more empirical content than an abstract term of pure mathematics or, for that matter, an undefined term of some abstract calculus.

But if substantive models require interpretation just as do calculi, then why have them in the first place? Why not just use an interpreted calculus as a purely computational device – a logical mill, as it were into which can be fed given facts in order to derive others? This manner of deductive ordering would surely be a form of systemization and, above all, would be unquestionably effective as a means of prediction. A calculus, K, with appropriate correspondence rules for relating it to observational subject-matter would be as logically sufficient for prediction and for the subsumptive ordering of known facts as would any model for K. Thus, any tendency to use substantive models instead of mathematical calculi for the systemization of scientific knowledge must rest on extra-logical factors. In this connection, even the most casual acquaintance with the historical development and *actual* use of scientific theory easily reveals that axiomatization is not the usual mode of exposition. More often it is a succession of substantive models without abstract calculi that mark the development of scientific theory and within which working scientists carry out their deductions for the systemization (prediction, explanation, etc.) of fact. And reasons for this are not hard to find. To begin with, substantive models are usually suggestive of an abundance of easily intuited possibilities for elaborating the consequences of a theory. This is because whatever assumptions they consist of are imbedded in a system of postulated subject-matter for which it is generally possible to envision relations that go beyond anything which the original assumptions state explicitly. Such relations then become a basis either for extending the model or for raising new experimental questions. Thus, the quantum model of the nuclear atom with its orbiting electrons suggested questions about why electrons do not all huddle about the nucleus in some orbit of lowest energy. It suggested the possibility of a spin for the electron, of a measurable magnetic moment for the atom due to its orbiting electrons, of overlapping orbitals and shared electrons to explain the covalent chemical bond and a great deal more. Such questions of course can be expressed entirely in terms of the formulations of mathematical quantum theory, but they seem to arise much more naturally and spontaneously in the substantive form of the theory.[42]

Secondly, substantive models are more readily assigned physical significance than are purely abstract calculi. Thus if quantum theory had first been presented as a piece of pure mathematics it would be difficult to imagine how correspondence rules could be introduced for coordinating experimental subject-matter with its formalisms. Thus, for example, when the formal theory is applied to a particle in a central field such as that of the hydrogen atom it yields a set of discrete values for the average (expectation) energy of the system. The substantive model with orbiting electrons, however, permits us to interpret expectation differences as probabilities that the electron will jump from one orbit to another. These "jumps" are then quite naturally coordinated with radiated energy, in turn associated with the observable intensity of a spectral line.

Finally, the theoretician finds more than visual concreteness in the substantive model. It provides a causal base for explaining phenomena. Abstract calculi are but logical algorisms for deriving observational consequences from observational premises. The substantive model, whether it occurs as an interpretation of some calculus or as a theory in its own right postulates the inner structures and theoretic entities, envisioned as the causal determinants of what happens visibly. Accordingly, a set of substantive theoretic assumptions, whether they be referred to as a "theory" or as a "model", are postulated not as mythical analogies which happen to give right answers, but as existential hypotheses about the actual nature of things whose truth implies the falsity of alternative theories. As is generally recognized the reliance on "intuitable" models has its hazards. They can hamper thinking and even lead it astray with inessential content. Nevertheless, one does not easily forgo the heuristic value and intuitive appeal of concrete imagery in thinking, and so, substantive models (theories) have been not only the antecedents to virtually all formalization in the genesis of empirical theory

[42] For example, the question of why electrons do not all cohabitate close to the nucleus can be abstractly considered as a question regarding the kind of wave functions which electrons are permitted to have.

In connection with this question W. Pauli postulated his famous Exclusion Principle which states that no two electrons can have the same four quantum numbers, each quadruplet of numbers representing a certain dynamical orientation of the electron in the atom and only certain numbers being allowed by the theory in the first place. The Exclusion Principle, however, can be equivalently and formally expressed by requiring that electrons have antisymmetric wave functions. See C. W. Sherwin, *op. cit*., pp. 224-229.

but more often than not, they have remained the sole constituents of such theory even at its more established and working stages. In such a case, there is available no *explicitly formulated* calculus which the substantive model interprets. There is in use only the substantive construction – "theory" or "model", as one may prefer to call it.

3. *Analogical Models*

To enhance the understanding of some subject-matter, S, investigators sometimes propose a set of postulates or model, M, characterizing another subject-matter, S′, different from S. In general, the model, M, is introduced because S, as characterized by some theory or set of other laws or even by a set of vaguely conceived principles, is too strange or too complex to be grasped readily. S′, on the other hand, is familiar and readily visualized. But though S′ is different from S, it resembles S in certain important respects. This may also be expressed by saying that S and S′ are analogues. Thus, in proposing M, one draws an analogy between S and S′. Consequently, we call M an *analogical model of S.* An example of such a model is the soap bubbles model of a multi-atomic ionic crystal. The inter-ionic forces may be computed by means of approximative methods based on quantum theory.[43] But by imagining a cluster of soap bubbles held together by some type of attractive force, yet held from collapsing by repulsive forces due to an elastic surface with pressure inside of it, one can visualize the forces at play in the crystal. But much more than this, the model suggests that the repulsive forces which keep the crystal from collapsing are not the same as those which are more easily calculated for diatomic molecules. The suggestion comes from imagining that the soap bubbles will be flattened on the interface between them. This will somewhat modify the shape and elastic properties of the remaining parts of the surface of the bubbles. As a result the bubbles will present a slightly different repulsive effect to neighboring bubbles. These differences may be approximately calculated by abstract mathematical methods of quantum theory.[44] Another example is Poincaré's well-known model of a Lobachevskian non-Euclidean universe.[45] This is a sphere of radius, R, whose temperature varies from a maximum at the center to absolute zero at the surface. At any point, r, the absolute temperature is proportional to R^2-r^2, and all bodies will expand

[43] See the Heitler-London method, John C. Slater, *op. cit.*, pp. 213-240.

[44] *Ibid.*, p. 235.

[45] H. Poincaré, *Science and Hypothesis*, New York, 1952, pp. 64-66.

or contract so that the dilation in their linear dimensions is proportional to their absolute temperature which is always that of their immediate environment. For observers in such a world the geometry of physical entities would satisfy the Lobachevskian axioms. Thus, the shortest distance between any two points within the sphere would not be a Euclidean straight line but an arc passing through the two points and perpendicular to the sphere's surface. Still other examples of analogical models arise when electrical systems are postulated for analyzing vibrating mechanical systems or when hydrodynamic models are used to illustrate the relations among the elements of ohmic electrical circuits. The anthropomorphisms which got into pre-scientific characterizations of natural processes may not all have been simply misguided speculations. The idea of nature "abhorring" a vacuum or "deploring" wastefulness or of a stone "wanting" to drop downward might well be taken for fictitious constructs, that is, for analogical models used – in the absence of any substantive theory – for enhancing one's grasp of the phenomena. Indeed, might one not see even in pre-systematic mythological accounts of the world an essentially allegorical mode of exposition strongly suggestive of the analogical model?

From our examples, note the following additional features of analogical models :

(1) The subject-matter, S′, which any analogical model, M, postulates may be of a kind found in the real world, such as an electrical analogue of a mechanically vibrating system. Or it may be an entirely fictitious construction such as Poincaré's thermo-metric world.

(2) Where the subject-matter under study, S, is characterized by a formulated theory, T, an analogical model, M, of S may also be an *interpretation* of T or, as one could also say, a model *for* T. This is rendered likely by structural correspondences between M, S, T, and S′. In such a case M serves not only to enhance our understanding of S but also to establish that T is just as consistent as M. Thus given that M is consistent, the consistency of T can also be asserted.

(3) M may enhance not only our understanding of S but also our understanding of theory, T, which is about S. If S is unfamiliar or complex, T may be similarly difficult to grasp. In such a case M may well turn out to be an interpretation of T and thus help fix our understanding of it.

The following diagram summarizes our categorization of scientific models. For convenience, we let M be any model, S the subject-matter which M is supposed to systemize or help understand, and S′ the subject-matter in terms of which M is framed.

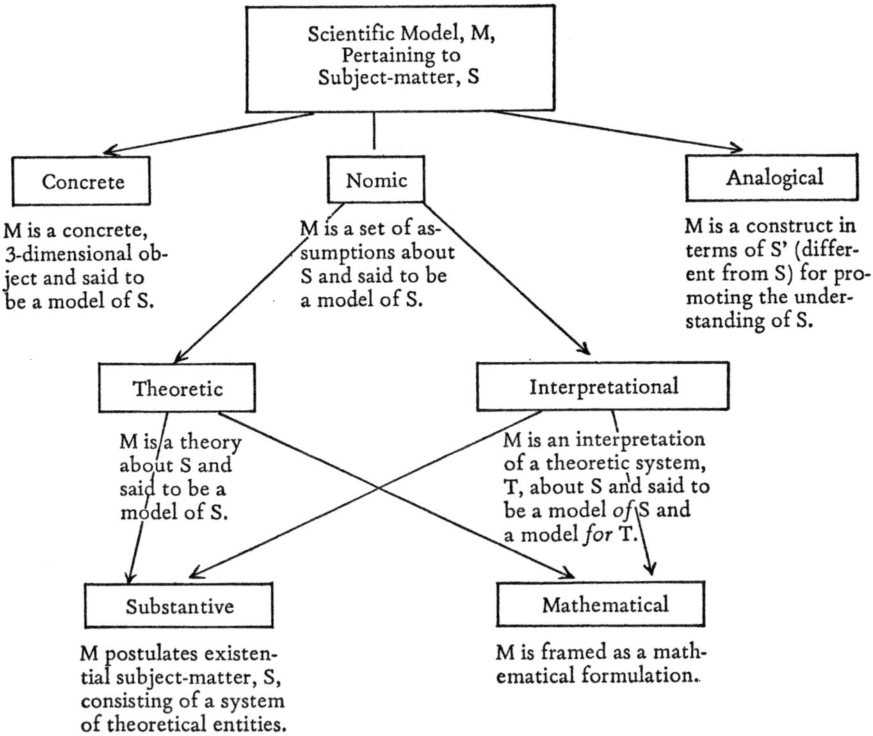

D. FORMALIZATION AND SCIENTIFIC THEORY

1. The formal or axiomatic representation of scientific theory consists, as we have noted, of (1) an abstract, axiomatized calculus, (2) an interpretation by means of correspondence rules that link the calculus with the observational domain and possibly (3) an interpretation, again, by means of correspondence rules, coordinating it with a *system* or *model* either of the substantive or mathematical sort, which in turn must be appropriately linked to the observational domain by still other correspondence rules. Today this mode for exposition is familiar even in the most unsettled of the sciences, some of which, despite their disquieted state, may be considered methodologically advanced in the sense that they give explicit attention to logical structure and, in particular, to the assumptions and conventions (postulates, definitions, rules) on which such structure stands. Even in the behavioral sciences, for example, where mathematical methods have been somewhat late in coming, the axiomatic method is not unfamiliar. And though such disciplines often labor under strenuous theoretical controversy,

some degree of formalization can help to locate the root of disagreement and thus provide a first step towards resolution.

2. The power of the axiomatic approach is considerable, indeed. It moves discourse towards broadly *four* systematic objectives, which may be labeled : *rigor*, *generality* (power, scope), *simplicity* (economy of structure) and (lacking a better term) *perspicuity*.

(1) A *rigorous* deductive argument or derivation is one which has been made explicitly and *exactly* in accordance with a system of valid rules and other assumptions. This demands that all such stipulations be given full and explicit recognition in advance of any derivation. The axiomatic mode of exposition provides the formal means for doing this. Thus, a specific set of postulates delineates the statemental assumptions of a system; syntactical rules give the manner of formulating and deriving expressions; semantical rules link the system to observational subject-matter or in some cases to other systems; and definitions introduce new symbols in place of more or less complex expressions already formulable in the system, thus providing the logical format for introducing derivative concepts into the theory. Concern for rigor and for the explicitly formal methods which it demands cuts down the likelihood of committing logical error. The chances of playing the game correctly, as it were, are greatly enhanced when all the rules are made explicit. In pure mathematics, where rigor is given far greater importance than in empirical science, the axiomatic method has become almost universally prevalent. The greater concern for formal rigor in mathematics than in empirical science is quite understandable. A derived result in mathematics has no empirical content and it is not therefore subject to observational test. In mathematics, therefore, logical error is the basic hazard and one for which there can be only logical precautions. The empirical scientist on the other hand, may often defer any concern for highly formal rigor. The truth of any result derived from empirical theory depends only on observational evidence. Hence, so long as a result, R, derived within theory, T, continues to agree with observational findings, the precise logical status of R within T need not be of immediate concern to the working scientist. Logical defects within T (even when these endanger consistency) or invalid steps (*nonsequiturs*) in derivations would *mainly* jeopardize explanation or other systemization but, ultimately, not the truth status of R. As a matter of historical fact, the careful rigorization of scientific theory, when it has occurred at all, has typically come only after it has had a long and successfully active career.[46]

[46] Interestingly enough the "logic" of the major branches of analysis in mathematics,

(2) *Generality* – sometimes called *power* – has come to our attention in connection with comparing laws (Chap. VI). In essence those considerations apply directly to the more specific case of theories. Comparing two theories with respect to generality or power is, broadly speaking, a matter of relating their respective domains of application with regard to proper subset inclusion. Thus, a theory in genetics, about all living things, would be more general than a theory in psychology about animal behavior, in the sense that the domain of the second is included within but does not exhaust – i.e., is *properly* included within – that of the first. Similarly, a theory of motion, generally, about physical bodies would be more general than one, specifically, about biological systems. For again the domain of the second is properly included within, i.e., is a *proper subset* of, the first. The domain of application, D, of a theory is the set of all individuals to which the theory's descriptive vocabulary applies. And since all the descriptive terms in a theory are either primitive or defined by descriptive primitives, this means that the power of a theory may be measured, in some way, by referring to the sets of individuals to which the descriptive primitive terms of the theory apply. In more standard semantical terms, these sets are called the *extensions* of the primitive terms. The *numbers* of individuals in the extensions of the primitive terms of a theory and, more particularly, their sum may well provide a basis for assigning a numerical measure to the generality or power of the theory. This, however, is still a highly problematic subject whose various technical aspects need not detain us further.[47] It is sufficient to note for our purposes that the high degree of explicit structural articulation which formalization provides makes it possible to speak more definitely of a theory's generality and perhaps, ultimately, of some comparative, numerical degree of such generality.

---

has also come late in their history. An example would be the infinitesimal calculus, unrigorized for nearly two centuries after its first applications. The most dramatic example is, of course, arithmetic whose axiomarization at the end of the 19th century comes tens of millenniums after its inceptions as a distinct discipline. But here, it must be recalled that all such branches of mathematics were applied disciplines whose principles *worked* successfully when applied to empirical subject-matter. Rigorous systemization of the principles of applied mathematics (and, until relatively recent times, all of mathematics could be classified as such) has, therefore, as in science, been a matter of either rare or deferred concern.

[47] In this general connection the reader may refer to such papers as the following : J. C. Kemeny, "Two Measures of Simplicity", *The Journal of Philosophy*, Vol. 52 (1955); L. Svenonius, "Definability and Simplicity", *Journal of Symbolic Logic*, Vol. 20 (1955); N. Goodman, "The Test of Simplicity", *Science*, Vol. 128 (1958).

But formalization offers still another dividend in connection with generality, in this case, an aspect of generality which it alone can provide. This is the generality which characterizes the uninterpreted calculus. As has been noted already, such a calculus, until interpreted, has no subject-matter at all. It is entirely without reference or, as one might say, semantically vacuous. Still it does have structural delimitations, so that not any subject-matter will do as an interpretation. All interpretations must share the same logical structure. Nevertheless, even within these formal limitations an uninterpreted calculus has, in general, an indefinite number of possible interpretations each of which could yield a distinctively different theory. Hence, the axiomatization of any theory provides a vastly general formulation. It provides a calculus which, though syntactically specific, is, so to speak, semantically "open", constituting, as it were, a logical matrix for generating theories or, as one might say, a *theory-form* of virtually inexhaustible potential from the viewpoint of theory construction and application. It features a form of abstraction and generality otherwise unparalleled in the history of systematic thought. Strictly speaking, the generality in point belongs not to a specific theory but to a theory component, i.e., to an abstract calculus. Nevertheless, it is an aspect of generality profoundly inherent in theory and made explicit by the formal mode of exposition. On the basis of a postulated set of formal features, it encompasses a *family* of possible theories.

(3) The *simplicity* of a theory is, roughly speaking, its economy of structure. And it is an undisputed canon of all scientific theorizing that such simplicity must be maximized without loss of any desired subsumptive scope. Theoretical compactness has been an objective on all levels of theorizing, even those which one might call pre-axiomatic. Some of the most powerful and encompassing theories in the history of empirical science, even though not formalized, or, at best, only partially so, have rested on dramatically few (often not more than two or three) factual assumptions. Axiomatization, however, in requiring that all assumptions be listed explicitly and exclusively, greatly strengthens the systematic concern for economy. Thus, a set of formally expressed axioms are more readily investigated for their independence. One postulate, for example, may be found to be a logical consequence of the others and so can be dropped from the system, thus reducing their number. The notion of logical compactness or economy of a system is, however, not easily explicated. One could not for example say that the fewer the axioms in a system the more compact it is. For obviously we can reduce any set of n distinct axioms $A_1, A_2 \ldots A_n$ to one single axiom consisting of the conjunction, A_1 and A_2 and ... A_n.

This might suggest that the simplicity of a theory be appraised on the basis of the number of statement conjuncts in its axioms. But this too will not work. For any conjunctive statement, S, can be equivalently expressed as a non-conjunctive one, S′, containing complex predicates which are defined in terms of the original predicates of S. This indicates that simplicity must be explicated not in terms of the number of axioms or of their more obvious structural features but in terms of a finer structure involving the basic descriptive predicates of a formalized theory. Thus, as it seems, the simplicity of a formal theory may be, essentially, the simplicity of its descriptive vocabulary. Some students of the subject working along these lines have made notable progress in getting at a sense of simplicity which may lead to appropriate comparative appraisals of this feature of formalized theories.

Structural simplicity and systemization are inseparable to the extent that we cannot have the second without the first. Indeed, the better part of what we mean by system in scientific discourse is the degree to which an extensive set of terms and principles has been reduced to – i.e., subsumed under or derived from – a few others taken as basic. Hence, choosing the simplest of possible hypotheses for covering a given body of fact is a lot more than a matter of free option; it is the rational strategy of systemization.

But the import of simplicity may go much beyond systematic considerations. The acceptability of any scientific theory in the sense of veridical status such as probability, confirmation, likelihood, etc., seems to hinge not only on the evidence immediately for or against theory but on such structural characteristics as its simplicity. Indeed, students of the subject are more and more of the opinion that no logic of induction for appraising theories can be adequate if it does not take their simplicity into account. A relevant manner of measuring the relative simplicity of any theory, therefore, seems necessary for assigning it any adequate measure of acceptability.[48]

[48] For a simple review of the problem of simplicity, the reader is invited to read R. S. Rudner's *Philospohy of Social Science*, Prentice-Hall, 1966, pp. 40-46, also his "An Introduction to Simplicity", *Philosophy of Science*, Vol. 28 (1961). Also in the same issue of *Philosophy of Science*, pp. 109-171, may be found : "The Weight of Simplicity in the Construction and Essaying of Scientific Theories" by Mario Bunge; "Safety, Strength, Simplicity" by Nelson Goodman; "Inductive Simplicity" by Robert Ackermann and "On Simplicity in Empirical Hypotheses" by Stephen Barker. N. Goodman's more technical article, "The Test of Simplicity" in Science, Vol. 128, has already been mentioned in a previous footnote in connection with the power of a system.

(4) Closely related to this aspect of generality in axiomatized theory is what we have called *perspicuity*. By this we mean the clear demarcation of formal from other aspects of a subject-matter which the axiomatic mode of exposition makes possible. The advantages of this are far-reaching. Separating the formal (syntactical) from the interpretational (semantical) features of a theory facilitates focusing on the logical structure of a theory in abstraction from its referential content. This could prove to be valuable in the construction of a new theory. In the attempt to systemize a class of phenomena, D, a theoretician may, as a matter of strategy, see some promise in endowing his theory with formal features similar to those of existing laws about phenomena other than D. (Maxwell and Kelvin made great use of such *formal analogies*, as they are sometimes called.) The implementation of such analogies will no doubt be a more deliberate and systematic process when a theory-to-be is first delineated as a formal calculus in which structural features are readily incorporated without the complicating concern for empirical significance. Maxwell and Kelvin did not employ the axiomatic method in any strictly formal sense of this, though the models which they eventually constructed for electrostatic potential (Kelvin) and the electric field (Maxwell) were genuine mathematical ones. Schrödinger, however, comes closer to theory construction by *formalization*. By implementing a purely formal analogy with the theory of standing waves on a vibrating string, he arrived at his famous wave equation whose solutions provide the formal machinery of a contemporary version of quantum theory. We have already noted the manner in which this was done in an earlier section.

Closely dependent on this advantage in the axiomatic method is another having to do with the explication of concepts. Consider, in general, a theory, T, about some class of entities states, or events, P. Such a theory serves not only as a means of deriving laws about P but also as an explication of the P-concept. Thus, in a very important sense, Newtonian mechanics tells us what mass and force are. Or perhaps, putting it more appropriately, it determines the mechanical concepts of mass and force. But, for all its apparent simplicity, this point bears a bit further consideration. The explication which a formalized theory may be said to provide for any of its constituent concepts occurs on two distinct levels. Again, these are the formal (or syntactical) and the interpretational (or semantical) phases of formulation. Asked to explicate a certain concept, P, one may respond by providing a theory, T, of P. This consists of a set of abstract axioms containing the term, 'P', together with an interpretation which, by means of correspondence rules, links P to some sort of subject-matter. The axiomatic mode of exposition thus makes it possible to divide the question of meaning clearly into two parts : the question of a concept's formal characteristics and that of its referential content. The first of these is answered by the abstract calculus (or its axioms), the second by an interpretation of this calculus. In the philosophical analysis of scientific concepts there can be a considerable advance in clarity if these two questions are kept distinct. An important contemporary issue in the philosophy of science, serving to illustrate this point, is the one over how to explicate the probability concept. It is a matter of virtually settled opinion that whatever the philosophical differences on how we are supposed to understand the term, 'probability', all probabilities must "have" certain formal characteristics as prescribed in the uninterpreted calculus of probabilities.[49] Philosophical disputation, therefore, on the meaning of 'probability' is not about the formal features of the term 'probability' but about what interpretation or semantical content to give it so as to transform the statement forms of the probability calculus into genuine probability statements. Hence, care to separate formal or syntactical considerations from interpretational or semantical ones in discussions about the meaning of probability can do much to clarify issues and even to avoid serious fallacies regarding the origin and calculation of probability metrics. In this connection, a not uncommon error is that of supposing that since the probability calculus sets forth the purely formal or logical feature common to all probability conceptions then any concept satisfying this calculus is a probability concept. Indeed, the probability calculus admits of an indefinite number of interpretations and of these only a mere few – most commonly *three* (though some argue for distinguishing a few more) – are considered to be explications of the notion of probability. In particular these are the limit-frequency (generally regarded as empirical) concept, the logical or *a priori* concept, and the rational conviction or subjective concept. As generally explicated and understood, each of these concepts corresponds

[49] The probability calculus can be most generally represented as a special part of set theory, in particular, the mathematics of additive, normalized sets. Probabilities are then formally regarded as measures of subsets, A_i, of some *basic* set, A, of individuals. A need be neither finite nor denumerable. All measures must then satisfy certain purely formal conditions. In particular, these are : (1) *additivity*, requiring that the joint measure of any two sets A_i, A_j, with no members in common, be the sum of the measures of A_i and A_j, respectively, and (2) *normality*, requiring that the measure of A be unity. These stipulations together with certain convenient definitions such as that of relative measure of a set, A_i, with respect to A_j (defined as the ratio of the measure of $A_i \cap A_j$ to the measure of A_j) suffice for deducing all the theorems traditionally belonging to the mathematical theory of probability.

to a radically different method of assigning probabilities to events or – what is the same thing – of verifying probability statements. At any rate, whether *three* or *two* or just *one* probability concept properly represent our use of the term, 'probability', in science and everyday life, what must be kept clear is that some such concept or interpretation of the probability calculus is indispensable for obtaining probabilities. The probability calculus or mathematical theory of probability is no more than a branch of pure mathematics and ultimately of logic for compounding or decomposing *given* metrics of a certain logical sort. It cannot yield any probabilities except on the basis of known probabilities. It is essentially a deductive system for computing what we might call *derived* probabilities on the basis of given, *initial* probabilities. Essentially, it provides the logical means of providing analytical transformations of given probability statements. Thus, what some have called mathematical probabilities are only deductive transformations involving undefined probability terms. Such transformations carry only the content that is put into them when probabilities determined according to one of the interpretations are put in for the undefined probability variable. Confusion has hampered discourse on the nature of probability and has even led to fallacious attempts at application of the probability calculus when it has been supposed that some of its theorems could *by themselves* suffice for determining the initial probabilities of actual events, in the sense of positive knowledge about such events.[50]

3. But despite the advantage of strict formalization, it has not been the dominant mode of theory construction although known since antiquity. One reason for this is undoubtedly that an axiomatized exposition is abstraction which naturally comes at an advanced rather than early stage of conceptualization. Axiomatization represents a distillation of what may be the most subtle characteristics of any subject-matter, namely, its formal attributes. It is no surprise, therefore, that a concept or subject-matter is, almost without exception, very well known before there is axiomatic theorization about it. As has already been noted, the axiomatization of even the most general and basic theories of empirical science has come only after they have had long and successful careers. In some cases, such as that of the Newtonian mechanics of a system of particles, rigorous axiomatization has been little more than a reconstructive, logistic exercise of interest

[50] It has already been noted in a previous footnote that the so-called theorem of Bayes on "inverse" probabilities – a purely mathematical theorem of the abstract probability calculus – has been one of the most notorious sources of such confusions. For an account of misuses of this theorem, see H. Reichenbach, *op. cit.*, pp. 93-96.

almost exclusively to philosophers and logicians rather than to physical scientists.[51]

Another reason for not using the abstract axiomatic method as a primary or initial mode of theory construction in empirical science has been one which we have already touched on in connection with the rise of substantive models. This is the traditional quest for causal explanations which, as has already been noted, has engendered the construction of theoretic models postulating underlying structures of subject-matter in order to explain the observable attributes of the subject-matter. Such (substantive) models supplemented by mathematical descriptions of their workings have constituted nearly all of what there has been to scientific theory until relatively recent times. Hence, from the viewpoint of causal explanation the strict formalization of theory has, in effect, been a methodological detour of only academic interest to the working scientist.

Contemporary science, however, has seen a reversed trend with new emphasis on mathematical models and a loss of centrality for causal explanation. As was noted in an earlier chapter on subsumptive inference, theoretic systemization in contemporary science is often a matter of abstract calculation, i.e., mathematical transformations of given data. Contemporary quantum theory is largely of this sort, along with some mathematical models in genetics and the behavioral sciences. Moreover, in some contemporary theories having formal structures that, in certain respects, differ radically from the mathematical representations of past theories, it is remarkably difficult to introduce suitable substantive models. The reason is the strangeness of the new theory and the fact that substantive models generally introduce underlying subject-matter and structure that can be visualized only in accordance with unsuitable traditional conceptions. Thus, despite the great value of substantive models in the application and even the extension of theory, their use is not without the hazard of hampering instead of facilitating the systemization of scientific fact. Substantive models generally lack the rigor of an explicitly axiomatized calculus. Consequently, the theoretic subject-matter which they postulate may be tacitly and fallaciously endowed with features leading to a misunderstanding of the theory and hence to unwarranted expectations regarding its consequences. For example, the interpreting of quantum mechanics as a "particle" theory, that is, in terms of a particle model, is by now a familiar source of misunderstandings. The trouble is that in using such a model, the sub-

[51] For a well-known textbook version of such a formalization within set-theory, see : P. Suppes, *op. cit.*, pp. 291-305.

atomic particles to which it refers, e.g., electrons or even neutrinos, are envisaged as particles having Newtonian properties such as determinate mass, position, and momentum. The result may be intellectual discomfort and a sense of paradox read into some of the theory's consequences or even into the "strange" results which the theory does not rule out. Examples of such perplexities are those which have attended "particles" (e.g., electrons) for which, according to theory, there cannot be simultaneously assigned a determinate position and momentum, or "particles" (e.g., neutrinos) which have no observable mass at all and sometimes behave in a manner similar to light, yet can interact with another particle as though they were bits of matter. The hindrance which substantive models pose to free-swinging trends in mathematical theory construction has led many contemporary scientists to distrust substantive theorizing or to recommend its total abandonment at the level of mature and stabilized theory. Be this as it may – empirical science, in actual practice, fashions its theoretical content (i.e., its nomic models) either in terms of *substantive* or *mathematical* (i.e., algorithmic, non-explanatory) constructions, though more recent science instantiates the latter to the greater extent. Where preferences are concerned some "purists" opt for the "intuition-free" flexibility of mathematical models, while other (perhaps, more traditionalist) commentators take substantive models, and the explanations they make possible, to be the only scientific constructions which are intellectually satisfying. But despite the fact that either one or the other form of theoretic elaboration may, in some ultimate sense, be logically dispensable, both modes of construction remain substantially represented in basic scientific theory. Indeed, it may well be the case that – as a matter of *psychological* necessity – neither mode of theorizing may be entirely excluded in the making of any growing science.

E. ANALOGICAL CONTENT IN THEORIES

We have already discussed analogy in connection with scientific discovery where analogizing with known or familiar material can be a major step in the creation of a new theory. We have also found that analogy is the explicit basis for one kind of scientific model, namely, the analogical model. Before concluding our concern with theories, it seems appropriate to return to the analogical aspects of scientific theory for one further consideration. The play of analogy as a strategy of discovery and, to some extent, in the use of analogical models (the latter being, strictly speaking, extrinsic to theory *proper*) leaves an unmistakable stamp on the actual form and content of established theory, itself. This manifestation – which one might call the *analogical content of theories* – takes two forms : The *first* is the striking similarity of formal structure which one finds between some theories about qualitatively diverse subject-matter within the same basic science. Such similarities are often traceable to formal analogies which were deliberately drawn, at the stage of theory creation, with already established theories. The *second* form of the manifestation concerns theoretical vocabulary, which is shot through with terminology and residual nuances that easily betray analogical origins. Undeniably, analogical content has much heuristic value in the application of theory and for purposes of its further development, in a certain sense, perhaps, enhancing the depth of "grasp" or, better still, appreciation of any theoretico-scientific interpretation of phenomena.[52] But such analogical content can also be misleading and pose real obstacles to understanding the role and significance of any given theory in the systemization of scientific fact. The difficulties tend to arise when analogies are overdrawn far beyond the limited resemblances which constituted their original bases. And most prone to the difficulties is the non-specialist confronted with, or looking in on, scientific discourse; although, as the history of science suggests, the professional himself can be similarly hampered.

Science offers an abundance of examples bearing on these considerations. Here are a few :

(1) The general notion of force in mechanics and its relation to induced motion no doubt springs analogically from the notion of muscular exertion in the moving of objects. Indeed, the term, 'force', goes back to the Latin, '*fortis*', which means strong. Instructive though this early association might have been, the mechanical concept of force in modern dynamics has gone a long way from its anthropocentric origins. And it is extremely doubtful that the old nuances would figure much in the thinking of any modern physicist, particularly since the German physicist, E. Mach, proposed formulations in which the term, 'force', is essentially eliminable by defining it in terms of mass and acceleration. But the extent to which the analogy was drawn no doubt had much to do with the stagnant state of ancient mechanics attributable, at least in part, to the failure to generalize the force concept so as to give more explicit and precise recognition to other manifestations of *mechanical* force such as friction and gravitation.

(2) The term, 'gas', first used by the Belgian chemist, Van Helmont (1644), was suggested by the Greek term, *chaos*, meaning formless or

[52] It should be noted here that we apply the term, 'content', in blanket fashion despite the purely structural nature of the *first* form of the manifestation in point.

disordered. Not only does this implicitly compare a gas to more palpable instances of matter in chaotic state but it also seems to hint strongly at the statistical content of the later kinetic theory of gases. Despite Van Helmont's coined word, however, gases, for the next hundred years of pneumatic chemistry, were called "airs" – e.g., Black's "fixed air" for carbon dioxide. This reflects a second analogy, this time, between gases (irrespective of their more specific properties) and atmospheric air. But the analogy was much too easily overdrawn in support of the phlogiston theory of combustion and, for a time, must have helped to obscure the possibility of inherently distinct gases, with dire consequences for the growth of chemistry and atomic physics.

(3) The familiar reference to atoms and cells as building-blocks of matter and living tissue, respectively, draws obvious but illuminating analogies with visible structures. But one must also recall here the various attempts to enshrine these building-block models in the notion of absolute, atomic indivisibility and in vitalistic doctrines of the physico-chemical irreducibility of "living" cells. From the viewpoint of later developments in nuclear physics and molecular biology these are plausibly regarded as retarding extensions of the analogy in point.

(4) The original form of the wave theory of light was an analogy with sound waves. The analogy, however, was carried too far by Newton who supposed that if light were a wave it could not travel in a straight line.[53] Both he and Huygens further stretched the analogy by failing to give sufficient consideration to the possibility that light, unlike sound, might be a transverse wave. Had this been done, they might have explained polarization – known to both of them – thus perhaps settling the wave-particle controversy about the nature of light more than a century sooner.[54]

(5) The early physics of heat drew heavily on supposed similarities of heat to fluids (Black and Watt). Sadi Carnot, who later came to regard heat as a particle motion, developed the still valid theory of the ideal heat engine utilizing the concept of caloric as a material fluid. What is more, broader theoretical formulations about heat flow were often mathematically patterned after known laws of hydrodynamics (Fourrier). But it took a Lord Kelvin to advance the generality of treatment by

[53] Light, unlike sound, is largely rectilinear because its frequency is enormously greater than that of sound, so that in ordinary cases (i.e., for ordinary obstacles) diffraction effects are often negligible.

[54] Newton did hint at something like transverse waves in connection with polarization, suggesting that "rays of light have several sides". See Isaac Newton, *Opticks* or *a Treatise of the Reflections, Refractions, Inflections and Colours of Light*, Book III, Queries 26-29.

recognizing the limitations of the analogy and introducing new analogical approaches drawn from electrostatic phenomena.

(6) In like fashion similarities of mathematical form (isomorphisms) with known laws of heat (Kelvin) and of incompressible (dielectrical) fluids (Maxwell) were postulated for articulating electrical theory. But though the reference to incompressible fluids is essential in the analogy, the comparison becomes troublesome when it is extended uncritically to include density and elasticity aspects of the fluid phenomena involved. Maxwell, in excluding specific reference to an ether from his formulation of electro magnetic theory, no doubt recognized the militations in the analogy. But later polemics about the "elasticity" and "density" of a supposed ether were, in part, consequences of an overdrawn analogy.

(7) The development of 20th-century atomic theory owes much to the now familiar analogy with cosmic solar systems. This analogy was a partial basis for a nuclear model of the atom with distant electrons orbiting around a compact central nucleus. Rutherford, whose work laid the first foundations of contemporary atomic theory, hesitated to accept this model because, as he construed it, it would not work. For, according to electromagnetic theory, such an atom would soon radiate all its energy and, contrary to fact, would run down. But conceived in this manner the analogy was too sweeping. It was Bohr who provided the appropriate deliminations with his quantum modifications of the model thus opening the way for the modern physics of the nuclear atom.

(8) As a more contemporary example we mention the important and dramatic analogy which led Schrödinger to the famous wave equation of mathematical quantum theory. This analogy was discussed in some detail in our sections on scientific discovery. It suffices to mention, here, that Schrödinger developed his central formulation by literally adopting some of the formalisms involved in the classical theory of vibrating strings. The analogizing is of an almost strictly formal sort. And the parameters of the resulting equation are given physical significance by means of deliberate stipulations which tie them to experimental ideas. Nevertheless, the analogy transfuses terms like 'wave' and 'resonance' from old to new contexts. And an incautious reading of such terms in quantum mechanical contexts can lead to confusion on just what sort of basic subject-matter quantum theory is actually about. Discourse relating to quantum phenomena seems to be particularly prone to the dangers of over-extending analogies originally drawn from classical contexts. Particularly noteworthy are such terms as 'particle', 'momentum', and even 'location', whose classical

nuances can seriously obscure their altered significance in quantum mechanical formulations.

A further point, it seems, should be made here with regard to analogies in scientific theory. As our examples suggest, such analogies may be grouped under two headings, theory-to-theory (or formal analogies) and subject-to-subject (or substantive analogies). Examples (6) and (8) illustrate the first kind, (5) is somewhat mixed and all the rest illustrate the second kind. In a theory-to-theory analogy, some of the structural features of a theory or other system of laws, T′, about a subject-matter, S′, are used for structuring a new theory, T, about some other subject-matter, S.

This is a well-known strategy in the extension of formal theory throughout many branches of systematic knowledge. Logicians, for example, have modeled their explications of the notion of confirmation on the abstract relations of the probability calculus. That is, among the various features stipulated for the concept, one is that it satisfy the abstract probability axioms. That this procedure need not have been followed is evidenced by the fact that a "non-probabilistic" account of confirmation has been recently proposed which does not adopt the formal features of the probability calculus.[55] In general, mathematics offers an abundance of examples in which new systems are constructed and extended in isomorphism with already known ones. Various branches of modern mathematics for example are often developed and extended in terms of an isomorphism or identity of structure between sets of various sorts (groups, subgroups, rings, etc.) for which defined operations are required to correspond in formal structure.

In the material sciences, theory-to-theory analogies constitute a major approach to theory construction. Our examples (6) and (8) are only two historically important illustrations of this; still others could be mentioned. Indeed, there are some very broad formal assumptions made tacitly by any scientist in first formulating a law, L. They require that L resemble all other known laws with respect to certain aspects of simplicity. One of these, for example, is that L does not change with time. It is sometimes expressed by saying that the time variable, 't', shall not occur explicitly in L.[56] In attempting to frame the laws of nature so as to satisfy this formal

[55] Cohen, J., "A Logic for Evidential Support", *op. cit.*, pp. 21-43, 105-126. *Vide supra*, footnote 17 ,Chapter II.

[56] As we have noted in earlier discussion, if time derivatives of any order were allowed in the mathematical expression of a law, then (except for some special functions of t) an explicit 't' could always be eliminated by a sufficient number of differentiations. The simplicity condition of which we speak, therefore, in order not to be trivial, requires a

condition, a broad theory-to-theory analogy has been drawn with all other known laws.

In subject-to-subject analogies a familiar domain of subject-matter, S′, whose characteristics are expressed by a set of known laws, is taken as a model for constructing a set of substantive assumptions, T, about subject-matter, S, different from S′. Thus, in Bohr's model of the atom, there are assumptions (T) made about the micro-system (S) which consists of a nuclear atom with orbiting electrons. The forces with which the electrons and nucleus interact are postulated on the basis of an analogy with the strong interactions of electrically charged *macro*-particles (S′). The assumed interaction therefore is that of an electrostatic Coulomb (inverse square) force holding the atomic electrons in orbit about the nucleus. The analogy is extended by introducing perturbations of the electronic orbits caused by electrostatic interaction between the orbiting electrons. Though the forces are electrical, the perturbation effects are calculated in identically the same manner as are planetary perturbations produced by gravitational interactions between orbiting planets. This time, an analogy has been drawn with planetary systems (S′). A number of analogies have been drawn in attempts to construct theoretic models of the atomic nucleus. One example is the Heisenberg notion of a nuclear exchange force between a neutron and a proton in which an electron is shared between them. This account was an attempt to explain the important phenomenon of *beta*-decay in which the shared electron is in some cases emitted as a *beta*-particle. It was modeled after the so-called "exchange forces" between two atoms due to their sharing a valence electron that tends to "divide" itself between them, thus producing covalent chemical bonding. Accordingly, the nucleus is envisaged very much as a miniature molecule in which protons and neutrons play a role analogous to that of atoms in a covalent molecule.[57] Another model of the nucleus is the shell-model in which it is assumed that the neutrons and protons in the nucleus form shells much like the orbital, electronic shells of the extra-nuclear atom.

In summary, one might say that both formal and substantive analogies have played a vast role in theory creation. This has resulted in the formal resemblances which established theories (particularly in the formal sciences) have to each other and in descriptive, theoretical vocabulary brought in by the analogies. As our considerations here and under discovery

restriction on the order of differential equations expressing the laws of nature. These are generally assumed to be no higher than second order.

[57] For Heisenberg's own account see Werner Heisenberg, "Uber den Ban Atomkerne I". *Zeitschrift für Physik* 77 and 78, 1932.

(Chapter III) point up, the role of analogy in the conceptual "leaps" required for most theoretical innovation seems to be crucial. Moreover, some analogical basis in more familiar material can be of inestimable value in the use and further expansion of any novel theory. Nevertheless, it is also true that analogical content in scientific theory calls for much caution. Indeed, overdrawn analogies, uncritically applied, can not only hamper scientific thinking but also lock it in fruitless paradox and conceptual dead-ends.

F. RECAPITULATION : WHAT IS A SCIENTIFIC THEORY?

1. The use of theory in the systemization of empirical knowledge, perhaps more than anything else, decisively separates science from common-sense levels of factual inquiry. And by 'theory' we mean here not an unconfirmed conjecture but a *set* of systemizing statements (hypotheses) differentiable from other law-like hypotheses by its "theoretic" vocabulary. This statement-set contains not only nomic universals but also existential sentences for introducing theoretical subject-matter. In ordinary scientific usage a set of theoretic assumptions is also loosely referred to as a *model* To cover this rather prevalent sense of the term, 'model', (that is, the sense in which it is synonymous with the term, 'theory') we have introduced the more specific term, 'theoretic model'.

2. We may group scientific theories under two broad headings : (1) those which consist merely of logical or mathematical apparatus for *calculating* observational effects from given ones, and (2) those which postulate underlying (theoretic) subject-matter (inner structure, entities, etc.) as a causal base for *explaining* observational effects. We call the first of these *mathematical* theories or models, the second *substantive* theories or models.

3. The most formal representation of mathematical models [58] is that of an abstract calculus whose descriptive terms are connected with experimental ideas by means of correspondence rules. Though such calculi, and their less formal counterparts, serve to systemize (e.g., predict, postdict, etc.) observational effects, they cannot properly be said to *explain* them in the sense in which "explaining" an effect tells us *why* the effect occurs.

4. It would be erroneous to assume that scientific theories are actually created as formal calculi together with appropriate rules for coordinating them with observational subject-matter. In actual practice scientific theories

[58] In this case it seems preferable to speak of a mathematical *model* since the term, 'mathematical theory', is often used for any theory whatever in which quantitative, mathematical relations are expressed.

are usually born as substantive models with more abstract versions coming as much later developments. But even where a theory is introduced at the outset as a mathematical model its vocabulary is such as to suggest both the manner of coordinating the theory with experimental ideas and the manner of providing for it a coherent substantive model or interpretation. Substantive interpretations can be of major value in applying and extending a scientific theory. They may also yield a sense of intellectual satisfaction in providing a causal (explanatory) base for the observational subject-matter which a theory systemizes. Much of contemporary scientific theory, however, especially in the advanced physical sciences, has proved to be difficult to interpret in the substantive mode. Available substantive models have been shown to be inadequate, in some cases, even hampering and misleading those who use them. The systemization of knowledge in contemporary theoretical science therefore presents a rather striking non-explanatory and, one might say, "a-causal" character.[59]

5. For the purposes of analysis, and quite apart from the actual manner in which theories come to be, there is great advantage in regarding any scientific theory as consisting, at least implicitly, of some abstract calculus or other and its appropriate interpretation. The interpretation of a calculus may either coordinate elements of the calculus with specific experimental ideas by means of correspondence rules (operational definitions, semantical rules, etc.), or it may coordinate the calculus with a system or *model*. Such a model may be either of the *mathematical* sort, transforming the original calculus into another (isomorphic) one, or it may be of the *substantive* sort, providing the existential and more or less "intuitable" flesh for the skeletal formalisms of the calculus. In either case it is a "system" having an explicitly indicated structure of its own. Substantive models, we may note, consist of what we have called *theoretic* subject-matter. By this we mean subject-matter which is not directly accessible in the sense that chairs and ships are.

6. To our account must be added mention of *concrete* and *analogical* models, neither of which we view as constituent parts of theories, but both of which are used to promote the understanding of subject-matter. The first of these are convenient three-dimensional representations – usually, either miniatures or enlargements – which may be manipulated concretely

[59] It seems preferable here to speak of the *a-causal* rather than the *non-causal* character of contemporary physical theory. The latter term is heavily burdened with metaphysical connotations and also brings to mind matters relating to whether a theory is deterministic or not, time-independent or not, etc. These are considerations which have no essential bearing on our characterizing some contemporary physical theory as "a-causal".

in place of less manageable prototypes. The second, unlike concrete models, are merely conceptualized or imagined ("intuited") representations of elusive subject-matter. Moreover, they are qualitatively distinct *analogues* rather than scaled copies of the original.

The following may be taken as a synoptic schema for representing the relations between abstract calculi substantive models and the observational domain.[60]

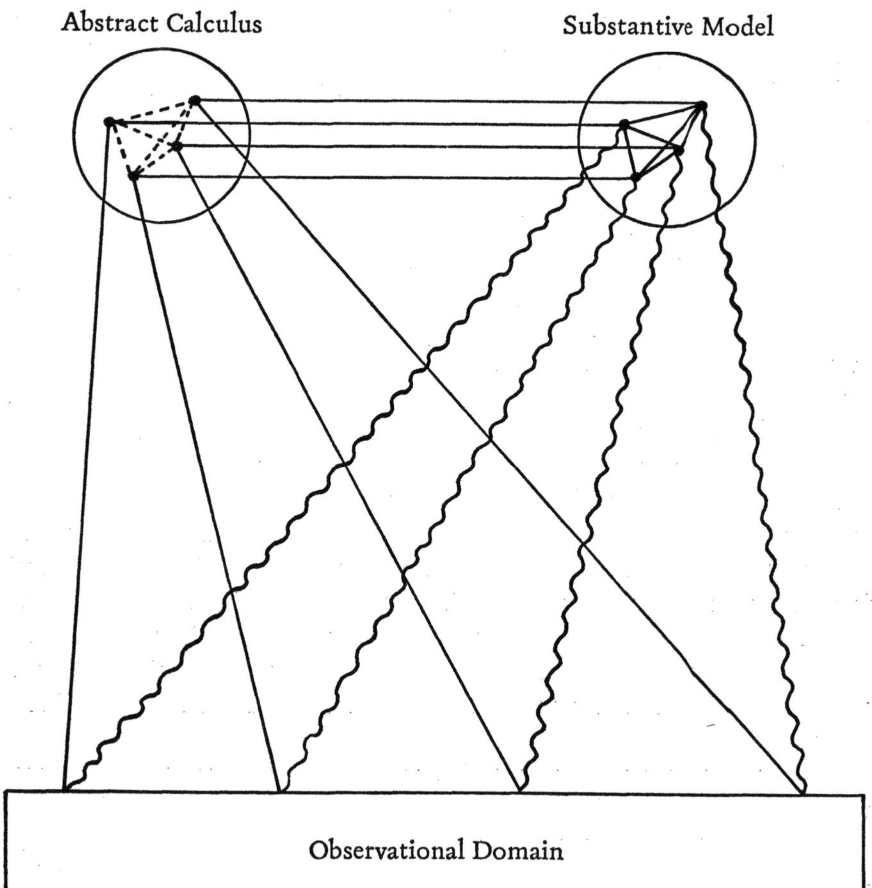

The dots represent postulated elements and entities of the abstract calculus and substantive model, respectively.

[60] Some of the features of this diagram are modeled after a schema by K. F. Schaffner, "The Watson-Crick Model and Reductionism", *The British Journal for the Philosophy of Science*, Vol. 20 (1967).

The horizontal lines represent correspondences between calculus and model.

The broken lines between dots represent formal relations between elements of the calculus.

The solid lines between dots represent structural (space-time) relations between entities of the model.

The oblique lines represent correspondences with the observational domain. Those stemming from the substantive model are wavy to distinguish them as causal.

VIII

GLIMPSES BEYOND

A. OVERVIEW

We have now examined various structural aspects of scientific knowledge, the major portion of our considerations having gone to science as *product* (rather than as process) of inquiry. Accordingly, our attention has been principally on *subsumption under hypotheses* as the basic mode of logical structure, with *nomic inference* (explanatory as well as non-explanatory) and *nomic statements* (laws, theories, models, etc.) as the central topics. The subsumptive systemization of "finished" science, however, is not the only subject which concerns the methodologist. The *growth of science* – and this includes both the *discovery* and *acceptance* of scientific hypotheses – has also been of interest to him even as far back as antiquity, though in very recent times, the prevailing hypothetico-deductive view of science has virtually removed discovery from the curriculum of the analytical philosopher, deeming it a matter for the historian, psychologist or sociologist rather than for any analytic discipline. The questions of acceptance (confirmation) and refutation (falsification) have, meanwhile, festered under the cover of unsettled opinion. Contemporary discussion in the philosophy of science, however, offers strong indications of renewed and expanding interest in the entire subject of scientific development or *ampliative science* as we have also called it. Opinion on how to characterize this exciting phase of scientific inquiry is lively and unsettled. But out of the dialectical exchanges some novel trends of thought are emerging which pose a root challenge to traditional methodologies.

Our own attention to ampliative science has, within the space of this work, been limited to the relatively brief and general considerations in Chapter III. For the sake of a better perspective, therefore, let us, in this concluding chapter, attempt to round off discussion with some glimpses of growing trends in the philosophy of scientific development.

B. CONVENTIONALISTIC TRENDS

After P. Duhem, H. Poincaré and, more recently, W. Quine, the philosophical analysis of empirical science has had to take into account a view which, in some important respects, represents a radical departure from traditional forms of empiricism in the philosophy of science. We refer here to the thesis known as *conventionalism*, which has various forms but, very generally speaking, claims that the acceptance of any scientific hypothesis is ultimately a matter of logically indifferent choice from among various possible alternatives.[1] In its most general aspects, the idea is not new and brings us back to Protagoras and his "man ... the measure of all things". Even Aristotle, who would hardly count as any sort of conventionalist, might, by some stretch of interpretation, be found suggestive along these lines with the idea that ontology (the study of being *qua* being) is intimately tied to logic and more generally language, some of the forms of which might be said to be matters of convention.[2] Modern conventionalism, however, owes something to Kant whose idea it was that how we "see" the world depends on the structure of the human understanding. Kant does not mean to suggest here that we have any choice in the matter (like Aristotle, he was no conventionalist); but his thesis does ultimately relativize any claim to scientific "truth", thus suggesting the general (Kant's term here might have been "transcendental") possibility of alternatives, as in the case that our minds had been differently structured.

But explicit conventionalism begins with more recent writers like the mathematician-philosopher H. Poincaré and the historian of science, P. Duhem, the latter's version of the view perhaps being the most sweeping.

An explicit account of this thesis might begin by recalling the elementary but suggestive point of logic (see Section E and *passim* in Chapter III) that no actual body of evidence can logically (i.e., deductively) imply any unlimited general hypothesis. It is hypotheses (given certain initial con-

[1] A. Grünbaum has drawn attention to features of Poincaré's thought which differentiate it from the Duhemian thesis and which are generally overlooked by commentators. If Grünbaum's reading of Poincaré is right, then it is the Duhemian rather than the Poincarean thesis which radically challenges a major empiricist tradition in a manner which we shall presently note and to which our considerations are essentially addressed.

See : A. Grünbaum, "Geometry, Chronometry and Empiricism", *Minnesota Studies in the Philosophy of Science*, edit. H. Feigl and G. Maxwell, vol. III, Minneapolis, 1962, pp. 493-506.

[2] This, however, is a most debatable point. One could, it seems, also argue (perhaps even more convincingly) that Aristotle's tying of language to ontology is an "objectification" of language rather than a "conventionalization" of ontology.

ditions) that imply data, and not the other way around. Indeed, for any body of evidence, one can, in general, find an unlimited class, K, of mutually incompatible hypotheses each of which can account for the evidence; so that, though one might be inclined to say that some body of evidence *confirms* an hypothesis (in the sense of *instantiating* it), he could not maintain that such evidence *necessitates* that hypothesis. This, however, entails that a given body of evidence could be said to confirm, in the sense of instantiating, not just a single hypothesis but any of an unlimited set, K.

Commonplace as these considerations might seem, they surely undermine any notion of evidential verification or *confirmation by instances*. For on given evidence, the adoption of any *one* hypothesis seems to remain a matter of logically indifferent choice from among an unlimited set (K) of alternatives. To note the full thrust of this consideration it must be recalled that the members of K are mutually incompatible, so that the truth of any one would entail the falsity of all the others. It would seem reasonable to suppose, therefore, that, for any member of K, any gain of confirmation (in some sense relating to "truth" or acceptability) implies some loss of such status for any other member. Hence, it would seem, the same body of evidence cannot confirm each member equally except in the case that such confirmation is *zero*.

There is indeed, here, a serious difficulty in the entire notion of confirmation by instances, a difficulty which suggests a *prima facie* case against any inductivistic philosophy of science, unless confirming instances give way to some other basis for inductive validation such as : simplicity, compatibility with existing context, etc.

But dire though these considerations might be for any justificationist view of hypotheses they are not quite as sweeping as those of conventionalism. They raise doubts only about confirmability but not about refutability. For, though the set of hypotheses fitting a given body of evidence is unlimited, there are, nevertheless, hypotheses (outside this set) which would not do at all. These are the hypotheses which are allegedly incompatible with the evidence and hence falsified by it. Indeed, this is the view of non-conventionalistic, hypothetico-deductive theorists (like K. Popper) who, though denying the confirmability of hypotheses, nevertheless maintain their refutability by unfavorable evidence.

We should pause to note here that, even in the face of dubious confirmability, the refutability of any hypothesis provides some guarantees of its factual content or *significance*. For if factual evidence can contradict an hypothesis then such an hypothesis must be "saying" something factual.

But refutability is what conventionalism denies. Its claim is that no scientific hypothesis, H, need ever be judged false in the face of unfavorable evidence. For it is not on the basis of H alone that we derive an observational consequence, O, but on the basis of H in conjunction with a cluster of tacit, assumptions, C. That is, *collateral* with H there is always a set of undecided hypotheses, C, which, as it were, can be "blamed" for the apparent failure of H. For greater explicitness on this point, consider the following :

Traditionally speaking, the falsification of any hypothesis, H, can be represented by a deductive inference called *modus tollens.* Letting O be an observational consequence of H, and assuming that O turns out to be false. i.e., not-O is true, we have :

$H \rightarrow O$ (read : "H implies O")
$\sim O$ (read : "not-O")

therefore, $\sim H$ (read : "not-H")

Conventionalism rejects the above schema on the grounds that it leaves out of account the collateral assumptions, C. The replacing schema then becomes :

$H \cdot C \rightarrow O$ (read : "the conjunction of H and C implies O")
$\sim O$ (read : "not-O")

therefore, $\sim (H \cdot C)$ (read : "not both H and C")

But $\sim (H \cdot C)$ is equivalent to $\sim H$ or $\sim C$. Thus, neither H nor O need be false, and we can have our pick. More specifically, let us now suppose that O′ materializes instead of O; then the conventionalist tells us that we may assume C to be false and replace it by some alternative collateral assumption, C′, i.e., a revision of C, such that :

$$H \cdot C' \rightarrow O'$$

Thus, let us suppose that according to some entertained hypothesis, H, a star should appear at position, P, but instead actually turns up at another position, P′. One, then, need not reject H; for it can be pointed out that light had been tacitly assumed to travel in a straight line and that such an assumption may not be true. Indeed, one may assume such a path for the light as would yield (given H) the position, P′, rather than P, for the displaced star. What is more, presently accepted principles about the rectilinearity of light are unavailing. They may, in fact, not hold for great distances, or for light passing close to great masses, or for any number of reasons pertaining to the particular case in point. Other examples are readily constructed : Thus, in making linear measurements, we assume that rulers do not shrink or distort in certain respects. In measuring masses with a gravitational balance we assume the same gravitational field for both arms of the balance, the irrelevance of color, of shape, and of other types of force fields that may be present, etc. Similarly, in testing the Newtonian laws of motion, we assume that we have correctly accounted for all forces acting and for the state of motion (re : relative acceleration) of our reference frame, for the behavior of our clocks and more. Again, a test of Copernican theory (*via* parallax) involves assumptions about terrestrial atmosphere, about the possible effect of motion on the mass of the object, even about the nature of perception with optical instruments, and much more.

Indeed, reflection strongly suggests that tacit assumptions either of a very broad underlying character (ontological, epistemological, perceptual, etc.) or more specific ones pertaining to experimental contexts (apparatus, fixed conditions, etc.) might *always* attend a "chosen" hypothesis and that, with logical impunity, any one of them can be denied to salvage that hypothesis. This seems to make the acceptance of any hypothesis a matter of logically free choice, that is to say, a matter of vacuous convention. For, if it is always possible to hold to an hypothesis, come what may, then that hypothesis cannot be about any matter of fact. Uncontradictable by "fact", it cannot be *about* fact. At best, it may be viewed as ultimately a device for fixing and relating concepts, that is to say, a *definition.*

This seems to be a somber and paradoxical result. The sciences – particularly the "natural" ones – are dramatically successful enterprises. But how does one account for this success if scientific laws and theories are nothing but definitions? Some writers have attempted to escape this consequence at least for physical science by maintaining that, once certain mathematical notions are fixed, in particular, the definition of geometric congruence, then the question of which geometry holds in physical description is a matter of empirical fact.[3] To this some might object that definitions like that of geometric congruence are themselves optional; so that this like any other *conditional* resolution of the matter is unsatisfactory. The opponent of conventionalism, however, could reply that the assumptions in point, optional though they may be, are, this time, purely *definitional* or semantical, rather than factual, in character. Hence, they merely establish significance for a body of discourse and neither add nor detract from its factual content. But if it be shown that the distinction between definitional

[3] See : H. Reichenbach, "The Philosophical Significance of the Theory of Relativity", ed. P. A. Schilpp, *Albert Einstein, Philosopher Scientist*, Evanston, Ill., pp. 286-88.

(analytic) and factual (synthetic) statements is open to question as Quine attempts to do, then even this sort of reply falls open to debate.[4]

Some writers have charged conventionalism with the error of confusing the formulations of pure mathematics which are arbitrary but without referential content with those of empirical science which make reference to existential subject-matter and whose acceptance is therefore subject to evidence.[5] But that this is a fair criticism of the most general claims of conventionalism is far from clear. Conventionalists, like Poincaré, do speak of the ease or "convenience" with which some hypotheses are applied, in contrast to others. But this would strongly suggest that interpreted rather than purely syntactical systems are being entertained.

Again, under pressure of the conventionalist thesis, some writers, including Poincaré himself, have made simplicity a key consideration in deciding which hypothesis to hold to.[6] At first regarded as a mere matter of preference, simplicity, as a structural feature of hypotheses, has come more and more to be regarded as inductively associated with their success and acceptance, and even with their objective plausibility. In our own references to simplicity we have considered the possibility of attributing to it this latter sort of validational import.[7] A thorough explication of the structural character of simplicity in the full context of scientific language is still to be accomplished in the contemporary philosophy of science.

We pause, now, for some further considerations on the conventionalist thesis. It will be recalled that the conventionalist's claim may be conveniently summed up in the following statement :

Given that $H \cdot C \rightarrow O$ and that O' turns up instead of O, then one can always find a C' (incompatible with C) such that $H \cdot C' \rightarrow O'$. Thus, C' as a replacement for C would "save" H from falsification.

We ask now : What sort of claim is this? More specifically – what does the conventionalist mean when he says that one can always find a C' which will save H? Does he mean merely that, *as a matter of circumstance*, a scientist can always count on finding a C'? Thus, is conventionalism maintaining that : the scientist's context of assumptions is always open-ended and variegated enough; and the scientist is always ingenious enough; and natural possibilities as he knows them are diversified enough – so that a

[4] W. Quine, "Two Dogmas of Empiricism", *From a Logical Point of View*, pp. 20-47.

[5] E. Nagel, *op. cit.*, pp. 199-202, 270-274.

[6] H. Poincaré, *The Foundations of Science*, tr. G. B. Halsted, The Science Press : New York, 1929, pp. 133-140.

[7] Nelson Goodman (even more strongly) characterizes simplicity as "a test of truth". See his "The Test of Simplicity", *op. cit.*, p. 1064.

C′ of unknown truth-value can always be ferreted out for saving H? Surely the conventionalist could not be offering any such matter of factual contingency as a methodological principle. For in that case his thesis would not be conclusive. Indeed, circumstances might not always be so accommodating as to satisfy the required contingencies. The scientist's context of assumptions might not be variegated enough, or his ingenuity not great enough or, again, nature, as he knows it, insufficiently diversified, for actually finding the required C′. The conventionalist thesis can be conclusive, and hence have methodological import, only if it is shown that the availability of C′ is a matter of logic rather than of factual contingency. C′ must have, we might say, "logical existence", that is, one based on logical evidence alone and not on the abilities of scientists or on any particular character of the world. This then makes the preservability of H a preservability *in logical principle* and one that is independent of all factual contingencies. Now, to say that C′ has "logical existence" is to say that its *formulability* under the stated conditions is guaranteed by logical considerations only and hence holds for all possible material circumstances. Another way of putting it would be to say that the notion of a C′, given the stated conditions, is not self-contradictory, and that, though one may, under actual conditions, sometimes fail to find a C′, its formulability nonetheless remains logically possible throughout.[8]

But the logical possibility of finding a C′ depends upon the conditions which C′ must satisfy. We have already noted that C′ must, in conjunction with H, yield O′. That is :

$$H \cdot C \rightarrow O'$$

But reflection easily indicates that other restrictions are in order. C′ must, so to speak, be scientifically admissible. Hence, restrictions, R, of the following sort also suggest themselves. Of C′ we would want to say that :

(1) It must not be trivial.[9]

(2) It must not introduce semantical assumptions (rules) that would alter the meaning (and possibly the truth status) of H.

(3) It must not be known to be false (or even highly disconfirmed). For

[8] There is nothing unusual about speaking of the "logical existence" of some formulation even though the formulation is itself unknown. Thus, mathematicians sometimes establish the (logical) "existence" of solutions to equations even without ever knowing specifically what these solutions are.

[9] C′, for example, cannot be logically equivalent to O′ or to $H \rightarrow O'$, so as to guarantee trivially that :

$$H \cdot C' \rightarrow O'$$

no account in terms of a false hypothesis would be considered scientifically appropriate.

(4) It must not be *ad hoc*.[10]

The conventionalist thesis may now be more explicitly stated to include conditions, R, as follows :

For any H and any C and any O and any O′ (incompatible with O); given that :

(I) $H \cdot C \rightarrow O$ and that O′ occurs instead

(II) Then as a matter of *logical possibility* one can always find a C′ such that :

$H \cdot C \rightarrow O'$

But no conventionalist has, on the basis of (I), explicitly demonstrated the possibility referred to in (II) : Indeed, it is difficult to see how (I), together with any set of logical considerations, can guarantee (II).[11]

Conventionalism evidently continues to be a highly debated thesis and one that still calls for essential clarifications.[12] But its influence on the philosophical analysis of science has been profound. Following Poincaré, contemporary methodologists are increasingly more inclined to speak of the success and acceptance of a scientific hypothesis rather than of its verification or falsification. The historian T. S. Kuhn, for example, sees the *great* innovations in science as a sequence of non-cumulative (i.e., "revolutionary") choices based on more than the interdicts of external evidence.[13] With available ingenuity, anomalies can always be resolved and counter-instances accommodated. The inconvenience, or the frustration of practical preferences, or the comparison of competing alternatives (with each other rather than with nature), and other such reasons are what

[10] Thus, it would obviously not do, in the face of unfavorable evidence, to let C′ be the hypothesis that some special factor is at work which perturbs our data in a given sort of experimental context but disappears *entirely* whenever we try to detect it in any other way.

[11] A criticism of conventionalism along these lines is developed by A. Grünbaum in his "The Duhemian Argument", *The Philosophy of Science*, 1960, Vol. 27, No. 1, pp. 76-77. Professor Grünbaum, however, points out not only that the conventionalist's conclusion (II) does not follow but also attempts to establish that this conclusion is *false*. This stronger claim, however, will not detain us. For a rejoinder to it see : P. K. Feyerabend, *Current Issues in the Philosophy of Science*, edit. H. Feigl and C. Maxwell, New York, 1961, pp. 155-161.

[12] For instructive polemics, see : discussion by P. K. Feyerabend and A. Grünbaum, *op. cit.*, edit. H. Feigl and C. Maxwell, New York, 1961, pp. 140-168.

[13] "... the act of judgment that leads scientists to reject a previously accepted theory is always based on more than a comparison of that theory with the world". T. S. Kuhn, *The Structure of Scientific Revolutions*, p. 77 and also Chaps. VIII and IX.

make an anomaly sufficiently irritating to set off a crisis and revolution in the scientific community. The process is viewed as essentially psychological and sociological rather than logical in kind, the underlying thesis being that no hypothesis need be abandoned because of evidence alone, provided we make the appropriate contextual adjustments.

Conventionalistic methodologies are the foci of contemporary philosophic contention particularly in connection with the ampliative phases of science. But, more constructively, they point to new emphases and fresh approaches in the philosophy of science. Even so, there is an important sense in which it seems doubtful that the conventionalist's critique conclusively dissevers scientific knowledge from its observational moorings. What conventionalism denies, explicitly, is not that there are experimental counter-instances, but that such counter-instances can logically compel the abandonment of any hypothesis. Hence, though counter-instances may not, by themselves, suffice for dislodging an hypothesis, they can be the occasions for revisions of one kind or another either in the primary hypothesis or somewhere else in the general body of knowledge. In Quine's words, "... our statements about the external world face the tribunal of sense experience not individually but only as a corporate body".[14] Some writers, as we have already noted, see in this the suggestion that simplicity and perhaps other structural criteria are more fundamental criteria of validation than has been traditionally recognized. Under such considerations, counter-evidence would have a more indirect methodological bearing on hypotheses. Rather than falsifying them, it would detract from their status by the extent of reshufflings and complexity of accommodations that might be required to rescue them.

Strictly speaking, then, this seems to leave open merely the question of how tenacious one should be concerning any scientific hypothesis, with theoretic centrality, conceptual familiarity, simplicity, and other such criteria as the operative guidelines – over-all simplicity perhaps being the major criterion.[15] Moreover, as has been already suggested (cf. Chap. III), some rationale seems possible for the use of such criteria in any scientific strategy. Indeed, one could speak of the scientific status of an hypothesis, H, in terms of the degree of "tenure", so to speak, which it has earned in the system. By this, one would mean the *tenacity* with which it would be held to in the face of evidence. The limiting case of such tenacity would

[14] W. Quine, "Two Dogmas", *op. cit.*, p. 41.

[15] "We must stop somewhere, and that science may be possible we must stop when we have found simplicity. This is the only ground on which we can rear the edifice of our generalizations". H. Poincaré, *Foundations*, p. 133.

be one in which H would be retained in the system, come what may. This extreme case is particularly interesting because, paradoxically, though it would represent maximum status for H, it would also spell a kind of scientific doom for it! Any H so shielded from all factual evidence cannot be saying anything factual at all and therefore collapses to a mere matter of option. One choosing to be more metaphysically minded about this might want to say that, in the extreme case, H is a *metaphysical* rather than an *empirical* hypothesis.[16] He could perhaps even speak, qualitatively, of conjugate *metaphysical* and *empirical* aspects of any scientific hypothesis, each of these corresponding, respectively, to the extent of "tenure" and "provisionality" enjoyed by that hypothesis. Some might see promise in such terminology of a link bridging the discontinuity left by positivists between empirical science and metaphysics.

C. INCOMMENSURABILITY; NON-REDUCTION AND NON-ACCUMULATION OF SCIENTIFIC KNOWLEDGE

Closely allied to the conventionalist critique is the contemporary challenge to the root notion of positivistic empiricism that all of science must be ultimately reducible to terms of familiar experience. In its early version this reducibility was envisaged as explicit translatability of all meaningful scientific statements into so-called observation-statements; and such translatability became embodied as a necessary condition for significance in the so-called *verifiability theory of meaning*. Under the pressure of serious difficulties, however, translatability was later abandoned for a less naive version of reductionism. The subtler idea was that, for any statement to be significant, it must be possible to say what sort of sensory evidence would count for and what sort would count against the likelihood of truth of the statement. In the case of a scientific theory, this means that some of the terms must be coordinated with (this is something less than translated into) an observational language whose content consists of the familiar notions of ordinary sense experience. This observational language is the ultimate source of semantical reference for the theory, which is only a vacuous calculus until interpreted in this fashion or until coordinated with another theory which has been so interpreted. Such an observational language is envisaged as context-independent. That is, it is assumed to be pre-existent to the theoretic framework concerned, and hence semantically

[16] A view along these lines is expressed by I. Lakatos, "Methodology of Scientific Research Programmes" in *Criticism and the Growth of Knowledge*, ed. I. Lakatos and A. Musgrave, Cambridge University Press : Cambridge, 1970, p. 184.

independent of it. Moreover, the language is presumably already understood. That is, it is already in use for communication and hence has an established and relatively invariant significance. It is all this – familiar meanings, theory-independence, and stable significance which presumably qualifies an observational language for adoption as a basis for interpretation of theory.

But that a theory derives its significance by the "taking on" of theory-free notions *via* coordination rules as required by the philosophy of the Vienna Circle has become subject to challenge. Kuhn urges that "no experiment can be conceived without a theory" and that "there can be no scientifically or empirically neutral system of language or concepts".[17] This means that data are "theory-laden" and that no experimental concept can be fully understood without the theoretical framework in which it is couched. Indeed, the counterpositivist would maintain that the tangle of theory and observational notions is so intimate that theories with *apparently* shared vocabularies are not semantically commensurable. And whenever a scientific revolution has changed a world view "what were ducks in the scientist's world before the revolution are rabbits afterwards".[18] In this connection, Feyerabend, for example, maintains that the notions of length, mass, and time in special relativity are very different from the "corresponding" ones in classical mechanics. Thus, relativistic length involves an element not present in its classical counterpart, namely, the velocity of the object concerned, relative to the reference system. Thus, distinct theories cannot be compared for content and "a judgment of verisimilitude cannot be made."[19] Accordingly, it is not required that theories be assigned significance by definitional or translational methods based on the myth of theory-independent observational languages. Their vocabularies can be learned much as in anthropological field work – not by translation but from scratch, in the very context of actual usage.[20]

[17] T. S. Kuhn, *op. cit.*, p. 87, 146. Note also the following passage : "... is sensory experience fixed and neutral? Are theories simply man-made interpretations of given data? The epistemological viewpoint that has most often guided Western philosophy for three centuries dictates an unequivocal, Yes! In the absence of a developed alternative, I find it impossible to relinquish entirely that viewpoint. Yet, it no longer functions effectively, and the attempts to make it do so through the introduction of neutral language of observations now seem to me hopeless". p. 126.

[18] T. S. Kuhn, *op. cit.*, p. 111.

[19] P. Feyerabend, "Consolations for the Specialist", *op. cit.*, ed. Lakatos and Musgrave, Cambridge, 1970, p. 222.

[20] "... Let us study the language of new theories not in the definition factories of the double language model, but in the company of those metaphysicians, experimenters,

These contentions have a deep bearing not only on the doctrine of verifiability but also on the positivistic programme which, as we noted in Chapter II, looks for the ideal of theory-to-theory reduction and the ultimate unity of science. In particular, the claim of incommensurability undercuts standard notions of verification; for these are based on the notion of theory-free evidence. But it also gravely undermines the possibility of systematic reduction of one science to another or even of one theory to another. Indeed, if taken seriously, it demands the abandonment of cherished "dreams" shared by both scientist and methodologist. We refer here to such matters as the "reputed" reduction of chemistry to physics, the "near" reduction of biology to chemistry (*ergo*, to physics) and the hopeful reduction of psychology to biology (*ergo*, to chemistry ... to physics). Incommensurability even reflects unfavorably on more circumscribed possibilities such as the reduction of certain particular theories to others, also known as derivations (explanations). Serious doubt is cast on the validity of such historic reductions as : the subsumption of antecedently known thermodynamic principles under later kinetic theory; the derivation of Newtonian mechanics from relativistic mechanics, as a limiting case; the derivation of optical principles from electromagnetic field theory; and other allegedly great systemizational achievements. Indeed, it is incommensurability bolstered by an underlying conventionalism which makes possible Kuhn's catastrophism in his account of scientific history. Science is for Kuhn not a cumulative enterprise but a vast graveyard of scientific theories, world views, metaphysical outlooks, conceptual frameworks, methodologies, etc. (He refers to any of these when they take on the dimensions of major scientific orientations as "paradigms".) One paradigm catastrophically replaces another which is impermeable to it, the succession representing neither the enlargement, nor improvement of knowledge. A given paradigm contains the rules of its own validation and its own elaborations. It is not only logically immune to counter-evidence (after Duhem and Poincaré), but neither can it be impugned or strengthened by any other paradigm. It can only be replaced by one.[21] For "differences between successive paradigms are both necessary and irreconcilable ... The normal-scientific tradition that emerges from a scientific revolution is not only

---

theoreticians, playwrights, courtesans, who have constructed new world views". *Ibid.*, p. 225.

[21] "In recent years ... a few historians of science have been finding it more and more difficult to fulfil the functions that the concept of development-by-accumulation assigns to them ... Perhaps science does not develop by the accumulation of individual discoveries and inventions". T. S. Kuhn, *op. cit.*, p. 2.

incompatible but often actually incommensurable with that which has gone before".[22]

These are deep-cutting reviews of reductionism and theory accretion, so long the touchstone of scientific progress and maturity of discipline. But suggestive and vigorous as the case against theory-independent observational bases and theory-to-theory reducibility may be, the doctrine of "incommensurability" of distinct theories is still largely an unexplicated claim and one that perhaps mitigates too drastically the distinction between the formal (syntactical) properties of theories and their referential (semantical) function. The two-language coordination thesis deep at the heart of positivistic empiricism does founder seriously as, for example, on the question of meaning of theoretical entities or on precisely what is to be understood by the notion of "coordinating" observational content with a theory. But the incommensurability thesis, as Kuhn admits, is still only an undeveloped alternative, calling for what promises to be very fruitful research in the philosophy of perception, learning and discovery.[23]

D. NON-METHODISM

The views we have just considered provide the framework for a trend of thought which one might call *non-methodism* – doubtlessly the most radical departure from traditional methodologies in contemporary philosophy of science. It challenges traditional methodologies with respect to both ampliative and subsumptive levels of scientific inquiry. With respect to ampliative science, i.e., *growth of science*, non-methodism rejects *inductivism*, or the view that factual evidence is the ultimate basis for accepting or discarding scientific theory. In this its point of departure is conventionalism, but the over-all critique is more sweeping. With respect to systemization, it diverges from some of the most entrenched standards of over-all structural unity.

Modern inductivism goes back to Hume who found no rationale for it and concluded that it was ultimately psychological in character. In our own times, Carnap has attempted to provide the needed "logic" of the "inductive relation", offering in place of Hume's *irrational* inductivism the most articulate elaboration of *rational* inductivism in contemporary methodology. His spectrum of "inductive methods" provides for numerical

[22] *Ibid.*, p. 103.

[23] "In the absence of a developed alternative, I find it impossible to relinquish entirely that viewpoint". His reference here is to the view that sensory experience is fixed and neutral. See T. S. Kuhn, *op. cit.*, p. 126.

degrees of confirmation (cf. our chap. III) and is predicated on the presumed rationality or "reasonableness" of these methods. On the other hand, others like Popper and I. Lakatos reject inductivism as a logic of justification but retain it (and rationality) as a doctrine of *falsification.*[24] More recently some writers like Kuhn and Polanyi[25] and, still more recently, P. K. Feyerabend decisively reject inductivism outright and, with it, rationality as methodological criteria of scientific development. To the basic question of whether there is a "logic" of scientific growth, these writers answer with a resounding "no". The patterns of scientific development are not those of rational appraisal, i.e., of logic and reason. Justificationism and falsificationism are both futile approaches in any correct account of science; and methodology ultimately reduces to psychology. Thus, in studying the progress of science, we study not a logical pattern but (as one writer puts it) the "Mind" of the individual scientist or of the scientific community.[26]

For Kuhn the development of knowledge is a succession of prevailing scientific orientations, or "paradigms" as he calls them, and their catastrophic overthrows. Each paradigm is dramatically dethroned following the crises and revolution caused by its successor when it is introduced as a promising competitor by some individual scientist or some larger segment of the scientific community.[27]

[24] Lakatos departs somewhat from the decisive falsificationism of Popper by maintaining that what he calls a "research programme" (a notion somewhat broader than that of theory) *is* and *should be* held to when refuted or even when shown to be inconsistent. Such "tenacity" should be exerciced at least to the point where deeper potentialities have been tapped from the theory, that is, until "degenerating problem-shifts" or "progressive problem-shifts" warrant (respectively) the rejection or maintenance of the programme. I. Lakatos, *op. cit.*, pp. 91-196.

[25] For Polanyi's account see his : *Personal Knowledge, Towards a Post-critical Philosophy*, 1958.

[26] Lakatos, I., *ibid.*, p. 179. In this passage Lakatos is giving an account not of his own views but of those of others.

[27] T. S. Kuhn, *op. cit.*, p. 6. Also see p. 8, where he says : "Competition between segments of the scientific community is the only historical process that ever results in the rejection of one previously accepted theory or in the adoption of another".

Precisely what Kuhn intends by the term, 'paradigm', is a matter of some conjecture. Thus, for example, Margaret Masterman tells us : "On my counting he uses 'paradigm' in not less than twenty-one different senses, ..., possibly more, not less". See her article, "The Nature of a Paradigm", *op. cit.*, ed. Lakatos and Musgrave, Cambridge, 1970, p. 61.

But, for our purposes, it suffices here to understand Kuhn's term as designating a major scientific orientation, construed broadly enough to include metaphysical beliefs, substantive theoretic "achievements", methodological rules, standards, equipment, technical approaches and abstract constructions, all marshalled for engaging in the

As we have already noted, for Kuhn, confrontation with evidence is not sufficient to mold any scientific commitment; so that, in this sense, the decision for change is not quite an inductive one.[28] Indeed, Kuhn maintains that the scientific community *does*, *should* and presumably *can*, with logical impunity, resist the encroachment of a new paradigm irrespective of all counter-evidence or of the occurrence of experimental anomalies. Indeed, it is this resistance to change that contributes heavily to the catastrophism in the eventual dislodging of one paradigm by another.[29] Feyerabend introduces the convenient label *principle of tenacity* for the methodological advice to stick to the more promising theory (and this could be the one which is presently in vogue) even if the actual difficulties it encounters are considerable.[30]

What should also be mentioned here, with respect to Kuhn's non-methodism, is a point we have already noted in the foregoing section on incommensurability. Recall that for Kuhn successive paradigms are mutually incompatible. Hence, no sequence of them can be viewed as an expansion or accretion of scientific knowledge.[31] Indeed incompatibility of paradigms is a necessary consequence of the inseparability of fact and

---

"puzzle solving" of daily or "normal" science. Examples of paradigms are Aristotle's account of terrestrial motion, Ptolemy's kinematics of the heavens, Galileo's relativistic analysis of motion, Maxwell's mathematization of the electromagnetic field, and Heisenberg's matrix mechanics.

For a detailed discussion of Kuhn's notion of a "paradigm" see M. Masterman's entire article, *ibid.*, pp. 59-89.

Kuhn gives his own examples of a paradigm; see : T. S. Kuhn, *op. cit.*, p. 23 *et passim.*

[28] "... the act of judgment to reject a previously accepted theory is always based on more than a comparison of that theory with the world". *Ibid.*, p. 77.

[29] Kuhn tells us to note first "what scientists never do when confronted by even severe and prolonged anomalies ... they do not renounce the paradigm that has led them into crisis. They do not treat anomalies as counter-instances ... No process yet disclosed by the historical study of scientific development at all resembles the methodological stereotype of falsification by direct comparison with nature". The defenders of a paradigm "... will do what we have already seen scientists doing when confronted by anomaly. They will devise numerous articulations and *ad hoc* modifications ... in order to eliminate any apparent conflict". *Ibid.*, pp. 77-78. For the details of Kuhn's position on this, see Chap. VIII, *op. cit.*, pp. 77-91.

[30] See P. K. Feyerabend, "Consolations for the Specialist" in *op. cit.*, ed. Lakatos and Musgrave, p. 203.

[31] "The transition from a paradigm in crisis to a new one from which a new tradition ... can emerge is far from a cumulative process, one achieved by an articulation or extension of the old paradigm. Rather it is a reconstruction of the field from new fundamentals, a reconstruction that changes the field's most elementary theoretical generalizations as well as many of its paradigm methods and applications". *Ibid.*, p. 85.

theory. This makes fact largely theory determined and – in the sense which we have already touched upon in the preceding section – incommensurable. "Scientific fact and theory are not categorically separable, except perhaps within a single tradition of normal scientific practice".[32] This last claim, that is, the incomparability of successive paradigms, has far-reaching significance for a methodology of science. We have already noted its adverse implications regarding the possibility of a unified science. But more generally, these implications undermine the possibility of broad-scoped simplification through theoretic unification. According to the deepest traditions in scientific methodology, overall scientific systemization (a special form of which is explanation) is the deductive subsumption of wide ranges of scientific knowledge under a few simple principles. The compactness which this affords not only provides unified understanding but, following Duhem and Poincaré, seems to provide the one remaining basis for assessment of theories. Indeed, we naturally tend to assign more scientific status to any system the more it can span and subsume with economy of assumption. The incommensurability of Kuhn's successive paradigms is incompatible with this far-reaching criterion of more traditional methodologies. Kuhn's account therefore is not merely anti-inductivistic. It is also – as one might put it – "non-systematistic" as regards the broadest aspects of subsumptive structure in science.

But, despite its overall non-methodism, Kuhn's account of science does leave room for a limited range of logical structure and method in the scientific enterprise. He distinguishes between "normal" and "extraordinary" science. Normal science is the day-to-day puzzle solving, mop-up and fill-in activity that goes on, sometimes over historically extended periods of time, under the conceptual framework of the prevailing paradigm. Within the *éclaircissements* of such a paradigm there can be progress.[33] Indeed, within the paradigm there may even be shades of falsificational (hence inductive) judgment. Thus, Kuhn admits that before scientists adopt a new viewpoint – "Nature itself must first undermine professional security by making prior achievements seem problematic".[34] Moreover, he offers some faint suggestions of over-all progress and hence cumulative structure even where extraordinary science (i.e., change of paradigm) is concerned :

"... when ... a new candidate for paradigm has been evoked, scientists will be reluctant to embrace it unless convinced that two all important

[32] *Ibid.*, p. 7.

[33] "... it is only during periods of normal science that progress seems both obvious and assured". T. S. Kuhn, *op. cit.*, p. 163.

[34] *Ibid.*, p. 169.

conditions are being met. First, the new candidate must seem to resolve some outstanding ... problem that can be met in no other way. Second, the new paradigm must promise to preserve a relatively large part of the problem solving activity that has accrued to science through its predecessors ... though new paradigms seldom or never possess all the capabilities of their predecessors, they usually preserve a great deal of the most concrete parts of past achievement and they always permit additional concrete problem solutions besides ... a community of scientific specialists will do all it can to ensure the continuing growth of the assembled data that it can treat with precision and detail".[35] Terms like 'professional security', 'evoked', and 'reluctant' are heavy with psychological and sociological content. Nevertheless, there are in these remarks underlying criteria that are "logical" or "rational" in character. They have to do with the pressure of evidence and the objective augmentation of knowledge. And, in the indicated sense, they moderate Kuhn's analysis. One last point having to do with moderation in Kuhn's non-methodism : He reminds us that the critical analysis of a prevailing paradigm in the face of crisis has played a major role in the progress of research. Thus, "the search for assumptions (even non-existent ones) can ... be an effective way to weaken the grip of a tradition upon the mind and to suggest the basis for a new one. It is no accident that the emergence of Newtonian physics in the seventeenth century and of relativity and quantum mechanics in the twentieth should have been both preceded and accompanied by fundamental philosophical analyses of the contemporary research tradition. Nor is it an accident that in both these periods the so-called thought experiment should have played so critical a role in the progress of research".[36] Admittedly, searching for the underlying assumptions in a tradition belongs more to philosophical analysis than to empirical strategies. Nevertheless, in recognizing the important place of critical analysis in the scientific process, Kuhn again grants an element of scientific rationality which moderates his non-methodism.

Kuhn's delineation of science represents a considerable departure from any standard conception of methodology, far surpassing in its divergence the conventionalism from which it springs. In reducing philosophy of science largely to psychology or sociology he has taken discourse about science virtually out of methodology, and we class him therefore as a non-methodist. But his approach as, we have noted, is qualified by unmistakable

[35] *Ibid.*, p. 169.
[36] *Ibid.*, p. 88.

elements of restraint. And these elements are shades of standard methodology which mark him as a moderate.

P. K. Feyerabend takes the non-methodist position to much farther limits, characterizing his own proposals as an "anarchistic methodology" whose only guide-line is the principle that "*anything goes*".[37] What is more, the abandonment of orthodox canons of method regarding scientific discovery and development is not a matter of option (taste, whim, ignorance or disattention); it is necessary! Violations of rule "... are necessary for progress". Thus, "... developments such as the Copernican Revolutions or the rise of atomism in antiquity and recently (kinetic theory; dispersion theory; stereochemistry; quantum theory) or the gradual emergence of the wave theory of light occurred either because some thinkers decided not to be bound by certain obvious methodological rules or because they unwittingly broke them. This liberal practice is not just a fact of the history of science ... It is reasonable *and absolutely necessary* for the growth of knowledge".[38] Basic to this position of course is the Duhem-Poincaré thesis that no amount of counter-evidence can logically dislodge any hypothesis. Accordingly, whether one bends to or resists the pressure of factual evidence is optional provided he makes appropriate adjustments elsewhere in the system. But this doctrine, if correct, merely frees methodology from inductivism as a logical constraint on the acceptance of hypotheses. Presumably, one might still choose to yield to unfavorable evidence and to hold to well-corroborated hypotheses. Feyerabend instead goes for more radical strategies. Not only does he allow proceeding either inductively or with Kuhnian *tenacity*, as the occasion may require, but by *counter-induction* as well. Thus, to Kuhn's principle of tenacity Feyerabend adds a principle of *proliferation* according to which it is also legitimate to do science by "... introducing, elaborating, and propagating hypotheses which are inconsistent either with well-established *theories* or with well-established *facts*".[39] Such a strategy, he counsels, is a necessary part of any critical empiricism. The assuming of a theory, T′, which is incompatible with some presently accepted theory, T, often helps unearth evidence for the test of T.[40] What is more the proliferation of alternatives even when the prevailing ones seem unblemished and, again, even in contradiction

[37] Feyerabend, P. K., "Against Method", *Minnesota Studies in the Philosophy of Science*, ed. M. Radner and S. Winokur, University of Minnesota Press : Minneapolis, 1970, p. 26.

[38] *Ibid.*, p. 22.

[39] *Ibid.*, p. 26.

[40] *Ibid.*, p. 27.

with known fact, is the rule rather than the exception in science. "... *not a single theory* ever *agrees with all the known facts in its domain*".[41] Moreover, the disagreement in point is not merely a matter of numerical error – random, systematic, substantive, etc. It is also a matter of qualitative discrepancy as for example in some of the difficulties and absurd consequences inherent in such established theories as geometric optics, classical electrodynamics and the classical electron theory of Lorentz.[42]

Indeed, the proliferation of counter-inductive hypotheses that clash with either data or existing theory is not only a matter of historical fact; it is essential. And "... science cannot exist without it". It is "... a legitimate and much needed *move* in the game of science". Indeed, counter-inductive hypotheses are the only external measures of criticism of a prevailing scientific tradition, however entrenched it may be. They are the needed measure sticks. And "... the first step in our criticism of customary concepts and customary reactions is to step outside the circle and invent a new conceptual system".[43] But, even more than this the proliferation of counter-hypotheses is what ultimately generates scientific content. It is dialectical interaction of incompatible and even incommensurable theories that leads to higher-level articulation and synthesis in scientific development.[44]

Thus the proposing of diversified viewpoints is the enrichment of science; it sharpens scientific criticism by comparative foundational analysis and yields scientific content by the dialectical synthesis of thesis and anti-thesis. Feyerabend's proposal is a pluralistic philosophy of science *in lieu* of the unacceptable monism of traditional methodologies.[45] Nor is Kuhn's one-at-a-time paradigm theory of science acceptable to Feyerabend. For it is essentially monistic and, as such, fails to account for the emergence of competing hypotheses. Indeed, Feyerabend denies the historical existence of Kuhn's one-paradigm normal science. "Such an enterprise is not only ill-conceived and non-existent, its defence is also incompatible with a humanitarian outlook".[46]

[41] *Ibid.*, p. 36.

[42] For more detail on these examples see P. K. Feyerabend, *op. cit.*, pp. 36-43.

[43] *Ibid.*, p. 45.

[44] Lakatos and Musgrave, *op. cit.*, p. 210. Feyerabend's debt to Hegelian dialect is obvious here. For his detailed acknowledgment to Hegel, see his : "Against Method", *op. cit.*, ed. Radner and Winokur, pp. 25-36.

[45] In this he decisively parts company with Kuhn who sees pluralism only in the transitional (crisis) stages of "extraordinary" scientific development during which several competing paradigms are under consideration by a scientific community that is generally resistant to change. What is more, "normal" science for Kuhn proceeds entirely by *tenacity* rather than by *proliferation.*

[46] P. K. Feyerabend, "Consolations for the Specialist", *op. cit.*, pp. 207-211.

But counter-induction and proliferation are not the only methodological liberalizations required by Feyerabend. We may also abandon practicality and the demand for experiential content.[47] Nor can we exclude the use of propagandistic machinations such as used by Galileo to produce receptivity in his contemporaries.[48] Another admissible device customarily excluded by methodologists is *ad hocness* which characterizes hypotheses invented merely to accommodate discrepancies, but which have no implications beyond the available evidence and, therefore, cannot be independently assessed. In defense of ad hocness, Feyerabend tells us that it has played a genuine role in the growth of knowledge and cites some of Galileo's use of *ad hoc* assumptions in the attempt to overthrow the old dynamics of falling bodies. What is more, *ad hoc* hypotheses are essential strategies :

It was good that he (Galileo) used them. Had he not been *ad hoc*, he would have been *ad hoc* anyway, but this time with respect to an older theory. Hence, as one cannot help being *ad hoc*, it is better to be *ad hoc* with respect to a new theory, for a new theory, like all new things, will give a feeling of freedom, excitement and progress. Galileo is to be applauded because he preferred protecting an interesting hypothesis to protecting a dull one".[49]

Non-methodism, particularly Feyerabend's version, presents an intriguing blend of considerations : valuational, historical and logical. It is a daring reconstruction that, at once, attempts to tell : (1) what would make science humane, interesting and worthwhile; (2) what, as a matter of fact, science has actually shown itself to be; and (3) what from the viewpoint of adequacy (efficacy, conceptual productivity, etc.) science must be.

In line with (1) Feyerabend urges that :

"... we want a methodology and a set of institutions which enable us to lose as little as possible of what we are capable of doing and which force us as little as possible to deviate from our natural inclinations ... It must be allowed to *retain* our ideas in the face of difficulties; and ... to introduce *new ideas* even if the popular views should appear to be fully justified and without blemish." (Both *tenacity* and *proliferation* are recommended here.)

Feyerabend's plea for freedom and pleasure through science could not but be engaging and refreshing. It is a departure from dehumanizing rigor for rigor's sake, and it contains powerful suggestions towards a blueprint for a "bridge" to the other culture (i.e., to the humanities). But this ideal, Feyerabend maintains, is in essential accord with science as we *actually* find it. [Point (2), above] In this connection he says :

47 "Against Method", *op. cit.*, ed. M. Radner and S. Winokur, p. 55.

48 *Ibid.*, p. 61.

49 *Ibid.*, p. 69.

"... actual science, or at least the part of actual science that is responsible for changes and for progress, is not very different from the ideal outline ... But this is a happy coincidence indeed! We are now in full agreement with our wishes as expressed above!" [50]

What is more [point (3), above], this policy is the one to be most highly recommended from the viewpoint of *sound* methodology. Its main strategies are *proliferation* and *tenacity* :

"Proliferation means that there is no need to suppress even the most outlandish product of the human brain. *Everyone may express his inclinations*, and science, conceived as a critical enterprise, will profit from such an activity. Tenacity : this means that one is encouraged not just to follow one's inclinations, but to develop them further, to raise them with the help of criticism (which involves a comparison with the existing alternatives) to a higher level of articulation *and thereby to raise their defense to a higher level of consciousness.* The interplay between proliferation and tenacity also amounts to a continuation, on a new level, of the biological development of the species, and it may even increase the tendency for useful *biological* mutation".[51]

There is much that is exciting, even irresistible, in these recommendations. But their three-faceted validity cannot be taken for granted as "happy coincidence". One could have little quarrel with the inviting and permissive character of Feyerabend's model of science and with the liberating overtones of its break with methodological law and order (which Feyerabend reads as "stuffiness"). He may be right that such a model would optimally promote human satisfaction and fulfillment – though this might best be decided as a thesis in empirical psychology. Moreover, one may even be able to pull several impressive examples from history that bear strong resemblances to the Feyerabend ideal. But, that such a model is conducive to sound science, in some sense of methodological adequacy, is at best merely a *suggestion* that urgently cries out for justification. Despite some suggestive epistemological and metaphysical references (which we shall touch on shortly), Feyerabend provides little antecedent analysis of any distinct notion of methodological soundness and of how his anarchistic model might achieve such soundness. Simply equating *sound* science, *ipso facto*, with *satisfying* and *psychologically wholesome* science would seem to beg an important question. Indeed, it would not be fair to suggest that this is Feyerabend's intent. For surely he recommends more than merely the replacing of old dogmas with new ones. Of course, the conventionalism underlying his views rules out any serious reference to traditional notions

[50] "Consolations for the Specialist", *op. cit.*, ed. Lakatos and Musgrave, pp. 209-210.

[51] *Ibid.*, p. 210.

of scientific truth even when such notions are expressed in terms no stronger than those of confirmation and probability. It might be supposed that he could still turn to other notions of scientific adequacy or soundness expressed, possibly, in terms of simplicity, degree of articulation, or mere systematic coherence. And, it would seem, some remnants of these notions might still serve him well. But counter-inductive proliferation and incommensurability take a heavy methodological toll here and virtually cut him off from even these relatively modest criteria.

Feyerabend shows unmistakable concern along these lines and eventually attempts to assure us with the following remarkable suggestion :

> We need not fear that the diminished concern for law and order in science and society that is entailed by anarchistic philosophies will lead to chaos. *The human nervous system is too well organized for that.* (Our own italics)

This passage makes a neuro-psychological claim of some importance to his thesis. Also, and somewhat paradoxically, it betrays a measure of continued commitment to system and organization as an ultimate objective of science. Similar indications may be seen in his use of terms like 'higher level of articulation' or 'knowledge and truth'.[52] But what makes Feyerabend so confident that the assumed integrity of the human nervous system (and what this is supposed to say about how such a nervous system relates to the world) will avert chaos despite the anarchy in anti-methodism?

Feyerabend does not delay too long in revealing the deeper doctrinal basis of his claims. This basis is found in the philosophy of Hegel, which provides a cosmological dialectic for assuring conformity of process-instrumentality with product-validity. It would be grossly unfair to object, here, that Feyerabend is naively confusing process with product and, correspondingly, the context of decision with that of validity (the first being causal-genetic; the second justificational). Indeed, it is precisely because these two aspects of concern are to be distinguished that Feyerabend attempts to relate them. He does this in what he takes to be a Hegelian manner, making references to portions of Hegel's metaphysics which, as he reads it, equates process with product. But if science is process, then it can have no fixed conclusion – no univocal framework. The "dialectical" processes of nature "... cannot be understood by attending to those elements in our subjectivity which are still in relative isolation and whose internal contradictions are not yet revealed. (Most of the customary concepts of

[52] As, for example, on p. 36 of his essay "Against Method", *op. cit.*, ed. Radner and Winokur.

science, mathematics, and especially the rigid categories used by our modern axiomaniacs are of this kind.) ... We must rather proceed dialectically, i.e., by the *interaction* of concept and fact". Thus, the scientific enterprise is criticism and discovery, but always in the context of antithetical alternatives. Science is product only insofar as it is also process.[53] Moreover, conceptual conformity and inductive consensus would not by themselves suffice to move this process onward. Theory-to-theory, theory-to-fact, and even fact-to-fact tension or outright contradiction are necessary. This also calls for dialectical oppositions even on the level of conscious method. For every stratagem there is a counter-stratagem; for every prohibition a permission. Indeed, there is no one method except opportunistic fluidity of approach and strategic change.

To buttress this view of the scientific process, Feyerabend calls upon the Hegelian principle that being and thinking are one. The concept "... not only mirrors a nature that 'lies entirely outside of it', it is not merely 'something subjective and accidental' ... it has the tendency to be the end result of the development of a specific thing, so that, finally the concept and this thing become one ... To sum up : Knowledge (and therefore inquiry) is part of nature and is subject to its general laws".[54] This portion of Hegel's philosophy provides Feyerabend with what he takes as a "cosmic" basis for the correctness of his model of science.

But despite the vigor and suggestiveness of this account it leaves much to be done in the way of clarification and substantiation. It is far from clear as to how any view identifying thought with being entails that "every product of the human brain", as Feyerabend tells us, "must be allowed". Indeed, even Feyerabend's sense of 'allowed' calls for elucidation. If his use of this term is such as to effect no more than a plea for socio-political freedom of expression or, perhaps, for open-mindedness in the initial stages of conjecture, then, all Hegelian metaphysics aside, his net recommendation could hardly meet with serious objection, though it is decidedly uninteresting. If, on the other hand, by "allowed" Feyerabend means *regarded as having scientific validity* (and this would seem to be the intended sense), then his recommendation becomes a central rule of method. And, quite apart from questions regarding its metaphysical basis, it raises serious questions. For without criteria on which to base acceptance, or to guide conjecture, this proposal threatens to inflate and ultimately dissolve the very idea of science. Indeed, if every product of cerebration is to *count* as science, then what

[53] *Ibid.*, p. 36.

[54] *Ibid.*, pp. 35-36. Parenthetical insert is our own.

distinctiveness is there *left* for science? How do we separate it from mathematics, logic, or philosophy? If these are vain differentiations – vestiges of an old distinction between factual and analytical inquiry – then where do we put poetry, myth, fancy and delusion? And how about error itself? How does one rule out its possibility where scientific knowledge is concerned without impugning factual content and ultimately every basis of certification? Following Feyerabend's Hegelianism, one might find it harmless to leave unchallenged the claim that error is an irreducible fact of life in the dialectic of process (and hence of being). But this is far from saying that as a necessary ingredient of process, eventually to be reconciled (*aufgehoben*) with the existing body of knowledge, error may go unrecognized as error and be granted full scientific parity with the *validated* products of inquiry.

Even as a strategic recommendation about the sort of working attitude or posture to assume in scientific inquiry (rather than as a strictly epistemological claim) Feyerabend's "everything goes" seems subject to serious practical reservations. To point up what we have in mind here, we note that some allegedly scientific claims can have great sociological import. But if no stabilized notion of validity is to figure in the acceptance of any hypothesis, then how, for example, does one deal with socially destructive claims which, under more standard methodologies, might have been examined for objective validity and appropriately rejected when found to be failing? Recent history is all too suggestive on how matters of fact, doctrinally immunized against discreditation, yet passing for legitimate science, could be espoused with resulting subjugation and demoralization of humans. Consider, in this general respect, the history of master-race ideologies and some recent claims about race and color-linked intelligence.[55]

Still another line of comment suggests itself at this point and this has to do with the Hegelian background to which Feyerabend makes frequent reference in developing his thesis against method. No doubt there is more than one way of reading and interpreting Hegel whose writings are amongst the most controversial in the annals of modern philosophy and whose

[55] The reader who has not already done so, is invited to acquaint himself with the theories of genetic inferiority propagated by W. Shockley, A. R. Jensen, R. Herrnstein and others. A short symposium of conflicting views is given in : K. Clerk, H. Eysenk, A. Jensen, J. Kagan, R. Light, D. McClelland and W. Shockley, "I.Q. and Race", *The Humanist*, January/February, 1972, pp. 4-19. For a brief review, see : Sandra Scarr-Salapatek, "Unknowns in the I.Q. Equation", *Science*, Vol. 174, Dec. 17, 1971, pp. 1223-1228. For the conclusions of R. Herrnstein see : "I.Q." in the *Atlantic*, Vol. 228, No. 3, September 1971, pp. 44-63.

thought has multiple aspects, some of which have been difficult for his followers to reconcile. Feyerabend himself is not systematically explicit on precisely what in Hegel he finds useful, though his debt to "dialectical method", according to which knowledge is advanced through "dialectical synthesis" or reconciliation of opposed theses, is fairly evident. Moreover, he obviously draws directly from the skepticism and "free aspect" of Hegelian philosophy. Indeed, non-methodism with its denial of traditional validation, its permissiveness of counter-evidence and counter-hypotheses, etc., seems to draw much inspiration from Hegelian thought. Of some particular concern to us here, however, is Feyerabend's focussing on what he calls Hegel's cosmology. The reference is to that phase of Hegel's thought which some would characterize as organicistic or holistic and which views the world as ultimately a totality in which all parts and aspects are *essentially* ("organically" or "internally") related. The basic sort of relation connecting things is, as Feyerabend tells us, "... not external. Every process, object, state, etc. actually contains part of the nature of every other process, object, state, etc." [56] In such a world, it would seem, there can be no "island" systems, no entities or aspects which can be properly understood apart from other entities and aspects and from the whole.

Now, it is not clear just how such a "cosmology" figures logically in Feyerabend's thinking. But even if this *portion* of the Hegelian metaphysics (as understood by Feyerabend) is only an inspiration rather than a systematic basis, it nevertheless offers a rather perplexing contrast with the *analytical* and abstractive tenor of scientific inquiry. Analysis and abstraction are root strategies of science, as we know it, in which only selected aspects of the world enter into generalizations, the rest being judged separable, negligible or irrelevant for the purposes on hand. Whatever the scientific purpose – the supposed irrelevance of virtually all aspects of the world except for a highly select few is basic to all abstractive analysis and underlies not only generalization from instances which focusses only on "isolated" invariances, but the techniques that go into the very process of experimentation which seeks to relate only a few parameters (pairs, triplets, etc. of them), neglecting as many of the remainder as possible.[57] Thus, on the face of it, at least, some of Feyerabend's avowed Hegelianism and the scientific outlook seem, in certain important respects, to be opposed. But Feyerabend provides no clear indication that there is any demonstrable systematic relation between his Hegelianism and his philosophy of science; and, indeed, the connection

[56] *Ibid.*, p. 32.

[57] For an interesting but somewhat technical account of some abstractive methods in science see : R. Lindsay and H. Margenau, *op. cit.*, pp. 36-55.

may be purely heuristic thus requiring no systematic account. But given the apparent incompatibilities an explicit attempt by Feyerabend to clarify the bearing of Hegelian metaphysics on his thesis against method would not seem to be inappropriate and could be philosophically very interesting.[58]

One last point can be added to our brief reflections. As a plea for more naturalness and humanism in science, Feyerabend deplores rigidity and narrowness in the education of scientists, maintaining that this maims their style of writing, their speech patterns, and their facilities for communication.[59] And undeniably there is truth in this claim. Surely there is much in the contemporary scientific posture that needs loosening and much looseness in the best examples of past science that calls for recognition and emulation. (My already strong convictions in these respects have grown considerably with the reading of Feyerabend's writings.) But, again, if Feyerabend is attempting to do more than merely to make the usual recommendations for naturalness of expression in all writing and communication, then he should offer an account which would have more specific bearing on the exacting linguistic and symbolic demands which all participants in the more mathematical sciences must meet. Let us, by all means, opt for more expression of pleasure, humor, hope and even disappointment in all scientific writing. But how does one humanize, or "loosen", or express with natural idiom, or mollify in any way the strict formalisms required in the elaboration of such scientific pieces as Maxwell's electrodynamics or Heisenberg's matrix mechanics? How much can we do with ordinary language to cut through the ineluctable formalism of a partial differential equation without verbal inundation. And could we be sure that linguistic informalities would not have to be so excessive as to detract from the artistic elegance, unity and mood of the treatise? To what extent can formal science (or its exposition) mate with comedy and pathos without fathering stylistic absurdity? Further reflection and perhaps even experiments along these lines should be fascinating ventures.

In summation : The contemporary trends of methodological thought which we have reviewed in this concluding chapter hint at the possibility

[58] It should perhaps be recalled here that, as is generally recognized, Hegel, in certain phases of his system, does give attention to natural science and to its methods of analysis and abstraction. But these, according to Hegel, are functions of the "finite understanding" in contrast to *synthesis* which is the higher and more "ultimate" process and which, as a function of "reason" and by way of dialectic, moves towards concreteness rather than abstraction. Feyerabend's references, however, remain largely inexplicit on these facets of Hegel's thought.

[59] P. K. Feyerabend, "Against Method", *op. cit.*, pp. 20, 95-98.

of salutary orientations *within* and *about* science. From the viewpoint of philosophy of science, they point to methodological alternatives that, in some important respects, diverge radically from standard viewpoints. In contrast to the approved and dominant themes, these trends focus attention largely on the ampliative phase of science (i.e., discovery and acceptance) rather than on subsumptive systemization (usually labeled "explanation") – though, as we have noted, they have profound implications even for the systemizational phases of science. In attempting to excise some of the most entrenched rules of method, non-justificationism, non-falsificationism and more generally non-methodism have generated vigorous objections many of which have yet to be met with pointed philosophical analysis. No doubt there is much more for the proponents of these new trends to say within the frameworks of their challenging claims. Indeed, Feyerabend himself cautions that we need a theory of error; [61] although, as if to minimize the need, he then adds the hopeful comment that "... we are dealing with well-trained scientists who are kept in place by an internal slave-master called 'professional conscience' and who are convinced that it is good and rewarding to attain, and then to forever keep, one's 'professional integrity' ".[61]

But these attempts at reformation are not merely refreshing and challenging. They cut, with deadly penetration, to the very heart of traditional criteria of rationality. Indeed, they seem gravid with philosophic payload and may well be the nuclei of important and radical developments in the coming years. In these respects our own account has been necessarily brief, at best affording only some glimpses of the general trends now taking form.

E. THE HISTORY AND PHILOSOPHY OF SCIENCE

As a closing consideration in our discussion it seems appropriate to give some attention to the long recognized distinction between two sorts of study *about* science, namely, *history* and *philosophy* of science. For methodologists – especially those in the analytical tradition – the differentiation is a clear and distinct one. One turns to the historian of science for a factual account of scientific development but to the philosopher of science for an analysis of any logical aspects in this development or in the body of knowledge which results. The historian functions ultimately in the

[60] *Ibid.*, p. 18.
[61] *Ibid.*, p. 20.

context of spatio-temporal and causal genesis bringing in sociological and other factors as required. The philosopher instead works in the context of validity and significance; and his method is essentially analytical. None of this, of course, means to deny that history and philosophy of science are closely allied disciplines. The alliance is widely acknowledged and the two disciplines are often paired under a joint academic rubric. But the association is viewed as a matter of convenient juxtaposition. The historian of science is depicted as borrowing philosophico-logical distinctions in laying bare the more systematic developments in science; the philosopher, in turn, is seen as going to past science for concrete examples in his search for logical pattern. And despite the profitable exchanges, the respective responsibilities of *historical* and *analytical* modes of inquiry remain essentially and profoundly distinct.

Without doubt, the differentiation of historical from logical modes of inquiry is fundamentally appropriate and reflects basic distinctions between genesis and structure, or between the context of causal development (which calls for explanation) and the context of validity (which calls for justification). In actual practice, however, this systematic differentiation can be emphasized and overdrawn to the point of a disservice. When applied rigidly for separating disciplines, it tends to distract from vital *interdisciplinary* aspects of concern. History of science, like *philosophy of science*, is in one sense *about* science, although itself potentially a science. Therefore, in its accounts, it would naturally have much occasion to employ meta-scientific notions (or, one might say, vocabulary appropriate to talk *about* science). At the same time, the philosopher can find in history a priceless store of concrete examples which provide not only instances or paradigm cases for primary scrutiny but also the grounds for establishing certain entertained *possibilities* on the basis of historical *actualities*. He might, for example, seek the concrete historical case needed for establishing the *possibility* of a rational strategy of discovery. Put to such use, history becomes, so to speak, a philosopher's touchstone and the proving ground of possibility. One can say therefore that the history and philosophy of science relate to each other not merely as convivial neighbors but as kindred disciplines. For they have more in common than a measure of practical technique and a general subject-matter. Notwithstanding their basic differences, their bond is one of fundamentally overlapping concern and method rather than of occasionally borrowing one another's vocabulary and results. It would, for example, be grotesquely unnatural for the historian to adopt any strict policy of "staying out" of the analytical phase of any account of science. Such a policy would have the awkward consequence that the more "rationality" there was in scientific activity the less there would be for the historian to do. And, even given that scientific development was only *partially* governed by logic-like strategies, any puritanical insistence on contextual separation of all causal-genetic from analytic exposition could so disrupt and devitalize the writing of human history as to mutilate it beyond recognition.

Standard methodologies in the analytic tradition would, of course, have no fundamental quarrel with these reflections. In actual practice, however, they have tended to over-emphasize the division, and neglect the basic aspects of kinship, between philosophical and historical study. The pat manner in which the hypothetico-deductive tradition dismisses the role of logical strategies in scientific discovery is symptomatic of this over-emphasis. Science as a *process* of inquiry belongs to history; science as the structured *result* of inquiry belongs to analytical philosophy. Hence, let all philosophers of science focus on product and leave process to the historian.

In opposition to this tradition, irrationalistic accounts such as Kuhn's and Feyerabend's tend to reduce the importance of the contrast between historical and philosophical approaches to science (though that distinction no doubt remains in principle). One reason for this is not only that such accounts see scientific development as driven by social forces other than commitment to logical method, but also that they cast doubt on the logical integrity of the product itself. This, then, makes all of science (i.e., both process and product) a chapter in the history of cultural dynamics, thus minimizing the concern for any *distinctive* role of analytic philosophy (i.e., methodological analysis) in study *about* science. Thus, for Kuhn scientific development is a series of revolutions in which it is not theories of greater validity, fundamentality, or generality that win out but "paradigms", incomparable with their predecessors, which, for "more than logical reasons", eventually win acceptance in the scientific community. Feyerabend similarly renounces all traditional notions of a logic of scientific development and sees the scientific process as a conceptual play of incompatible and "incommensurable" alternatives which must be proliferated and among which "the most outlandish products of the human mind" also hold scientific rank. But more than this – and here is perhaps a second reason for the lessening of the history-*versus*-philosophy of science distinction – for Feyerabend, the conceptual play of alternatives is a "necessary" one; it is part of, and mirrors, the underlying dialectic of nature itself. Thus, *apart from these natural dynamics*, there are no valid prescriptions to be made, so that in the final analysis, *autonomous* logical method vanishes on all levels of science and all we find is history. The task of any study *about* science therefore cannot be more than that of unraveling the dialectical processes of inquiry as they actually occur. If so, it seems to matter little whether we call the study of such processes philosophy (i.e., methodology) or history; the "two" modes of inquiry are ultimately *one*!

We have seen, however, that despite their philosophically provocative and suggestive power, such non-methodistic accounts of science, especially in their more sweeping forms, are subject to serious critical review. But quite apart from the question of their over-all defensibility and the dubiousness (or at least obscurity) of their underlying metaphysics, such approaches tend to direct fresh attention to the roles of both history and philosophy in discourse *about* science. To some extent, they divert philosophic attention from science as *logically stabilized product* to science as *dynamical process* and from scientific thought as *rigor-bound method* to scientific thought as *free* and *imaginative exploration*. On the face of them, these emphases seem to preempt the philosopher's role in any study about science and to grant it all to the historian, thus reinforcing traditional separatist views of the two disciplines. But the effect could well be of another sort. By reaffirming the aspect of process and likening it to product – albeit on darkly metaphysical grounds – the non-methodistic account could help balance out the excesses of the hypothetico-deductive tradition which radically divorces product from process in order to maintain that only the former can be a legitimate concern of analytical philosophy. The notable irreverence of non-methodism for such hard and fast cleavages could well make it the *auguria oblativa* of broader and more unified study about science – study which can exclude neither historical nor philosophical considerations, in the very best sense of the underlying distinction.

INDEX

GPSR Compliance

The European Union's (EU) General Product Safety Regulation (GPSR) is a set of rules that requires consumer products to be safe and our obligations to ensure this.

If you have any concerns about our products, you can contact us on ProductSafety@springernature.com

In case Publisher is established outside the EU, the EU authorized representative is:

Springer Nature Customer Service Center GmbH
Europaplatz 3
69115 Heidelberg, Germany

Batch number: 10607222

Printed by Printforce, the Netherlands